Elements of
Systems Analysis

Elements of
Systems Analysis
Third Edition

Marvin Gore
Mt. San Antonio College

John Stubbe
Mt. San Antonio College

Wm. C. Brown Co. Publishers
Dubuque, Iowa

Copyright © 1975, 1979, 1983 by
Wm. C. Brown Company Publishers. All
rights reserved

Library of Congress Catalog Card Number
82–71568

ISBN 0–697–08169–9

Printed in the United States of America
10   9   8   7

Contents

2

3

4

5

6

7

13

Preface

In fewer than three decades, computers have evolved from cumbersome scientific machines with few commercial applications to indispensable tools of such awesome proportions that they affect almost every aspect of modern business—and of our personal lives. Throughout the 1980s, computer systems, large and small, will continue to proliferate. Furthermore, the potential for *cost-effective* uses of computers and for new computer applications will increase. As computer performance increases, the cost for that performance will decrease. The greatest challenge of the 1980s is to increase business and industry productivity using the full potential of computer-related technologies. *Elements of System Analysis* focuses upon the methods of systems analysis and the skills needed to meet this productivity challenge.

Elements of Systems Analysis is a significant update of the successful text, *Elements of Systems Analysis for Business Data Processing*. The authors recognize that there are few shortcuts to good system design practice, and emphasis is retained on the structured, life-cycle process for the design of usable computer-based systems—the focal point of the previous editions. Through its focus upon the system development life cycle, and its presentation of both conventional and newer structured systems analysis methods, the text meets the requirements of the model Computer Information Systems curriculum (CIS–4) endorsed by the Data Processing Management Association.

This edition extends the applications scope of systems analysis beyond the business data processing activities of the 1970s to the much broader information systems environment of the 1980s, which includes microcomputers, communications networks, automated office systems, software applications packages, and distributed data processing systems. Within this environment, persons whose training and experience includes systems analysis will become the information resource managers of the future.

Training in systems analysis is important for all business students because of the vast number of options becoming available for the implementation of information systems. Accordingly, the tools and techniques described here are current. The examples and case studies in the text and student workbooks are topical, representing the information system needs of modern business. To illustrate, the case study presented in the text reflects a distributed data processing environment with communications network and remote data processing options.

Elements of Systems Analysis includes a thorough introduction to the "top-down" process for systems analysis. This process is based upon the four life-cycle phases of business information systems: study, design, development, and operation. The life-cycle methodology emphasizes the integration of systems analysis activities, management reviews, and documentation. Examples of documentation—which is concurrent with the entire life-cycle process—are a feature of the text. A case study is introduced in the study phase, and an example of a completed study phase report is presented at the end of that phase. This report is then expanded upon in the design and development phases to provide design phase and development phase reports as examples of cumulative documentation. Current system design techniques, such as data flow diagrams, decision trees, hierarchy and IPO charts, structured walk-throughs, and top-down computer program development and testing are presented as life-cycle events at appropriate points in the study, design, and development phases.

Elements of Systems Analysis is written in clear terms the business student can understand. Divided into six major units, the text is organized so that it can be used in both introductory and advanced courses in systems analysis. The first two units introduce background concepts and acquaint the student with the basic tools and techniques of systems analysis. The four units that follow apply these tools to each of the four life-cycle phases.

Unit 1, Information Systems and the Systems Analyst, provides the background concepts most important to the profession of systems analysis. Chief among these are the life-cycle concept and the perception of a business as an information system and the recognition of information as a resource to be managed.

Unit 2, Tools and Techniques of Systems Analysis, introduces skills that are the working tools of systems analysis. These include coding, forms design, charting, and written and verbal communications. These skills are introduced at this point, instead of at the time that they are encountered in the life-cycle process, for two reasons: (1) to reinforce learning by allowing the student the opportunity to apply material with which he or she already has become familiar; and (2) to anticipate learning digressions that might interfere with the ability of the student to perceive the dynamics and continuity of the life-cycle process. A particularly important tool introduced in this unit is the project plan.

Unit 3, The Study Phase, introduces the first of the four life-cycle phases. This unit prepares the student to perform activities essential to identifying a computer-based business information system problem and recommending a solution. These activities include performing an initial investigation, defining system performance, conducting a feasibility analysis, and preparing a study phase report. Important features of this unit are the introduction of a continuing case study and an exhibit of a completed study phase report.

Unit 4, The Design Phase, teaches the student to perform basic computer-based system design tasks. These tasks include general system design, output design, input design, and file design, including an introduction to data base management systems. The continuing case study illustrates important design techniques and leads to an exhibit of a completed design phase report.

Unit 5, The Development Phase, acquaints the student with activities that must be undertaken to develop a system from a completed design. Two principal topics discussed are preparing for implementation and computer program development. Again, the case study is used to provide continuity and illustrative examples, including an exhibit of a development phase report.

Unit 6, The Operation Phase, makes the student aware of the operating environment of computer-based business information systems. Changeover, routine operation, performance evaluation, and management of change are described. The life-cycle process is reviewed within the context of the management of change, and the importance of management control and of documentation is reemphasized.

Important features of *Elements of Systems Analysis* are:

1. A preview of chapter and student learning objectives at the beginning of each chapter.
2. Presentation of **key terms** and their definitions at the beginning of each chapter and highlighting of these terms in boldface type in the text. Important *new terms* are indicated by italic type.
3. Questions for discussion and a summary of main points at the end of each chapter.
4. Careful adherence to the life-cycle "roadmap." Systems analysis activities, cumulative documentation, and critical management reviews are integrated throughout the four phases of the life cycle.
5. Inclusion of a detailed, continuous case study that provides examples, enabling the reader to follow the life-cycle "roadmap" for his or her own future applications. Particularly important are the examples of performance definition, feasibility analysis, and the study, design, and development phase reports.
6. A reinforcement learning process. Skills are introduced early in the text and then repeated and applied in greater detail in later chapters.
7. The availability of an integrated learning package for the use of those instructors who may wish to teach a course oriented toward measurable student performance. The learning package consists of the text, an instructor's manual, alternate student workbooks, and solution's manual for the workbooks. The second alternate workbook and its solution's manual are to be published in the spring of 1984. Each workbook contains assignments keyed to a case study that parallels the example case study in the text. The instructional supplement contains unit and chapter goals. In addition, for each chapter it contains:
   a. Student performance objectives
   b. Key points, indexed to page and figure number (transparency masters are provided for all important figures)
   c. Answers to "For Discussion" end-of-chapter questions
   d. Chapter quizzes and answers—two sets per chapter
8. Usefulness as a continuing reference. The text can be used not only for an introductory systems analysis course (or courses), but also as a guide for student projects of the type that are the "capstone" of many information systems curricula.

9. A first course leading to a career in systems analysis. Career opportunities for individuals skilled in systems analysis are extensive throughout modern businesses, and these opportunities are projected to increase. Systems analysts will be responsible for the orderly solution of business information system problems, including the effective use of computers. Systems analysts will become the information resource managers of the future.

10. Appropriateness to the needs of data processing majors as well as other business students. Programmers are not merely "coders." To develop effective programs, they must understand the system design process. Systems analysis has been, and will continue to be, the professional growth path most often open to programmers. To be able to take advantage of opportunities for promotion, programmers must prepare themselves through both experience and education.

No matter how completely a problem is defined to a programmer, additional decisions must be made based on an analysis of alternatives and their consequences. Systems analysis provides the methodology and tools for making such decisions.

11. Usefulness in educating another group of individuals—the largest group of all— users of computer-based business systems. A user may be defined as an individual who must provide input data to, or use output information produced by, a computer-based business system. Almost all business students will become users, since the trend of the 1980s is toward more end-user involvement. Factors creating this involvement are:

   a. Increasing use of distributed data processing systems to bring computer power to the location where the information is needed
   b. Development of "user-friendly" languages, including data base management systems with extensive query and report writing/display capabilities
   c. Availability of application packages and software-design services as alternatives to the use of in-house resources for system development.

In an even broader sense, a user is any individual who is affected to a significant degree by a computer-based information system. We need only switch on our TV set or go to the nearest bank or department store to realize that we all are information systems users. Familiarity with data processing and systems analysis principles will help us to become literate users of these systems.

We would like to acknowledge the constructive evaluation and suggestions of the following people who reviewed the manuscript for this third edition: George L. Miller, *Seattle Community College;* James R. Necessary, *Ball State University;* John A. Sharp, *Southern Illinois University at Edwardsville;* Norman E. Sondak, *San Diego State University;* and Mohan R. Tanniru, *University of Wisconsin at Madison.*

Marvin Gore
John Stubbe

Walnut, California

Elements of
Systems Analysis

Information Systems and the Systems Analyst

Unit 1

1
Information and Its Management

Preview

Computers have increased greatly in numbers and performance and decreased in cost since they were first used for business data processing in the early 1950s. Between 1950 and 1980 we progressed through three information eras: the Early Era, the Growing Era, and the Refining Era. Each era contributed to the evolution of business information systems and led to the identification of systems analysis as a structured method for solving buisness information system problems. We now have entered a fourth era, the Maturing Era, an age in which information is increasingly being acknowledged as a vital corporate resource. As the person who can create usable and productive business information systems, the systems analyst is assuming responsibility for information resource management.

Objectives

1. You will gain an understanding of important business information system concepts and will acquire an associated vocabulary.
2. You will learn the principal characteristics of each of the four information eras.
3. You will understand the reasons why difficulties were encountered in using computers successfully to solve business information systems problems during the first three eras.
4. You will learn why the Maturing Era, which will experience a continual explosion in information resources, may be called "the era of the systems analyst."

Key Terms

system   A combination of resources working together to convert inputs into outputs.

information   Refined data; a useful system output.

business information system   A system that uses resources to convert data into information needed to accomplish the purpose of the business.

systems analysis   A structured process for identifying and solving problems.

life-cycle methodology   A four-phase, systems analysis process for solving business information system problems.

systems analyst   An individual who performs systems analysis during any, or all, of the life-cycle phases of a business information system.

distributed data processing (DDP)   The ability to locate processing power wherever it is needed.

usability   The value of an information system as perceived by its principal users.

Business Information Systems

Systems Concepts

At present, computers and their applications significantly affect almost every aspect of modern business. Yet, by the end of the 1980s, industrial and financial institutions will be dominated by information systems with capabilities far beyond those visible today. Therefore, the study of systems analysis, which is a structured process for designing and developing effective business information systems, is becoming increasingly important. Certain concepts and definitions are necessary to our study of business information systems and to our understanding of the importance of systems analysis. Among these are: system, information, and business information system.

Generally we think of a system as being a group of related parts that work together as a unit. The definition of a system that we will use is based upon this idea; however, our definition is more specific to the study of business information systems. Hence, we define a **system** as a combination of resources working together to convert inputs into outputs. The conversion process is depicted by the symbols in figure 1.1. As this figure shows, the *resources* used by a system include personnel, facilities, material, and equipment. Usually we consider the inputs to a system to be unrefined, or raw, data. The outputs are refined, or useful, data that are called **information.** Information that is the output of one system may become an input to another system, since it must be further refined for other uses. Examples of this are decision support systems (DSS), which use summary information and projections to assist managers in planning and decision making.

A **business information system,** then, is a system that uses resources to convert data into information needed to accomplish the purposes of the business. An example of a business information system is a retail store system that converts sales transaction data into information used in the preparation of customer billings, inventory management, and calculation of profit and loss. Another example is a financial system that enables customers of a bank to enter personal identification data into a terminal in order to make deposits, transfer funds between accounts, or make cash withdrawals.

Computer hardware and software are examples of information system resources. *Hardware* refers to the physical components of a computer system, such as input, storage, processing, and output devices. Hardware is an equipment resource. *Software* is the collection of programs that facilitates the use of a computer. Typically, software resides on a magnetic medium, and it is an example of a material resource.

Business information systems usually are made up of smaller components that are, themselves, systems. These are called *subsystems,* since they are parts of the larger system. Figure 1.2 illustrates this concept for a product-oriented enterprise. Major systems usually present in such enterprises include marketing, finance, product development, and administration. As this figure also indicates, these major systems are composed of subordinate systems, or subsystems, which perform specific functions. One example is a marketing system that contains sales and distribution subsystems. Here an input to a sales system is typically a sales order, and an output is a shipping order. These "subsystems" are often conventionally referred to as "systems." Similarly, we often hear the term computer system (or computer data processing system) used to refer to the entire business information system. In this text we will consider the computer system (in other words, the computer hardware and supporting software) to be

Figure 1.1

A system definition. A system is a combination of resources working together to convert inputs into outputs. In a business system, which converts data into information, resources include personnel, facilities, materials, and equipment.

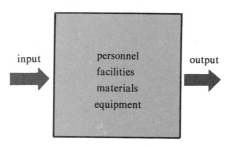

Figure 1.2

Systems and subsystems. Business information systems are usually large systems, themselves composed of small systems, called subsystems. A marketing system (a), a product development system (b), a finance system (c), and an administration system (d) are typical major systems that are made up of subsystems.

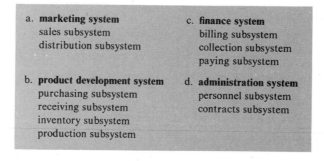

a. **marketing system**
 sales subsystem
 distribution subsystem

c. **finance system**
 billing subsystem
 collection subsystem
 paying subsystem

b. **product development system**
 purchasing subsystem
 receiving subsystem
 inventory subsystem
 production subsystem

d. **administration system**
 personnel subsystem
 contracts subsystem

a resource. It is an important resource that may support one or more business information systems. It is not, however, the business information system itself. In most such systems many functions that are outside of the computer "system" also rely upon personnel, facilities, material, and equipment resources.

Systems Analysis and the Systems Analyst

Systems analysis is a general term that refers to an orderly, structured process for identifying and solving problems. In this text we are concerned with business information system problems. We call the systems analysis process the **life-cycle methodology,** since it relates to four significant phases in the life cycle of all business information systems: study, design, development, and operation. The life-cycle methodology is introduced in the next chapter, and the four phases are covered in later chapters. Analysis implies the process of breaking something down into its parts so that the whole may be understood. The definition of systems analysis includes not only the process of analysis, but also that of synthesis, which is the process of putting parts together to form a new whole.

A **systems analyst** is an individual who performs systems analysis during any, or all, of the life-cycle phases of a business information system. The systems analyst not only analyzes business information system problems, but also synthesizes new systems to solve those problems or to meet other information needs.

The Productivity Challenge

Opportunity and Challenge

Computers have indeed created a revolution in the information resources available for increasing business productivity. As an example, for many years large capital investments have been made to apply computers to routine manufacturing tasks. Robot factory workers are becoming commonplace in modern industries such as automotive and electronics assembly plants, where large quantities of similar products are fabricated. Attention now is being focused upon improving the productivity of the office, which has changed little for many decades. The office of the future will be automated in a number of ways that will rely heavily upon computer technology.

A computer, however, is a sophisticated tool that must be applied with great skill if it is to be used effectively. Computer hardware has undergone spectacular improvements in performance, at the same time decreasing in cost and size. Although computer software developments tend to lag behind hardware advances, software effectiveness has also improved and will continue to do so in the decades to come. Nonetheless, hardware and software alone do not make a computer application successful. The ability to analyze business problems and to manage the development of complex, computer-based business information systems, in an environment where information resources are proliferating, are the challenges that accompany the opportunities of the future.

The early part of the 1980s is a good vantage point from which to review the past and to identify some of the major characteristics of computer hardware and software and their business applications that ushered in the information age. Four eras are shown in figure 1.3. These eras and the approximate intervals that they represent are:

Early Era	1940–1955
Growing Era	1955–1965
Refining Era	1965–1980
Maturing Era	1980–

The first three eras represent periods during which very important changes in computers and their commercial applications occurred. They summarize the past, which is a prologue to the future, represented by the fourth era. This era, the Maturing Era, will be an information age. Information will be acknowledged as a resource that must be managed in order to maintain and to improve the productivity of businesses.

Figure 1.3
The Information Eras and computer system performance. Computer system performance per dollar cost has increased throughout each information era. The Maturing Era, or Information Age will profit from information resource management from 1980 until at least the end of the century.

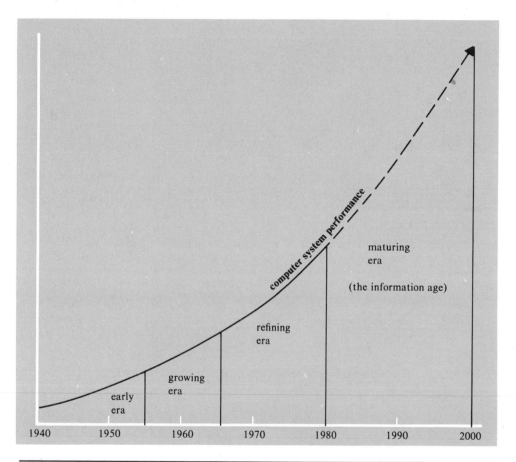

The Prologue

The Early Era: 1940–1955

The *Early Era* began with the development of the stored-program digital computer, so-called because it could store not only data, but also the sequence of instructions that controlled the processing operations to be performed upon the stored data. Characteristically, the hardware components of computers of this era were large. Part (a) of figure 1.4 shows a typical vacuum tube, which was the principal component of computers of the Early Era.

Figure 1.4
Progress in miniaturization of computer components. Three stages in the miniaturization of computer components extend from (a) vacuum tubes (Early Era) to (b) transistors (Growing Era) to (c) integrated circuits (Refining Era). The microelectronics techniques of the Maturing Era will continue to increase the complexity and reduce the size of electronic building blocks.

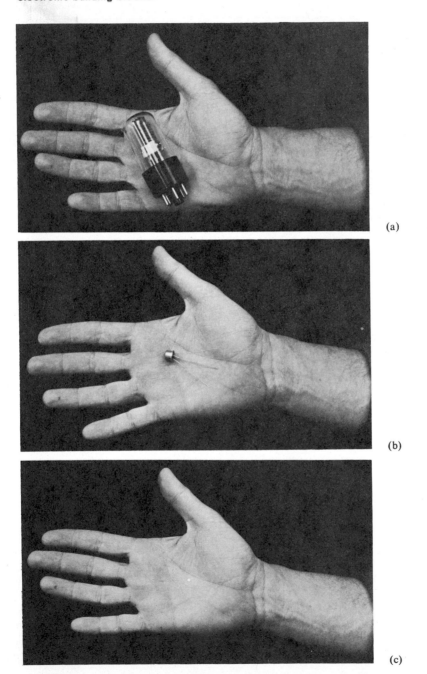

(a)

(b)

(c)

During this era, software was primitive and programming was both difficult and tedious. All characters had to be coded in machine language, which was based upon the binary digits "0" and "1." Throughout most of the Early Era digital computers were used for scientific computation. Almost all business data processing was performed on electro-mechanical equipment, called unit record machines. Punched cards, coded to represent physical units of data, were carried from machine to machine for specialized operations, such as sorting, reproducing, collating, calculating, and printing.

However, toward the end of the Early Era the UNIVAC, an acronym for UNI-Versal Automatic Computer, was introduced by the Remington Rand Corporation. UNIVAC was initially used by the Census Bureau and was the first digital computer to be produced in quantity for commercial use. Shortly thereafter IBM and other manufacturers introduced competitive computers, and business uses for these machines began to grow rapidly. However, procedures for the effective use of digital computers to solve business problems were not in evidence during this era.

The Growing Era: 1955–1965

The *Growing Era* produced many important improvements in computer hardware. The most significant hardware development involved replacement of the bulky vacuum tube with the transistor. In fact, the transistor was developed at the Bell Telephone Laboratories in 1948; however, the use of transistors in computers did not become widespread until the Growing Era. As figure 1.4b shows, transistors were much smaller than vacuum tubes. They also were much more reliable, and they were used in large numbers to make special purpose computers—some to solve scientific problems and others to perform business data processing. At the close of the Growing Era, IBM introduced its System 360 (for each of the degrees of the compass) series of computers. These were general purpose computers, and they could be used for both scientific and business applications. Thereafter, the large majority of computers produced were designed to be general purpose machines.

Significant software progress was also made, particularly in the development of procedural languages, which used English-like symbols instead of "0's" and "1's." These languages greatly assisted programmers in writing computer programs and in communicating with the users of those programs. Most of them are in use today. For example, FORTRAN (FORmula TRANslation) is a procedural language used extensively for mathematical computation. FORTRAN may be used for commercial applications; however, a more suitable language, COBOL (COmmon Business Oriented Language) was developed for commercial data processing. Statements such as MULTIPLY HOURS BY RATE GIVING GROSS-PAY can be written in COBOL. Clearly, this type of statement is meaningful and easy to remember. Another business-oriented language developed during this period was called RPG, which stands for Report Program Generator. It is useful in preparing printed reports, which were and remain a principal output of most business systems.

Throughout most of the Growing Era, the primary method for processing data, such as sales transactions, was to collect records in "batches" and transport them to a computer location where data were entered onto punched cards for processing at a later time. This method of data processing was called batch processing. Also, through-out this era, difficulties were encountered in developing all but the most simple business applications. Not surprisingly, the principal causes of these difficulties were:

1. Lack of management acceptance. Computers were not accepted by managers. At best they were tolerated.
2. Lack of managerial knowledge. The new technology was not understood. There was a reluctance to apply standard management techniques because of the computer "mystique."
3. Lack of standards. There were no historical standards by which computing efforts could be measured.
4. Lack of stability. The computer sciences, equipment, languages, and applications were undergoing continual change.

The Refining Era: 1965–1980

The *Refining Era* was an explosive period in the history of computers and data pro-cessing. It began with the dominance of large, central-site computer installations and ended with "a computer revolution within the computer revolution." This internal rev-olution came about because of the electronics industry's ability to miniaturize the elec-tronic building blocks of computers. Individual transistors, which had replaced vacuum tubes, were, in turn, replaced by *integrated circuits,* called "IC's." The IC's were the result of an evolving microelectronics technology capable of forming thousands of tiny transistors and other circuit elements upon small wafers of silicon, called "chips." Fig-ure 1.4c shows a typical chip, which is about one-fourth the size of a postage stamp. IC's made possible order of magnitude improvements in the performance and relia-bility of digital computers at greatly reduced costs. By the end of the 1970s, large-scale integrated circuit (LSI) technology (microelectronics) made it possible to place an entire computer (that is, central processing unit and memory) upon a single silicon chip. Figure 1.5 shows a single-chip computer with more processing power than a com-puter of the mid-1960s, which used several desk-size components.

The IC technology not only made possible very powerful central-site computers, but also led to widespread growth in *distributed data processing* (DDP), which is the ability to locate processing power wherever it is needed. With the development of com-munications networks, distributed systems could be located at users' locations and could operate independently or be linked to a central-site computer.

Another major development in the application of microelectronics was the pro-duction of low-cost computers, called *microcomputers,* for personal computing and for small business applications. Computer systems of all sizes began to proliferate as com-puting power became available at a fraction of its cost less than a decade ago. It soon became difficult to classify computer systems because of overlapping capabilities and costs. The selection of an appropriate computer system became increasingly applica-tion dependent, thus adding to the complexity of business information system design and development. Computer systems were commonly described as micro, mini, midi, or maxi. Two broad classifications of computers, which we will also use, were small and large. When we refer to "small" computer systems, we will mean "micro" or "mini"

Figure 1.5
Computer on a chip. Progress in miniaturizing computer circuits made it possible to place electronic circuits for the central processing unit and the memory of an entire computer upon a single, small chip of silicon.

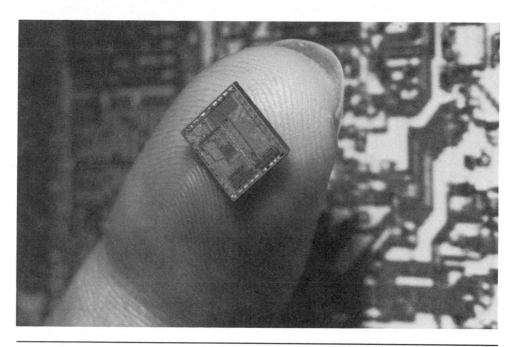

machines. When we refer to "large" systems, we will mean "midi" or "maxi" computers. Figure 1.6 distinguishes between computer systems on the basis of approximate cost. Figure 1.7 depicts the dramatic increases projected in the number of small and large computer systems in the United States. By the mid-1970s it became evident that, in the future, the number of small computers would by far exceed the number of large systems.

Software developments also were significant, particularly for large computer systems. These included advanced operating systems, which were collections of general purpose programs designed to assist programmers in debugging and testing applications programs, in handling repetitive input and output operations, and in working with a variety of devices. Nonetheless, throughout most of the Refining Era, major difficulties continued to be experienced in developing business information systems. The principal causes of these difficulties were: (1) underestimating the complexity of many business information systems; (2) the inability of computer programmers and users of concept programs to communicate in a common language; and (3) lack of a structured procedure, or methodology, for developing complex systems.

Figure 1.6
Small and large computer systems. Four common classifications of computer systems are micro, mini, midi, and maxi. They represent overlapping ranges in computer system performance and cost.

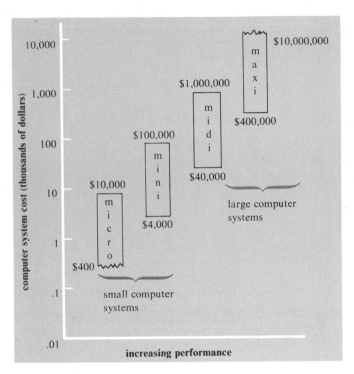

Toward the end of the Refining Era, however, greater success was achieved with business information systems. Three factors were responsible. First, there was a growing management dissatisfaction with the return on computer investment, forcing management education and participation. Second, systems analysis techniques that had been developed in high-technology industries for the management of large computer-related projects became more widely available. Third, a higher place in the corporate organization was created for computer-based activities because of their interaction with all business operations. As a result, individuals with both a computer and an applications background began to emerge and assume responsibility for the analysis, design, and development of business information systems. These persons were called systems analysts, and by the end of the era the profession of systems analysis was well established.

Figure 1.7
Growth trends in numbers of computer systems. As shown by the logarithmic (power of ten) scale, the numbers of computer systems will increase throughout the Maturing Era, with the number of small systems greatly exceeding the number of large systems.

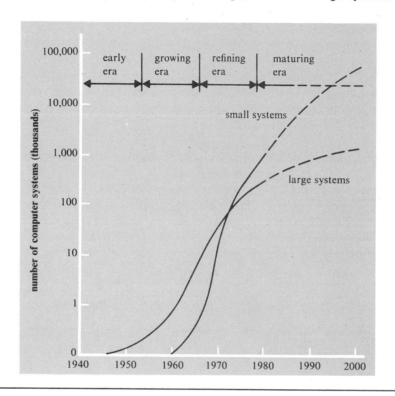

Information Resource Management

The Maturing Era: 1980–

The explosive growth in computing power experienced in the 1970s is but the fore-runner to even more extensive changes that will occur throughout the 1980s and beyond. Considerations that will be of paramount importance to business information systems throughout the Maturing Era include:

1. *Productivity*   Although the cost of information processing is decreasing, labor and management costs are increasing.
2. *Involvement*   Users of computer-related information systems will become more involved in deciding how their information needs are to be met.
3. *Usability*   **Usability,** which is the value of information systems as perceived by their principal users, will become the measure of the success or failure of such systems.

4. *Decisions*   Decision makers at all levels will rely more and more on computer-generated information.
5. *Services*   Information service organizations will assume broader responsibilities in order to meet both local and remote information needs in distributed data-processing environments.
6. *Communications*   The expansion of communications capabilities and services will be a phenomenon of the Maturing Era, supporting the convergence of data processing and office automation.
7. *Life cycle*   The life-cycle methodology for the management of business information system projects will become part of the training and experience of all persons involved with the design and development of business information systems.

There already is considerable evidence that hardware improvements will continue throughout the Maturing Era. Large scale integrated (LSI) circuits will be further reduced in size. VLSI (very large scale integrated) and VLSII (very large scale integrated, indeed!) circuits will become commonplace as advances in mircoelectronics technology continue to reduce the cost and increase the performance of computer hardware. Advanced memory techniques, such as the use of electronic circuits operating at temperatures close to absolute zero, will even further increase computer processing speeds. A multiplicity of devices will become available for processing data and manipulating information in many forms.

Software developments that enhance information system usability and contribute to improved productivity will be particularly important during this era. Some examples of this type of software, which were introduced during the Refining Era and which will continue to mature, are:

1. Advanced data base management systems
2. Nonprocedural languages
3. Applications packages
4. Structured software development aids

An important characteristic of integrated information systems is that their input data and output information needs do not necessarily conform to organizational boundaries. Rather, these systems rely on the concept that data need be recorded only once and that, thereafter, they can be accessed and used for many purposes. Such data, when captured and stored on a suitable device, such as a magnetic disk, are said to comprise a "data base." This data base, then, may be shared by independent computer applications programs. Special purpose computer programs, designed to create, access, and update data bases, are called *data base management systems (DBMS)*. Future data base management systems will provide extensive capabilities of this type for a multiplicity of users.

The availability of *nonprocedural* languages will increase. These are called "user-friendly" languages since they do not require the user to be knowledgeable about computer program logic or procedures. Instead, they are written to assist users to apply programs, often providing directions and prompting in English. Query languages, which are associated with DBMS, are examples of nonprocedural languages.

The market for well-designed, documented, and supported *applications packages* will continue to grow because these are general purpose programs that can enable many organizations to acquire information systems quickly and to reduce their investment in and dependence on large staffs of programmers. Also, many structured software development aids will become available. These are computer programs written to assist systems analysts in performing and documenting all of the activities associated with structured, life-cycle processes for computer-based business information system design and development.

Information: A Corporate Resource

All of the foregoing considerations emphasize the growing awareness that information is a vital corporate resource and that, as such, it must be managed as skillfully as any other important resource. The importance of managing information gathering, processing, storage, and dissemination resources in increasingly complex business environments was recognized by the Association for Systems Management (ASM), which adopted a program, called "Project 80s." This project identified four general classes of information technology: data, text, voice, and image, which would require management by systems professionals. Figure 1.8 illustrates how the data processing, automated office, and communications disciplines are merging and creating an overall information resource management need.

Three important applications areas that illustrate the complexity of future computer-based systems and the need for information resource management are: distributed data processing, the automated office, and decision support systems. **Distributed data processing** systems are computer-based information systems that place data processing power wherever it is needed. Users' demands for DDP systems are increasing because these systems provide flexibility and quick response for local operations. Also, through user involvement, they tend to increase the usability of the information generated. However, the expense of distributing the managerial and technical resources required to support DDP must be balanced against the benefits.

Extensive *office automation* will be one result of the convergence of the data processing, word processing, and communications technologies. To the data processing-oriented concepts of input, process, storage, and output, we need add only the communications-oriented concept of distribution to visualize the functions that will be performed by the automated office. Word processing machines, with extensive text editing capabilities, will become small computers, and storage will assume some of the characteristics of data bases. Input will be by means of keyboards, optical character readers (OCR), and, in some instances, by voice. Output, called reprographics, will involve all of the reproduction and document handling tasks that extend from the word processor to the point at which the document is to be distributed. Distribution methods will include facsimile transmission, telecommunications, and electronic mail. Figure 1.9 depicts an information network that might be typical of many corporate offices by the late 1980s.

In the 1970s, much attention was focused upon *management information system (MIS)* development. These systems, which refined information developed to meet operational needs, were designed to provide managers with information needed for effective planning, control, and decision making. In the 1980s, this emphasis will continue

Figure 1.8
Convergence of information technologies. The need for information resource management is emphasized by the convergence of data, text, voice, and image technologies in business applications of data processing, word processing, and communications.

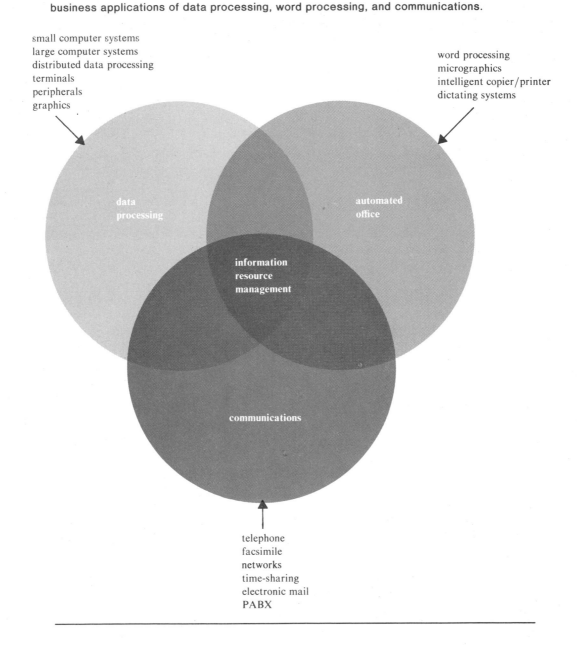

small computer systems
large computer systems
distributed data processing
terminals
peripherals
graphics

word processing
micrographics
intelligent copier/printer
dictating systems

data processing

automated office

information resource management

communications

telephone
facsimile
networks
time-sharing
electronic mail
PABX

Figure 1.9
An automated office of the 1980s. The office of the future will be highly automated.
Typically, a corporate headquarters of the late 1980s will include executive work
stations, word processing/data entry stations, data storage (micrographics) stations,
data reproduction stations (intelligent copiers), and centralized data processing services.
The "hub" of the office will be the corporate communications system—linked to a
satellite communications network—which will provide for teleconferencing, electronic
mail, and communications with remote sites, some of which may be offices in a briefcase.

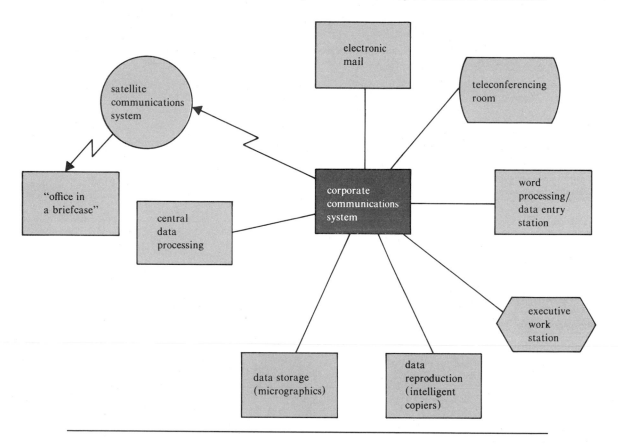

in order to improve the productivity of managers of all levels. These systems will be
designed to provide timely information for decisions, which, typically, might relate to
allocation of capital, new product development, acquisition analysis, and market strategy. Collectively, these sophisticated systems will be called *decision support systems
(DSS),* and their development will be a challenge to systems professionals and to the
managers of corporate information service organizations.

Information Resource Management and the Systems Analyst
Two major purposes of our discussion of the four information eras were (1) to emphasize the growing dependency of businesses upon computer-related information systems, and (2) to establish the background for our study of systems analysis, which is

a means of developing usable and productive information systems. Each of the information eras that we described was dominated by particular technologies. The technologies of each era, often greatly enhanced, carried forward to subsequent eras. Each also added an additional dimension to the use of machines by humans to solve business problems. The Early Era focused upon hardware design and was an era in which business applications of computers were very limited. The Growing Era introduced many important programming tools, and these significantly improved the ability of humans to communicate with machines. However, throughout this era communication between programmers and users was difficult and, as a consequence, the usability of many business information systems was poor. During the Refining Era, structured techniques for managing information system projects were introduced. The value of systems analysis was recognized, and an important role began to emerge for the systems analyst.

Throughout the Maturing Era, the role of the systems analyst will become even more important as the contention for information resources increases and as productivity is emphasized. In the Maturing Era, the major tasks of the data processing industry will not be the design of more effective computers or computer programs. Rather, it will be to improve (1) the quality of information services as perceived by the users of such services; (2) the productivity of end-users; and (3) the management of data processing systems and applications efforts. *Information resource management* will thus be a career path for the systems analyst trained in the life-cycle methodology.

Summary

We are entering an information age, one in which the management of the information resources of a corporation will be of vital importance. Business information systems are systems that use these resources to convert data into information in order to improve productivity. Business information systems usually are composed of smaller systems, called subsystems. Computer hardware and software are important resources that support business information systems and subsystems.

Systems analysis is a general term that refers to an orderly, structured process for solving problems. This process, when applied to business information systems, is called the life-cycle methodology. Four phases—study, design, development, and operation—make up the life cycle of computer-related business systems. A systems analyst is a person who performs systems analysis during any or all of the life-cycle phases. The systems analyst not only analyzes business information system problems, but also synthesizes new systems to solve these problems.

The four information eras are: the Early Era (1940–1955), the Growing Era (1955–1965), the Refining Era (1965–1980), and the Maturing Era (1980–   ). The Early Era concentrated on hardware, and human-machine communication was very difficult. The Growing Era improved this communication through the introduction of English-like programming languages; however, techniques for managing computer-related projects were lacking. During the Refining Era, explosive growth occurred in the development of large (midi and maxi) and small (micro and mini) computer systems and in their applications. Developments in microelectronics technology contributed significantly to this growth. Throughout most of the Refining Era, in spite of a proliferation of applications, difficulties were encountered in using computers to solve business problems. However, toward the end of this era, a structured systems analysis process—called the life-cycle methodology—came into increasing use as a means of developing usable business information systems. Structured techniques for the analysis, design, and development of computer-related information systems will be enhanced in the Maturing Era. These techniques will be used to develop information systems in applications areas such as distributed data processing, the automated office, and management-decision support. This will be an era in which information will be acknowledged as an important corporate resource. The systems analyst will assume an important role in managing the information resources of the corporation.

For Review

system
information
hardware
subsystems
life-cycle methodology
Early Era
Refining Era
integrated circuit (IC)
data base management
  systems

applications package
automated office
decision support
  system
usability
resource
**business information
  system**
software
systems analysis

systems analyst
Growing Era
Maturing Era
microcomputer
non-procedural
  language
**distributed data
  processing**
management
  information system
information resource
  management

For Discussion

1. Define: system, information, and business information system. Give an example of each.
2. What is the difference between a system and a subsystem?
3. What type of resource is computer hardware? Software?
4. Distinguish between systems analysis and the life-cycle methodology.
5. Use the terms "analysis" and "synthesis" to describe what a systems analyst does.
6. Describe, briefly, the principal characteristics of the following eras as they relate to business information systems:

   Early Era
   Growing Era
   Refining Era
   Maturing Era

7. Why have difficulties been encountered in developing computer-based systems in the past? What factors lead to greater success in the latter part of the Refining Era?
8. What is the productivity challenge of the Maturing Era?
9. Why may the systems analyst become the life-cycle manager of the Maturing Era?
10. Discuss the importance of "usability" as it applies to computer-based business information systems.

2
Life Cycle of the Business Information System

Preview

Systems are dynamic. They are characterized by a life cycle that has four distinct phases: study, design, development, and operation. Three important activities must take place concurrently throughout all phases of the life cycle of a computer-related business information system. These are (1) performance of tasks; (2) user and management reviews; and (3) cumulative documentation. Relationships between the life-cycle phases and the ongoing activities are explained by a means of a life-cycle flowchart.

Objectives

1. You will be able to visualize the life cycle and to understand how the concept is applied to the management of business information system projects.
2. You will learn the principal characteristics of each of the four life-cycle phases.
3. You will become familiar with the life-cycle flowchart and will understand why a systems analyst often can be referred to as a life-cycle manager.

Key Terms

life cycle   A concept that describes the relationships between the four phases of a business information system.

life-cycle phase   One of the four life-cycle phases: study, design, development, and operation.

study phase   The life-cycle phase in which a problem is defined and a system is recommended as a solution.

design phase   The life-cycle phase in which the detailed design of the recommended system is specified.

development phase   The life-cycle phase in which the system is constructed according to the design phase specification.

operation phase   The life-cycle phase in which the system is installed, operated, and maintained.

baseline specification   A reference for system maintenance and change. The principal baseline specifications are the performance specification, the design specification, and the system specification.

systems analysis   The performance, management, and documentation of the activities related to the four phases of the life cycle of a business information system.

systems analyst   A person who performs systems analysis during any, or all, of the life-cycle phases of a business information system. A life-cycle manager.

The Life-Cycle Concept

Systems are created by a dynamic process that moves through a series of stages, or phases. The concept of a **life cycle** has evolved to describe the relationship between these phases. This concept includes not only forward (in time) motion, but also the possibility of having to return, that is, cycle back to an activity previously considered completed. This cycling back, or *feedback,* may occur as the result of the failure of the system to meet a performance objective or as the result of changes in or redefinition of system objectives.

Hardware and Software End-Products

Practical techniques for the management of relatively complex projects have been developed from the life-cycle concept. However, in the past, most such projects have had as their objective the creation of hardware end-products. These *hardware end-products* are described primarily by their physical, for example, electrical or mechanical, attributes, which can be observed and measured as the hardware moves through the several stages of its development. Examples of complex systems that have been developed by the application of life-cycle techniques are rockets, communications networks, computer systems, and spacecraft.

The computer-based business system also contains hardware components; however, its most significant characteristic is a *software end-product.* Software may be defined as a collection of programs or routines that facilitates the use of a computer. This definition includes operating systems, which facilitate the general use of computers, and application programs, which are written to solve specific problems. The latter is the end-product associated with a computer-based business information system. Software, in contrast with hardware, does not possess attributes that can readily be observed and measured from concept to end-product. The software end-product is information. Although it may be stored or printed on a physical medium, such as a magnetic disk, a reel of tape, or a sheet of paper, information is transient and fragile compared with hardware.

Many of the past difficulties in developing effective computer-based business systems stemmed not only from belated efforts to apply management controls, but also from failure to recognize that techniques applicable to the development of hardware end-products could not be applied without modification to the development of software end-products. However, as a result of experience gained from large government and commercial software projects in the latter part of the 1960s and throughout the 1970s, the concept of life-cycle management was adapted to fit the development of computer-based business systems.

The key to modifying the life-cycle concept for the management of software projects was the recognition that, although supporting documentation *accompanies* a physical product throughout its development, documentation *is* the software product. This distinction is made clear in figure 2.1.

Figure 2.1

Distinction between hardware and software end-products. As contrasted with hardware, for which both documentation and a product are visible throughout the product development cycle, the visible evidence of a software product is documentation, without which there is neither progress nor product.

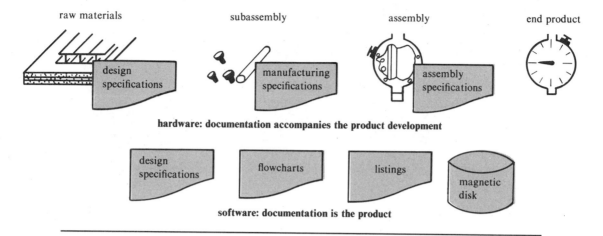

raw materials   subassembly   assembly   end product

design specifications   manufacturing specifications   assembly specifications

hardware: documentation accompanies the product development

design specifications   flowcharts   listings   magnetic disk

software: documentation is the product

Life-Cycle Phases and the Life-Cycle Manager

Like most systems, the life cycle of a computer-based system exhibits distinct phases. These are:

1. The study phase
2. The design phase
3. The development phase
4. The operation phase

All of the activities associated with each **life-cycle phase** must be performed, managed, and documented. Hence, we now define **systems analysis** as the performance, management, and documentation of the activities related to the four life-cycle phases of a computer-based business system. This is an extension of the general definition, given in the previous chapter, of systems analysis as a structured process for identifying and solving problems. Similarly, we now can identify the **systems analyst** as the individual who is responsible for the performance of systems analysis for all, or a portion of, the phases of the life cycle of a business system. The analyst is, in effect, a **life-cycle manager.**

The Life-Cycle Activities

The Life-Cycle Flowchart

The principal focus of this book is on the operations associated with performing, reviewing, and documenting all the life-cycle activities of a computer-based business system. In order to view these life-cycle activities in perspective, let us consider figure 2.2.

Figure 2.2
The life cycle of a computer-based business information system. Performance of tasks, cumulative documentation, and review by user-management are activities that parallel forward progress through the life-cycle phases: study, design, development, and operation. As the feedback paths indicate, return to an earlier point in the life-cycle process may occur at any time.

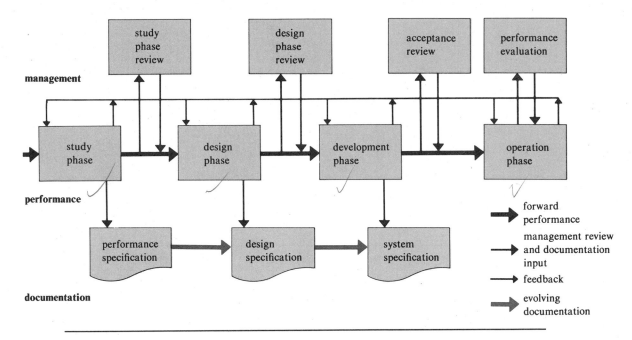

This figure is a pictorial overview with which we will become increasingly familiar; we will refer to it to mark our progress through this book. The pictorial representation used in figure 2.2 is called a flowchart. *Flowcharts* use predefined symbols to represent operations or information flow in a business system. The flowchart symbols used in figure 2.2 will be defined in context as we discuss performance, management, and documentation of the life-cycle activities.

Performance of the Life-Cycle Tasks
The life-cycle methodology for developing complex systems is a modular, top-down procedure. In the study phase, modules that describe the major functions to be performed by the system are developed. The procedure is called *top-down* because in successive phases the major modules are expanded into additional, increasingly detailed, modules. Powerful graphic tools have been developed to structure the top-down design and development of systems. We shall learn about these as we study the design and development phase activities in detail. For the present, we can summarize the principal tasks associated with each of the phases of the life cycle of a computer-based business system as follows:

The Study Phase

This is the phase during which a problem is identified, alternate system solutions are studied, and recommendations are made about committing the resources required to design the system. Tasks performed in the **study phase** are grossly analogous to (1) determining that a shelter from the elements is needed, and (2) deciding that a two-bedroom house is a more appropriate shelter than a palace, a cave, or other possible selections.

The Design Phase

In this phase the detailed design of the system selected in the study phase is accomplished. This is analogous to drawing the plans for the two-bedroom home decided on in the study phase. In the case of a computer-based business system, **design phase** activities include the allocation of resources to equipment tasks, personnel tasks, and computer program tasks. In the design phase, the technical specifications are prepared for the performance of all allocated tasks.

The Development Phase

This is the phase in which the computer-based system is constructed from the specifications prepared in the design phase. Equipment is acquired and installed during the **development phase.** All necessary procedures, manuals, software specifications, and other documentation are completed. The staff is trained, and the complete system is tested for operational readiness. This is analogous to the actual construction of our two-bedroom house from the plans prepared in its design phase.

The Operation Phase

In this phase, the new system is installed or there is a changeover from the old system to the new system. The new system is operated and maintained. Its performance is reviewed, and changes in it are managed. The **operation phase** is analogous to moving into and living in the house that we have built. If we have performed the activities of the preceding phases adequately, the roof should not leak.

The symbol used in figure 2.2 to represent the life-cycle phases is called a "process" symbol. Processes are similar to systems because processes are operations that convert, or transform, inputs into outputs. Thus, the study phase consists of all the processes required to produce an output that is identified on the flowchart of figure 2.2 as a performance specification. The performance specification is represented by the symbol for a "document."

Review of the Life-Cycle Activities

Management review of the life-cycle activities may occur at any time. However, the conclusion of each phase is a natural time for a major management review. The major management reviews are shown by the process symbols at the top of figure 2.2. These are formal scheduled reviews that must occur before a phase can be considered complete. They are essential to a structured interaction between the system analyst and the user, ensuring user involvement at critical decision points. Three types of decisions can be forthcoming at each review: (1) proceed to the next phase; (2) cancel the project; or (3) redo certain parts of a previous phase.

Activities that are redone must be reviewed before the project can proceed to a subsequent phase. Management review often is the mechanism that triggers "cycling back" (feedback) to an earlier state in the life cycle, to remedy performance deficiencies, or to respond to changes in requirements. Each successful review is a renewal of management commitment to the project.

Documentation of the Computer-Based Business Information System

The accumulation of documentation parallels the life-cycle performance and management review activities. Documentation is not a task accomplished as a "wind-up" activity; rather, it is continuous and cumulative. *Cumulative documentation* is implied by the colored lines and arrowheads of figure 2.2. The most essential documents are called **baseline specifications** (that is, specifications to which changes can be referred). There are three baseline specifications:

1. *Performance specification*   Completed at the conclusion of the study phase, and describing in the language of the user exactly what the system is to do. It is a "design to" specification.
2. *Design specification*   Completed at the conclusion of the design phase, and describing in the language of the programmer (and others employed in actually constructing the system) how to develop the system. It is a "build to" specification.
3. *System specification*   Completed at the conclusion of the development phase and containing all of the critical system documentation. It is the basis for all manuals and procedures, and it is an "as built" specification.

The design specification evolves from the performance specification, and the system specification evolves from the design specification. And, since these documents are the only measurable evidence that progress is being made toward the creation of a useful software end-product, it is not possible to manage the life-cycle process without them. Thus, documentation is not only the "visible" software end-product, but also the key to the successful management of the life cycle of computer-based business systems.

Summary

The life-cycle concept evolved to describe not only the four phases through which systems move, but also the relationships between those phases. The distinction between hardware and software end-products led to the identification of documentation as the visible evidence of a software end-product, such as a business information system.

The life-cycle phases of a computer-related business information system are:

1. The study phase, in which a problem is identified, alternate solutions studied, and system recommendations made
2. The design phase, in which detailed design of the system selected in the study phase takes place
3. The development phase, in which the system is constructed from the design specified in the design phase

4. The operation phase, in which the developed system is installed, operated, and maintained

At the end of each phase a review by involved managers and users is required before proceeding to the next phase. Documentation parallels the performance of the life-cycle tasks; it is cumulative throughout the life-cycle phases. The most essential documents are called baseline documents, and they are:

1. The performance specification, created during the study phase
2. The design specification, created during the design phase
3. The system specification, created during the development phase

In terms of these concepts, we can now define systems analysis as the performance, management, and documentation of the life-cycle phases of a computer-based business information system. The systems analyst can be considered to be not only an information resource manager, but also a life-cycle manager.

For Review

life cycle	**study phase**	**performance**
feedback	**design phase**	**specification**
hardware end-product	**development phase**	**design specification**
software end-product	**operation phase**	**system specification**
flowchart	cumulative	**systems analysis**
top-down	documentation	**systems analyst**
life-cycle phase	**baseline specification**	life-cycle manager

For Discussion

1. Describe the life-cycle concept.
2. Distinguish between hardware and software end-products and explain how this distinction relates to the development of computer-based business information systems.
3. Name and describe each of the four phases of the life cycle of a business information system.
4. How is systems analysis defined in this chapter?
5. Why is the life-cycle methodology a top-down approach to system design and development?
6. What is the meaning of the term "baseline document"? Name and describe the three baseline documents.
7. What does "cumulative documentation" mean? Why is it an important life-cycle activity?
8. Who participates in the review of the life-cycle activities?
9. Why can the systems analyst be called a life-cycle manager?
10. Why can the systems analyst often be considered to be an information resource manager as well as a life-cycle manager?

3
The Role of the
Systems Analyst

Preview

The role of the systems professional has changed as business information systems have evolved from manual systems to modern, computer-oriented systems. Computer-related systems of the 1970s are being broadened in scope by the emerging technologies of the 1980s. This broadening will increase the importance of systems analysis and will elevate the reporting level of the systems group within the corporate organization. Through a study of business organization charts, we can better understand the functions of systems analysis and the requirements for successful careers in the profession.

Objectives

1. You will learn about the evolution of business information systems and the reasons for the transition from traditional computer-oriented systems to nontraditional systems.
2. You will be able to understand and to prepare business organization charts and organization function lists.
3. You will learn the major functions performed by the systems group and where that group is placed on the organization chart.
4. You will become familiar with the personal qualifications, education, and experience required for a successful career in systems analysis.

Key Terms

manual system   A system designed around a pattern of manual operations.

traditional computer-oriented system   A computer-oriented system with data processing as the focal point.

nontraditional computer-oriented system   Computer-oriented system broader in scope than data processing.

organization chart   A flowchart that identifies organizational elements of a business and displays areas of responsibility and lines of authority.

organization function list   A document that describes the major activities performed by each organization shown on an organization chart.

The Evolution of Computer-Oriented Information Systems

Manual Systems

Throughout the *Early Era* (1940–1955), business systems were **manual systems** designed around a pattern of manual operations performed by people who were the principal system resources. Although unit record equipment was used to perform business data processing operations, these applications, which were primarily financial, were designed to emulate a sequence of manual tasks. The systems analysts of that era were called systems and procedures persons, and most of them had limited contact with the electromechanical, punched-card machines of the era. Also, as contrasted with the modern business environment, the systems effort spent upon data processing applications was relatively small.

Computer-Oriented Systems

During the *Growing Era* (1955–1964), after computers were introduced for business data processing in the mid–1950s, computer-oriented systems began to grow in number. These were systems in which the capabilities of computers and humans were blended to take advantage of the unique capabilities of each. As a result, major changes began to occur in the emphasis of systems activities and in the education and training required for entry into the profession of systems analysis.

The design and development of effective computer-oriented systems proved to involve far more than getting the computer to perform a series of operations in the same manner that a human would. It was learned that, although they could perform many repetitive tasks, computers could do much more. Thus, throughout this era, a reorientation of systems thinking was required in order to take advantage of the special capabilities of computers. Computer-oriented systems were found to differ from manual systems in at least four important capabilities:

1. *Accuracy*   There is a greater potential for accuracy. Once the data is entered correctly into the system in a machine-readable format, it is not necessary to reenter it again. This reduces the chances for error by reducing the number of times humans are involved. Of course, the computer system is vulnerable to the entry of invalid data.
2. *Data collection and communication*   Methods for collecting and communicating data are faster and more efficient. Modern computers allow data communication networks to be established to collect data and to respond to inquiries. An airline reservation system is an example of a network of data communication sites tied into a central data processing center. A reservation clerk may inquire about space availability on any flight and receive an almost immediate response. Seat reservations can be made and confirmed while the customer waits. The communication sites may be nearby or remote.
3. *Data storage*   Another way in which manual and human-machine systems differ is in the quantities of data that can be stored and accessed. In computer-oriented systems, data is kept in master files, usually magnetic, in a machine-readable format. A collection of related files forms a data base. Data bases allow for the centralized storage of data, thereby eliminating the need for multiplicity of redundant files. Data base systems require special measures to prevent their contamination by bad data.

4. *Speed of response* Speed of response—the time required for information to become available—can be greatly increased by use of computers. Retail operations provide an example of the capabilities of fast-response systems. Specially designed cash registers not only perform the functions of traditional cash registers, but also serve as data terminals. As terminals, they become part of a "real-time" system that can perform such functions as credit checks, search for lost credit cards, accounting for cash and credit transactions, and inventory management. These systems are able to make information available as needed rather than after the close of an accounting period of perhaps weeks or months.

Traditional and Nontraditional Systems

The Refining Era (1965–1980) witnessed the introduction of life-cycle techniques for designing and developing effective computer-oriented business information systems. Initially, these systems were designed to meet the financial needs of corporations; however, by the end of the era they also supported nonfinancial applications, such as marketing, production, and general administration. Although it was not unusual to see the power of computers wasted on systems that were "manual" in concept, the transition to a data processing orientation was widespread by the end of the era. In 1981 the Association for Systems Management completed a survey of the major activities of systems groups within major industries.[1] The results are shown in figure 3.1. Although all of the activities listed showed significant participation, two activities were identified by over 80 percent of the respondants. These were: (1) data processing, including analysis, programming, and management, and (2) methods and systems analysis. Data processing showed the most significant gain on the basis of comparisons with previous surveys.

In the information-technology dominated systems world of the *Maturing Era* (1980–   ), a more important consideration than "manual" versus "computer" is emerging. This consideration is the transition from **traditional computer-oriented business information systems** to **nontraditional systems.** Data processing was the focal point of the traditional systems of the 1970s. The nontraditional systems of the 1980s are broader in scope, make use of an expanding array of information resources, and demand more of the systems professional. An example is the emergence of the *automated office,* which involves the combination of data processing, word processing, and communications technologies. Other examples are systems that rely extensively on distributed data processing and systems that must be flexible enough to assist managers to meet not only defined information needs, but also to explore solutions to new needs as they arise.

The Functions of Systems Analysis

The basic functions performed by systems analysts have not changed as a result of the changes in technology and in business information needs. These functions are:

1. To analyze business systems with problems and to design new or modified systems to solve those problems

---

1. *Profile of the Systems Professional* (Cleveland: Association for Systems Management Bookshelf Series, 1981).

Figure 3.1
Major systems activities. Data processing and systems analysis are the activities of systems professionals identified by most businesses. These activities will continue to increase in importance because of the continuing need for effective computer-related business information systems.

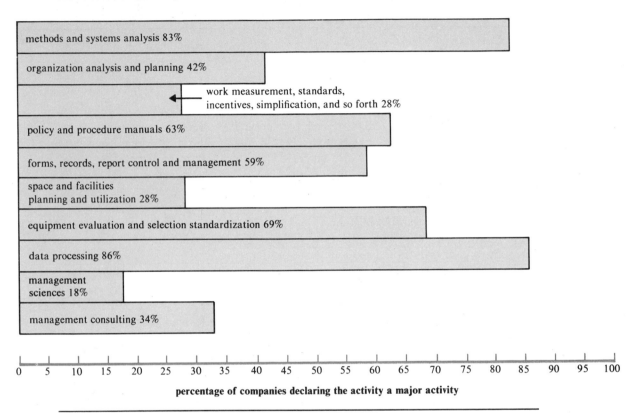

methods and systems analysis 83%

organization analysis and planning 42%

work measurement, standards, incentives, simplification, and so forth 28%

policy and procedure manuals 63%

forms, records, report control and management 59%

space and facilities planning and utilization 28%

equipment evaluation and selection standardization 69%

data processing 86%

management sciences 18%

management consulting 34%

0 5 10 15 20 25 30 35 40 45 50 55 60 65 70 75 80 85 90 95 100

percentage of companies declaring the activity a major activity

2. To develop business systems to meet new information or operational needs
3. To prepare and maintain manuals to communicate company policies and procedures
4. To design the various business forms used to collect data and to distribute information
5. To perform records management, including the distribution and use of reports
6. To participate in the selection of information processing equipment and to establish standards for equipment selection
7. To prepare and maintain business organization charts

Although the basic functions of systems analysis have not changed as a result of growth in the information technologies, the environment in which they are performed has changed greatly. Because of the complexity and far-reaching impact of modern business information systems, there is a greater emphasis on planning in order to anticipate needs, rather than to simply react as these needs are discovered. Also, greater emphasis is placed upon the ability of systems professionals to work effectively with both end-users and information processing specialists. This is because many activities of systems analysts involve them as members—often leaders—of project teams responsible for developing new information systems. Also, analysts are often required to monitor and maintain existing systems to ensure that they remain current. Two results of these kinds of emphasis are a higher corporate reporting level for systems analysis and enhanced career opportunities.

Among the previously listed responsibilities of the systems organization is the preparation and maintenance of business oragnization charts. Through an awareness of the business organization chart and its uses we can better understand the position of the systems professional in the corporation and the requirements for successful careers in systems analysis. Therefore, we will describe the role of the systems analyst in the context of business organization charts.

The Business Organization Chart

Organization Chart Responsibility and Use

The owners of a business decide on the organizational management and structure that best meet their goals and objectives. The owners, or their legal representatives, the board of directors, hire managers and other employees. Each employee from the president on down occupies a position that must be defined with respect to those of all other members of the organization. This definition is necessary to control activities and channel the flow of information within the business organization.

The **organization chart** is a flowchart that identifies the organizational elements of a business and displays areas of responsibility and lines of authority. It is the responsibility of top management to define and to update the organization chart. However, a continuing effort is required to prepare organization charts and modify them as organizational plans are altered to cope with changes in the business environment. Hence, the responsibility for the preparation and maintenance of organization charts usually is assigned to the systems staff. This is an important systems activity because it stimulates management to keep its organizational plan up to date.

Management has many uses for current organization charts. These include (1) reviewing functions performed by major elements of the company; (2) aligning the corporate structure with business opportunities; and (3) comparing salaries, authority, and organizational size at equivalent and subordinate levels.

As useful as they may be to management, organization charts are essential to the systems analyst. In all aspects of work the systems analyst deals with individuals who have a specific position in the organization. Therefore, the analyst must understand the organization chart—not only what is printed on it, but also the personalities who are behind it. Without an understanding of the latter, the analyst may be unaware of real but unwritten authority.

Figure 3.2
Basic superior-subordinate relationship. Most business organization charts are structured as a series of superior-subordinate relationships. Rectangles identify positions and functions; lines and arrowheads indicate the downward flow of authority and the upward flow of responsibility.

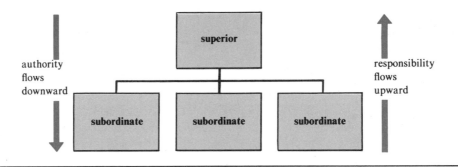

Organization Chart Structure

The structure of the typical modern business organization chart has evolved from concepts of authority and responsibility handed down from the governments of Greece and Rome and from the feudal period of history. These concepts have been formalized in more recent times by religious and military organizations as a series of *superior-subordinate relationships,* as illustrated in figure 3.2. The standard means for presenting an organization chart is a flowchart that uses lines to connect rectangles identifying individuals and functions. As is indicated in figure 3.2, authority can be delegated downward, but responsibility must flow upward. When expanded, this type of organization chart displays vertical overlapping group linkage. Figure 3.3 is a segment of a typical organization chart for a modern business enterprise. It clearly depicts the superior-subordinate relationship. It also displays the characteristic aspect of this type of organization—the expansion into successively subordinate levels, using the connecting elements as *link pins.* Thus, to the extent that the organization is shown, the vice-presidents, the directors, and the managers are link pins.

There is considerable evidence that an organization like the one in figure 3.3 functions more effectively if its operating characteristics are participative rather than authoritarian. For example, organizations in which departmental goals are established by group participation tend to have superior motivation and to outperform those in which goals are rigidly set by "orders from above." Also, in most organizations there are meaningful requirements for horizontal as well as vertical linkages. One area in which such a requirement is strikingly evident is systems analysis. Systems analysts work with information systems, and these systems often flow across organizational boundaries. Horizontal linkage permits systems activities to cross organization chart boundaries. This need is particularly accentuated in the case of computer-based business systems, since many of these generate outputs for the use of more than one organizational group and derive their inputs from many different groups.

Figure 3.3

Organization chart. Organization charts can be expanded to display the linkages between successive levels. Positions for which both a superior and subordinate reporting relationship are shown are called link pins. On this chart, the link pin positions are vice-president, director, and management.

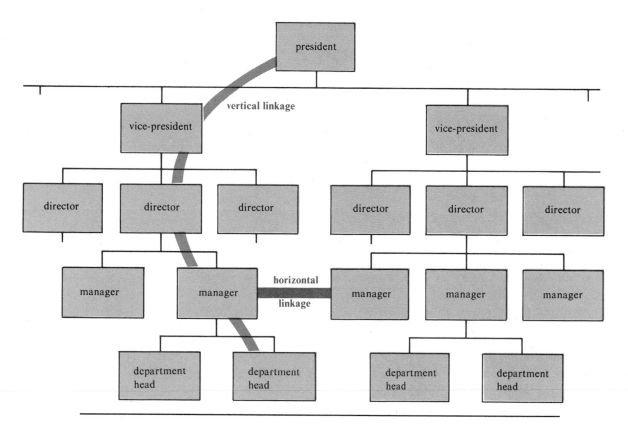

Organization Chart Guidelines

There are no universal standards for the construction of organization charts. However, certain general principles apply. Above all, it is important when considering what to put on a chart and what to leave off to realize that the organization chart is the picture of the company seen not only by its management and employees, but also by its general business environment. Therefore, this chart should properly reflect to important vendors, customers, and agencies the picture the company wishes them to see. With this in mind, some general guidelines for organization charts are suggested:

1. *Layout*   The layout of the chart should be attractive. The picture should be made up of rectangles and lines. It should be centered, with margins and white space selected to make the chart pleasing to the eye. The structure of the chart should be symmetrical. As figure 3.4 illustrates, there is more than one way of

Figure 3.4
Organization chart symmetry. There are many ways of laying out symmetrical organization charts. The three methods shown are equivalent, since the superior-subordinate relationships are the same.

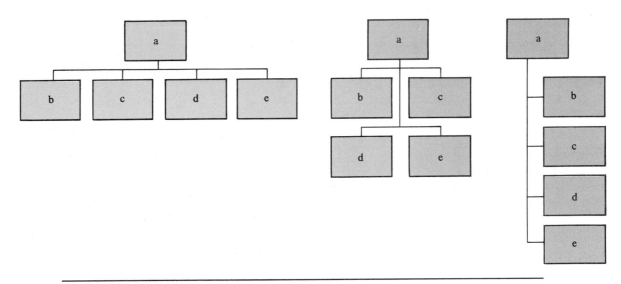

displaying equivalent relationships. In the three sections of this figure the same relationship is shown in different ways. In all cases B, C, D, and E are at the same level and report to A. A general guideline is to make balanced use of the space available.

2. *Title and approvals*   The organization chart should have a meaningful title. A standard position should be provided for approvals, date, and other identifying information.

3. *Scope*   One organization chart giving an overview of the organization's main elements is required. The chart shown in figure 3.5 is an example of an overview organization chart. This chart illustrates the use of the "dashed line" convention to represent a *staff position* as contrasted with a *line position*. The latter is a direct authority relationship, and the former is a service relationship. Some executives—the President of the United States for example—have large staffs. In the example of figure 3.5 the staff person is a full-time employee of the corporation. In other instances, the individuals may be consultants, "on-call," or members of the board of directors. For these cases, one convention is to also represent the position rectangle with dashed lines.

The information services organization appears as a line organization in figure 3.5. However, because it has no direct authority over the other line organizations that it services, the information services organization is also a type of staff activity. This organization does derive authority indirectly from the specifics of its assignments however, and from a line of command that commonly leads to the president through a senior executive, such as a vice-president of information services.

Figure 3.5

Overview organization chart. The overview organization chart is one that provides a long-range picture of the corporation. The information presented in an overview organization chart should be limited, with detail left to lower level charts.

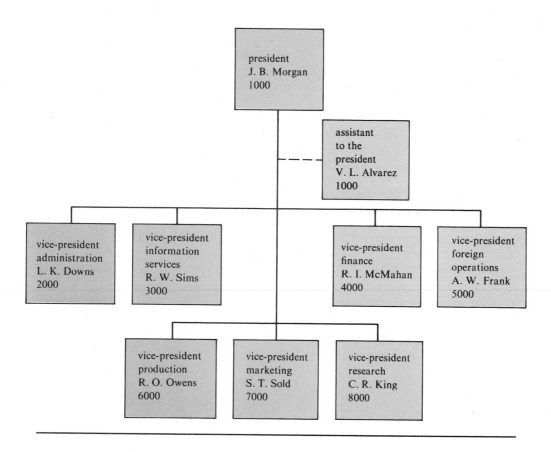

Figure 3.5 also illustrates the use of a special area for title, approvals, and date. In this case the only approval required is that of the president. Usually, lower level organization charts bear the signature of the highest level person shown on the chart and the signature of that person's supervisor.

The detail on an overview organization chart should fit easily on an 8½ by 11-inch page, viewed from the 11-inch edge. This type of layout is common. The amount of information presented on an organization chart should not exceed that which can be conveniently shown on a standard page. The overview picture of the entire organization is like a long-range photograph. Some closer-range pictures are required also.

A medium-range picture should be prepared for each of the organizational elements shown on the overview organization chart. Figure 3.6 is an example of a medium-range organizational picture.

Some close-up pictures also may be required for lower level organizational elements. For example, organization charts often are prepared in detail by managers responsible for particular projects or activities. Such charts are necessary for assigning and monitoring tasks. However, the systems group should exercise restraint with respect to the number of organization charts they prepare, because these charts must be maintained. In a dynamic company, this maintenance can be both costly and time consuming.

Restraint should be exercised also in deciding on the amount of detail to be presented on an organization chart. For example, figure 3.6 shows only significant individual staff positions, such as the assistant to the person occupying the key position on the organization chart.

4. *Organization chart distribution*   The overview organization chart should be distributed to top management and to all operating officers of the business. It should be available to customers and to employees who express an interest in the general organization of the company. Normally, new employees are provided with a copy of the top view organization chart as part of their indoctrination. Other organization charts should be distributed to individuals who have their responsibilities shown on the chart, to their superiors, and to any other persons, such as systems analysts, who have a legitimate need for the information. Of course, a file should be maintained of all current and past organization charts. These should be available to the president and to other authorized persons upon request.

5. *Information provided*   Each organizational rectangle on the chart should contain a title with functional significance (for example, vice-president, information services), the name of the individual in that position, and an identifying organization number. Figures 3.5 and 3.6 show typical formats.

Other kinds of information, such as salaries and number of individuals supervised, can be added to charts for the use of managers who wish to review their organizational plan. This type of information usually is confidential and is not for general distribution. The organization chart identifies the major functions of the organizations shown. Additional details are supplied by supporting documents known as organization function lists.

Figure 3.6
Information services organization chart. The information services organization chart is an example of a medium-range picture for a major element of a corporation. Two principal functions shown are corporate systems and data processing, the directors of which report to a corporate vice-president.

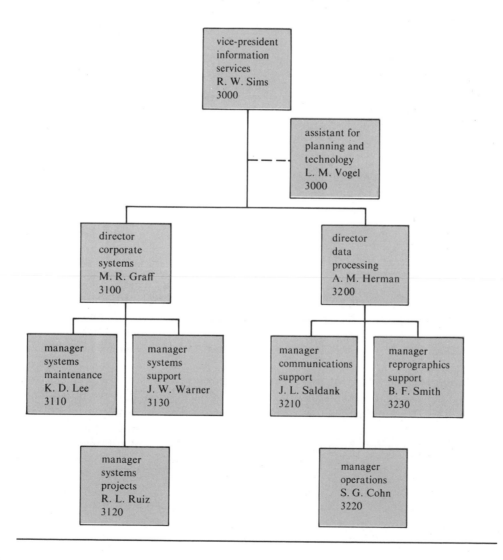

Figure 3.7
Organization function list. An organization function list provides information about the major activities performed by an organization. A separate organization function list is prepared for each element shown on an organization chart.

> Organization Function List
> Systems Support Department
> Department Number: 3130
> Date: 6/15/XX
> Approved:
>
> *J.W. Warner*
> J. W. Warner
> *M. Graff*
> M. R. Graff

1. designs and controls forms

2. manages and retains records

3. performs work measurement studies

4. prepares and maintains organization charts

5. analyzes reports

6. writes policies and procedures

The Organization Function List

An **organization function list** is a document prepared for each organization shown on an organization chart to describe the specific major activities performed by that organization. It is keyed to the organization chart by use of the organizational titles and numbers shown on the organization chart. Figure 3.7 is a function list for the systems support department of the information services organization shown in figure 3.6. Note that the functions are described briefly in the present tense using action verbs (*controls, performs,* and so forth).

Systems analysts who understand the organization chart and its associated function list are better equipped to improve the efficiency of business systems for which they are responsible. Analysts can use the organization chart as a means of increasing their knowledge of operational processes, job responsibilities, and information flow.

Figure 3.8
Reporting trends for systems groups. With increasing frequency, systems groups report to a high level, nonfinancial executive. This trend is due both to the application of computer-based information systems to many business areas other than accounting, and to the recognition of the importance of information resource management.

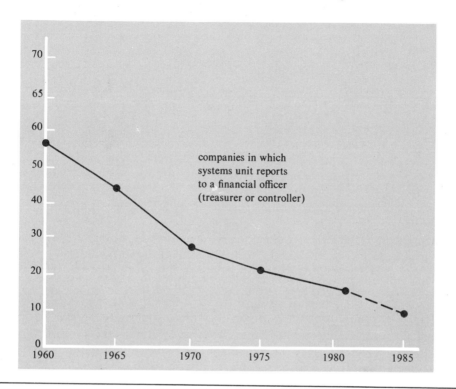

Careers in Systems Analysis

Opportunities and Compensation

The current outlook for careers in systems analysis is bright. The Department of Labor estimates that there will be 400,000 systems analysts in 1990, which is more than a 100 percent increase from 1978. Although the systems profession occupation is demanding, persons skilled in information resource management are needed, and there are opportunities for advancement. An important factor regarding this profession is the reporting level of the systems analysis group; another is the compensation relative to other careers in information services.

As we have mentioned, one of the most significant changes in systems work is the organizational level to which the systems group reports. With increasing frequency, this group is found to report to a high-level, nonfinancial executive. Figure 3.8 exhibits this trend.[2] Figure 3.9 demonstrates the increasing recognition of the importance of

---

2. Ibid

Figure 3.9
Reporting level of systems groups. An Association of Systems Management survey shows that 65 percent of the systems groups in the responding companies report to corporate headquarters, with 88 percent reporting at the division level or above.

systems group	
reporting level	*percent of total*
corporate headquarters	65
plant	2
division	21
department	12

the systems analysis function, with 88 percent of the groups operating at the division level, or above.[3]

Increasingly, the senior systems executive appears on the organization chart as a vice-president who reports directly to the president of the corporation. Typical titles for this person are vice-president of information systems, vice-president of information resources, and vice-president of information services. Figure 3.6 displayed an information services organization of the type that will be required to manage the information resources of the 1980s. Salaries paid to information system professionals tend to be highest for those positions requiring the most creativity and responsibility. Salaries for many jobs correlate to the point in the life cycle of a system where individuals begin to apply their skills. Figure 3.10 illustrates this relationship.

The systems analysis staff typically is headed by a director of corporate systems who reports directly to the vice-president of information services. The director of corporate systems is at the same organizational level as the directors of the user areas of the company and has horizontal linkages to them. This individual also has a vertical linkage to top management.

Personal Qualifications

What are the personal attributes of a successful systems analyst? In order to accomplish the task of analyzing current and potential systems, the systems analyst often forms a team of technical specialists. These specialists are selected from the operating personnel of the department involved with the system. They are familiar with the day-by-day operational problems and can add to the knowledge of the systems analyst. In order to manage this team, the analyst must be a leader.

The systems analyst must be able to secure cooperation. Because assignments often are of a staff or advisory nature, the systems analyst cannot force people in other organizational areas to provide necessary information about their business systems or

---

3. Ibid

Figure 3.10
Salaries and the computer-based system life cycle. Systems analysis is a career growth path for information professionals. Salaries are highest for systems analysts, since their creative skills and acquired experience are initially applied in the study phase, where decisions are made affecting the entire life-cycle process. Also, systems analysts have major responsibilities throughout all of the life-cycle phases.

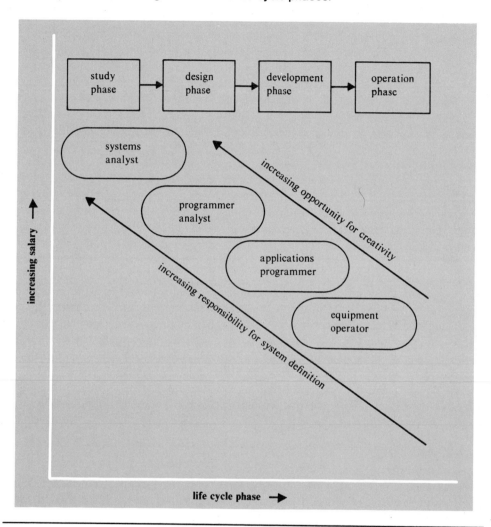

their problems. Many people whom the analyst contacts feel threatened by probing and questioning and are reluctant to cooperate. The systems analyst must, therefore, have skills in human relations.

The systems analyst must also be creative. An important part of the job is to modify existing systems or to design new systems to eliminate problems that users are having. An almost limitless number of combinations of system components exists to produce desirable systems. Some will provide better solutions to the problem than others. Using limited resources, such as time and money, the systems analyst must design a system that is the best solution possible. Within these constraints, the systems analyst must have enough creative imagination to develop and then evaluate feasible system solution candidates.

The systems analyst also must be able to speak and write well. After the analyst has defined the problem and selected a "best" system, management must be convinced that it is a good problem solution. Management also must be sold on the idea that the business should spend resources to design, develop, and put the system into operation. Selling management on the idea includes preparing the effective written reports and verbal presentations. If management approves the project, the analyst must also convince the system users that the system will work. The new system will benefit the organization only if it is accepted and used. Unless analysts have the ability to sell ideas to others through the written and spoken word, all their work may be of little value.

Educational Background

As computer-based business systems continue to increase in scope and complexity, the background required for systems analysis becomes more important and more demanding. Most systems analysts now entering the field have a bachelor's degree or higher. The most common college majors for systems analysts are business administration, accounting, computer science, and data processing. Obviously, the major fields of study should be in areas that will contribute to the understanding of a computer-based business system.

A formal education is only the beginning of an educational background. Many companies hold in-house seminars and executive training programs to provide a specific knowledge of how their company operates and to broaden the backgrounds of their professional personnel. Often there are opportunities to attend courses at local colleges or classes sponsored by data processing vendors. Also, there are several professional organizations, such as the Association for Systems Management and the Data Processing Management Association, which provide an opportunity for a free exchange of ideas. These associations also hold seminars designed to help their members discover new approaches to solving problems.

Work Experience

Formal and informal education are very important contributors to the systems analyst's know-how, but there is no substitute for "having been there." Work experience separates theory from practice, the things that work from those that do not work. Anyone entering the field of systems analysis will have to serve an apprenticeship period to get on-the-job experience. When starting to work on projects, the new analyst should undertake small assignments, usually under the supervision of senior analysts. In this way, the analyst can gain experience, build personal confidence, and earn management's confidence.

Senior positions in systems analysis are management level positions. Most analysts get into systems analysis through promotion from other jobs. The most common "other job" has been computer programming. There are several reasons for selecting programmers for promotion to systems analysis. First, they are familiar with many existing systems. Second, they know how to work with computers and speak the "language" of the computer. Third, as programmers, they have shown an ability for analysis and for thinking in a logical manner.

Programmers, of course, are not the only candidates for a systems analyst's position. Anyone who has demonstrated analytical ability and has a detailed knowledge of particular systems or their components is a potential candidate.

Summary

Throughout the successive information eras business systems have evolved from manual to computer-oriented systems. Four important advantages of computer-oriented systems over manual systems are: (1) accuracy; (2) data collection and communication; (3) data storage; and (4) speed of response. Data processing was the focal point of the "traditional" computer-oriented information systems of the 1970s; the "non-traditional systems of the 1980s will include additional information resources, such as extensive communications and word processing elements.

The primary tasks performed by systems analysts have not changed as businesses have become increasingly computer oriented. However, the involvement of systems analysts with computer-related activities has increased greatly. This has led to increased emphasis upon planning and upon the ability of analysts who work effectively with end-users and with information processing specialists. Also, as a result of increased responsibilities, the systems group now reports at a higher level in most organizations.

The organization chart is a flowchart that identifies the organizational elements of a business; it displays areas of responsibility and lines of authority. The organization chart structure has evolved from historic concepts of authority and responsibility and is based upon a series of superior-subordinate line relationships. There also are staff positions without direct, line authority. The systems group, which is engaged in activities that cut across organizational boundaries, has staff characteristics. However, the systems group reports to a senior line executive, frequently a vice-president of information services. This is a result of a transition from reporting to a financial executive, as was the case in the past when primary emphasis was upon accounting systems. The current activities of systems groups are much broader than financial systems, encompassing areas such as marketing, production, and research. As a consequence, career opportunities for systems analysts have been enhanced greatly. Persons with an appropriate background in business and with computer-related education and experience may anticipate success as systems professionals.

For Review

**traditional computer-
  oriented system**
**nontraditional
  computer-oriented
  system**
organization chart
**organization function
  list**

manual system
superior-subordinate
  relationship
staff position
line position
link pin

automated office
Maturing Era
Early Era
Growing Era
Refining Era

For Discussion

1. Describe four ways in which computer-oriented systems differ from manual systems.
2. What are the differences between traditional computer-oriented business information systems and nontraditional systems?
3. Identify the principal functions performed by the systems group.
4. What are the principal purposes of organization charts?
5. Who is responsible for defining the business organization chart? Who is responsible for preparing and maintaining the chart?
6. Relate the flow of authority and responsibility to the superior-subordinate relationship.
7. What is the difference between an authoritative and a participative organization?
8. What is an organization function list? How does it relate to the organization chart?
9. What is the trend in reporting level for the systems group? Why do fewer systems groups report to financial executives?
10. Describe the qualifications for a career as a systems analyst.

4
The Business as an
Information System

Preview

Businesses may be considered to be systems of systems, since they are made up of major units that are themselves systems. Information-oriented and product-oriented information flowcharts can be drawn to exhibit the relationships among these systems. These information flowcharts are useful to the systems analyst, but at the same time they are complex due to the amount of information that must be generated to meet the needs of a business.

Business information needs exist at four levels. In addition to an operational level, there are three management information levels: supervisory, tactical, and strategic. All of the latter rely on feedback and control to assist managers in decision making. Many top-down, structured techniques are available for business information system design and development. The life-cycle methodology encompasses all of these techniques. Because the application of this methodology requires management skills, the systems analyst often acts as a life-cycle manager.

Objectives

1. You will become aware of the principal characteristics of businesses as information systems.
2. You will learn how systems analysts use flowcharts to describe information flow.
3. You will be able to distinguish between the four information levels in a corporation.
4. You will become aware of the importance of feedback and control.
5. You will be introduced to current information system design techniques.

Key Terms

goal   A broadly stated purpose of a business.

objective   A concrete, specific accomplishment necessary to the achievement of a goal.

product flow   Flow of raw materials into finished goods.

information flow   The network of administrative and operational documentation.

information generator   A business information need, either external or internal in origin.

feedback   The process of comparing an actual output with a desired output for the purpose of improving the performance of a system.

control   The actions taken to bring the difference between an actual output and a desired output within an acceptable range.

MIS   Management information system.

DSS   Decision support system.

DBMS   Data base management system.

A Business: A System of Systems

Business Goals and Objectives

In chapter 1 we defined a business information system to be a system that uses resources to convert data into information in order to accomplish the purposes of the business. We classified these resources as personnel, facilities, material, and equipment. Also, we introduced the concept of system and subsystems. In this chapter, we will add to this foundation. We will consider the entire *business* to be an information system.

The purposes of a business can be defined in terms of goals and objectives. A **goal** is a very broadly stated purpose. Examples are the goal of making profit and the goal of educating students. **Objectives,** on the other hand, are concrete and specific accomplishments necessary to the achievement of goals. For example, an automobile manufacturer must have as an objective the production of a competitive product in order to achieve a profit goal; a college must have as an objective relevant curricula in order to achieve its educational goal. Major objectives are composed of lower-order objectives. Accordingly, before a car can be made, subassemblies must be produced and, before that, proper tools must be designed. Goals are relatively long term, and objectives are relatively short term.

Most business enterprises fall into one of two broad categories: production or service. Examples of *production enterprises* are manufacturing, farming, construction, and agriculture. *Service enterprises* include transportation, communications, medicine, and education. Each enterprise, whether production or service, has its particular goals and objectives supporting it. Certain system concepts are applicable to all business enterprises. Systems analysts must be familiar with them to understand how businesses are structured and organized. This chapter presents important concepts that will sharpen your perceptions of business systems.

A System of Systems

As we learned in chapter 1, business systems are composed of smaller elements, which also are systems. These systems transform or convert inputs into outputs. The transformation process was shown in figure 1.1. It also is shown in figure 4.1. Additionally, figure 4.1 introduces the idea of constraints. A business functions within a set of constraints, which generally it cannot alter significantly. Examples of such constraints are federal laws, social environment, total market, raw materials limitations, and scientific principles. The resources and constraints that affect the transformation of inputs into outputs also are shown in figure 4.1.

Principal functional systems associated with most product-oriented enterprises are shown in the flowchart of figure 4.2. In order to emphasize the relationship between these functional systems, management and administrative systems, although present in all enterprises, are not included in figure 4.2. Customers, employees, and vendors, indicated by dashed-line rectangles, are shown to provide continuity to the illustration. The nine basic functional systems shown in this figure are:

1. *Purchasing*   Procuring from the vendors the goods and materials needed by the business.
2. *Receiving*   Inspecting and accepting delivered goods and materials.
3. *Inventory*   Storing the received goods and materials.

Figure 4.1

The system environment. The ability of a system to convert data into useful information is affected by limitations on resources and by environmental constraints that are external to the system.

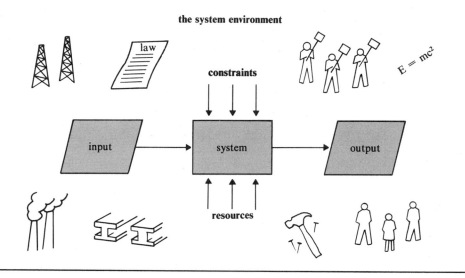

Figure 4.2

Functional business systems. Nine principal systems are associated with most product-oriented businesses: purchasing, receiving, inventory, production, sales, distribution, billing, collection, and paying. Persons who interact with these systems include customers, vendors, and employees.

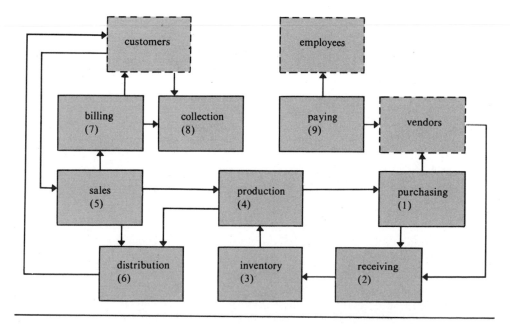

4. *Production*   Designing and manufacturing the goods to be sold.
5. *Sales*   Marketing the goods produced.
6. *Distribution*   Supplying the customer with the goods sold from a produced-goods inventory.
7. *Billing*   Sending statements of the amount owed to customers.
8. *Collection*   Receiving payments from customers.
9. *Paying*   Making payments to those whom the business owes money, such as vendors and employees.

Each of these functional systems produces one or more outputs in the form of products or documents. These outputs establish the relationship of each system to other systems and to the business as a whole. Since systems are assigned their own necessary resources they are, to a degree, relatively independent elements of a business. This is why we have defined a *system* as a combination of resources (personnel, materials, facilities, and equipment) working together to convert inputs into outputs. Again, we note that the definition of a system is similar to the definition of a business. However, whereas a system produces outputs, a business integrates the outputs of its component systems to accomplish objectives and to achieve goals. Often a system is thought of in terms of its mechanics, such as the methods, techniques, and procedures by which it achieves its purpose. The above definition of a system, by inference, includes such mechanics. For example, a distribution system is defined to include not only its functions, but also the associated written methods and procedures for processing shipping orders.

Because systems are its major elements, a business may be considered to be a *"system of systems."* This concept is illustrated in figure 4.3, which depicts the nine operational systems and a central complex of management and administration systems. By extension of this concept, most systems may be considered to be composed of *subsystems,* with the same transformation characteristics as systems. For example, as is shown in the first part of figure 4.4, a production system may be composed of subsystems such as engineering, production planning, manufacturing, and quality control. In practice, the distinction between systems and subsystems is fluid. It depends on the size, field of enterprise, objectives, and goals of the particular business. Thus, the other example in figure 4.4 shows the billing, collecting, and disbursing systems of figure 4.3 assembled as subsystems of an accounting system.

The dependence of the detailed characteristics of systems and subsystems on the unique goals and objectives of a business reveals an essential characteristic of most business systems. This is that systems and subsystems are not readily transferred from one industry to another. It is even difficult for systems used by one business to be used by another business within the same industry. The inability to pick up and use someone else's system without modification has been one of the most costly lessons learned in the short history of modern computer-based business systems. It is the reason why many predesigned and "packaged" systems have not been entirely successful. In recent years, however, much progress has been made in developing applications packages that are either industry-specific or general purpose. A hospital accounting system is an example of the former. A project management system, related to the life-cycle methodology, is an example of the latter. With the market for application packages already identified as a major software growth area for the 1980s, the "make" or "buy" recommendation will become an important responsibility for systems analysts.

Figure 4.3

The business: A system of systems. A business may be considered to be a large system composed of smaller, operational systems. This "system of systems" is integrated, through management and administration, to meet its objectives and goals.

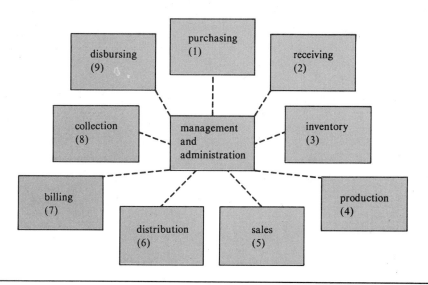

Figure 4.4

Systems and subsystems. Most large systems are made up of subsystems that also transform inputs into outputs. Two examples of such large systems are a production system and an accounting system.

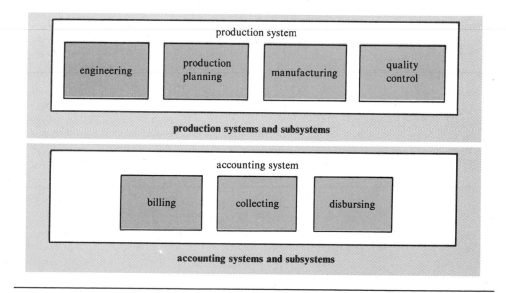

The Business Information Structure

Product Flow and Information Flow

The organization chart, which we studied in chapter 3, is only a one-dimensional picture of a business. Behind it there is a constant flow of information-oriented and product-oriented activities. All these activities involve individuals with differing levels of responsibility and authority. In order to examine the information structure of a business, let us start by distinguishing between information flow and product flow. We will use a manufacturing company as an example.

Product flow, which is relatively easy to visualize, is the flow of raw materials into subassemblies, then into assemblies, and finally into finished goods. **Information flow** consists of the creation and movement of the administrative and operational documentation necessary for product flow. Information flow is more difficult to visualize than product flow because its physical manifestation is a vast network of data carriers, forms, or electronic communications. Yet it is this network that the analyst must understand. Although their actions must be governed by the physical reality of the goods that the company is producing, systems analysts deal primarily with the creation and management of documentation. Therefore, it is necessary that the information flow be known in those segments of the business for which the analyst has assigned responsibilities.

To bring information flow into sharper focus, we can redraw figure 4.2, which depicts *functional business systems,* to emphasize the distinction between product flow and information flow. This is done in figure 4.5. In this figure, the heavy flow lines trace the product flow path. The lighter lines indicate paths by which information flows among the nine major functional business systems.

If we identify the principal documents associated with these information flow paths, the relative complexity of information flow can be depicted. This complexity is illustrated in figure 4.6. The documents shown in figure 4.6 can be defined, in context, as follows:

1. The *purchase order* is prepared by Purchasing, which sends the original to the vendor, retains a copy, and sends a copy to Receiving.
2. When the materials ordered arrive, Receiving verifies the order against its copy of the *purchase order,* inspects the material, and informs the Purchasing department of its arrival and acceptance by means of a *receiving report.* The material is transferred to Inventory accompanied by an *inventory transfer.*
3. By means of a *purchase requisition,* Inventory requests Purchasing to order those materials that are not on hand in sufficient quantity.
4. Production designs and develops the product. The components that are built in-house are combined with the components or subassemblies that are procured out-of-house. Production uses a *material requisition* to request needed materials from Inventory. Inventory notifies Production of the availability of the requisitioned materials by returning a copy of the *material requisition.*
5. Sales contacts the customer, sells the product, and prepares the *sales order.* The customer is provided a copy of the *sales order.* A copy of the sales order, entitled *sales notice,* is sent to Billing and to Production. An additional copy, the *shipping order,* is sent to Distribution.

Figure 4.5

Product flow and information flow. Product flowlines trace the flow from raw materials to end-product. Information flowlines not only parallel product flow, they also link functional systems not directly related to production.

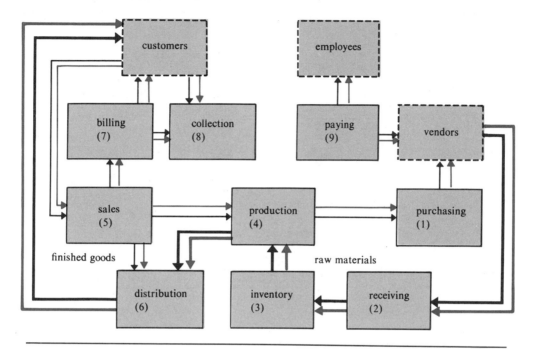

6. Distribution receives the finished goods from Production accompanied by a *warehouse transfer notice.* Distribution ships the product to the customer and informs Sales by means of a *shipping notice.*
7. Billing prepares and mails the *customer invoice.*
8. Collection receives *customer payments* from the customer and sends updated information to Billing by means of a *customer payment notice.*
9. Paying makes payments to vendors by means of a *vendor check.* This check is prepared after the vendor has submitted a *vendor invoice* and after that invoice has been verified and forwarded by Purchasing. Paying also distributes *paychecks* to employees. The amounts of the *paychecks* are based upon *time tickets* submitted by employees.

As complicated as the network shown in figure 4.6 may appear, it is a necessary oversimplification of the real volume of information flow in a typical corporation. Every major functional system is composed of complex subsystems, each of which has its documentation needs. In most enterprises, the production of printed reports (hardcopy) is reduced through use of visual displays, or screen images (softcopy). Usually hardcopy is available on demand, and it is convenient to use the document symbol, as in figure 4.6, to display information flow. However, where appropriate, other symbols, such as that for a screen display, could be used to represent data carriers.

Figure 4.6

Principal information flow documents. Information flow documents illustrate the complexity of the network of information carriers in a business. Information carriers may be "softcopy," for example, a CRT display, as well as "hardcopy" documents shown here.

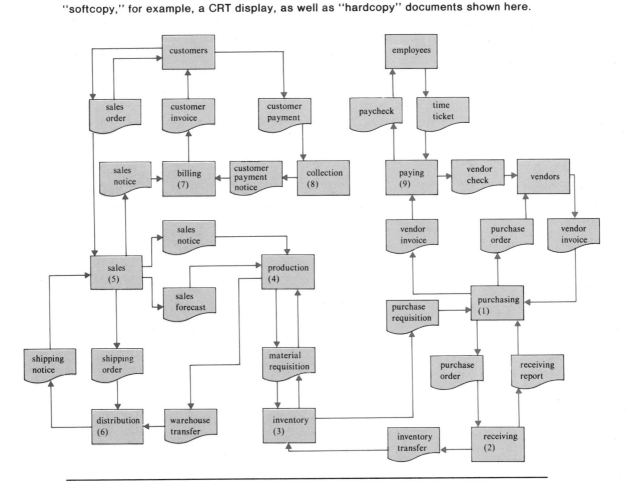

Information flow exists for service enterprises as well as for product enterprises, and it exists in different forms at different reporting levels within an organization. We will describe the information needs and uses at these levels; however, there are two other factors that add to the complexity of the information network: (1) there are external as well as internal generators of information; (2) information must be reported with different emphasis and formats according to the needs of different levels of management. Each of these factors requires discussion.

Information Generators

A company must develop information systems to meet not only its internal reporting needs, but also the external reporting needs that arise from its general business environment. Figure 4.7 distinguishes between the two information needs that act as **information generators.** The internal information needs are represented by the nine

Figure 4.7
Information generators. The information that a company must provide to meet its business needs is generated not only by its internal reporting requirements, but also by its external reporting environment.

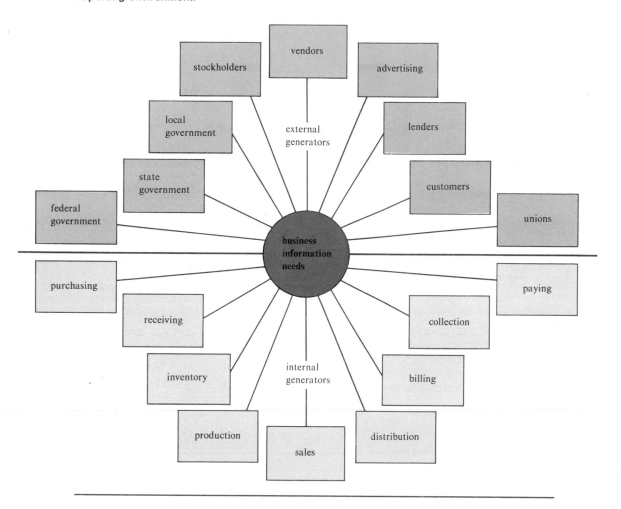

functional business systems discussed previously. Externally generated needs are represented by agencies such as federal, state, and local governments, shareholders, vendors, advertising, lenders, unions, and customers.

Not all the information generated externally is useful to the corporation. That which is useful is most likely to be of use to the upper levels of management. Also, in its "raw" form not all the internally generated information is useful at all levels within the corporation. Different types of reports must be prepared to meet the information needs of each level of user. Because of these differing user needs, we must identify levels of information systems.

Figure 4.8
Information system levels. Information systems meet not only operational needs, but also three levels of management needs. Both horizontal and vertical integration exists among all four information levels.

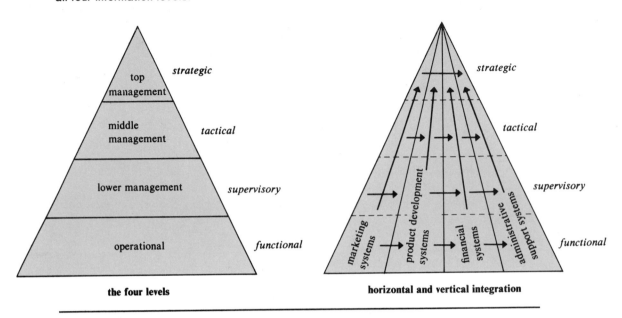

the four levels   **horizontal and vertical integration**

Information System Levels

Figure 4.8 depicts the four levels of information systems that exist in a typical business of moderate to large size. These are (1) operational; (2) lower management; (3) middle management; and (4) top management.

At the operational level, routine production or clerical operations are performed. Operational systems provide little feedback directly to the employee. For example, the materials clerk receives a material requisition, fills the requisition, and files a report of action taken. A supervisor evaluates the employee's performance. However, records of transactions occurring at the operational level constitute data that, when collected, organized, and processed, becomes the basis for higher level management actions.

Lower management performs supervisory functions that are short term relative to the higher levels of management. They deal with day-to-day job scheduling, checking the results of operations, and taking the necessary corrective actions.

Middle management functions are tactical in nature. This level is responsible for allocating and controlling the resources necessary to accomplish objectives that support the strategic goals of the business. Planning is performed; authority is delegated to the supervisory level; and performance is measured.

Top management functions are strategic. They include establishment of the goals of the business, long-range planning, new market and product development, mergers and acquisitions, and major policy decisions. Appropriate authority is delegated to middle management.

Figure 4.8 also shows that there are both horizontal and vertical information system structures. In this figure, for illustrative purposes, the nine functional systems are collected, vertically, as major systems. Marketing systems include Sales and Distribution; product development systems include Purchasing, Receiving, Inventory, and Production. Financial systems include Billing, Collection, and Paying. Support operations, such as personnel and contracts, are grouped as administrative support systems.

Horizontal integration may occur within or between major systems. An example of horizontal integration, internal to a major system, is the combination of Purchasing, Receiving, and Inventory (within product development) into a procurement system on the basis of a shared data base. An example of horizontal integration, between major systems (in this case between finance and administrative support), is a personnel-payroll system based on employee-related data elements common to both Personnel and Payroll.

The following is an example of a possible vertical integration of an information system within Production:

1. Machine assignment and job time reporting—functional
2. Machine scheduling—supervisory
3. Make (in own shop) or buy (from vendor) decision—middle management
4. New product decision—top management

All management-level positions require that decisions be made. These decisions range from the routine to the complex. However, all management decisions involve two elements:

1. A process that includes objectives, measurement of performance against objectives, and corrective action
2. The availability of appropriate information on which to base decisions

The first of these elements establishes management as a feedback and control process. The second leads to distinctions between the characteristics of information required at each management level. Feedback and control and management-level information characteristics are described in the next sections.

Feedback and Control

Feedback and control are essential to the design of any management system. **Feedback** is the process of comparing an actual output with a desired output for the purpose of improving the performance of a system. **Control** is the actions taken to bring the difference between an actual output and a desired output within an acceptable range. These concepts are illustrated in figure 4.9. Part A depicts a system in which there is no feedback and control. This system does transform an input into an output; however, because it lacks feedback, it is called an open loop system.

From part A of figure 4.9, it is evident that the output depends solely on the characteristics of the input and of the system. If the output is not satisfactory, there is no provision for modifying either the input or the system. Operating such a system is analogous to driving a car while blindfolded. If, however, the driver compares the

Figure 4.9
Feedback and control. (a) An open loop system provides no opportunity for management control. (b) A closed loop system provides managers with both a comparison of actual and desired outputs (feedback) and a means of taking action to bring them into correspondence (control).

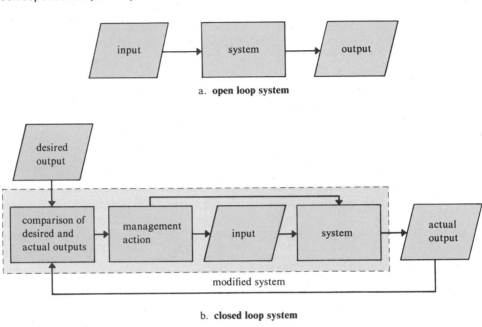

a. **open loop system**

b. **closed loop system**

position of the car with the location of the white line in the middle of the road, corrective action is possible. In this case, the feedback is visual, and the corrective action is initiated by manually turning the steering wheel. Because there is feedback, this type of a system is called a closed loop system. The elements of a closed loop system are shown in part B of figure 4.9. This flowchart demonstrates that the comparison of desired and actual outputs results in management action, which may modify the inputs, the system, or both. The result is a modified system that will produce altered outputs. Many operational systems are open loop systems. All true management systems are closed loop systems.

Because management information systems are feedback systems, they are more complex to design than are open loop operational systems. Feedback systems are effective only if they can respond quickly enough for the necessary corrective action to take place when needed. Otherwise, the time lapse may be so great that belated management action only makes the situation worse. Such systems are said to be unstable. Instability may also result if information is fed back at such a rate that it cannot be absorbed or if management action is premature.

Management Information Systems and Decision Support Systems
The term management information system is frequently encountered in business systems literature. Usually capitalized, a **Management Information System** (also called

Figure 4.10

Management decision levels: Information use and requirements. Information uses and, therefore, information requirements, differ at each management decision level. At the higher decision levels more emphasis is placed upon external information, long-term trends, and "what if" considerations.

MANAGEMENT LEVEL	INFORMATION USE	INFORMATION REQUIREMENTS

TOP MANAGEMENT	1. GOAL SETTING
	2. LONG-RANGE PLANS
	3. STRATEGY
	4. RETURN ON INVESTMENT
		1. EXTERNAL INFORMATION, e.g.
		   Competitor actions
		   Government actions
		   New markets
		   Resource availability
		2. INTERNAL INFORMATION, e.g.
		   Financial reports
		   Key exception reports
		3. LONG-TERM TRENDS
		4. "WHAT IF" INFORMATION

MIDDLE
 	
MANAGEMENT
	1. OBJECTIVES DEFINITION
	2. MEDIUM-RANGE PLANS
	3. TACTICS
		1. INTERNAL INFORMATION, e.g.
		   Financial reports
		   Exception reports
		2. SHORT-TERM TRENDS
		3. SOME "WHAT IF" INFORMATION

LOWER
 	
MANAGEMENT
	1. OBJECTIVES ATTAINMENT
	2. SHORT-RANGE PLANS
	3. SUPERVISION
		1. INTERNAL INFORMATION, e.g.
		   Recent historical information
		   Detailed operational reports
		   Appropriate exception reports

an **MIS**) is an information system that displays these two characteristics: (1) at least one level of vertical integration; and (2) **feedback** and **control.** A term that will be heard with increasing frequency in the 1980s is **decision support system (DSS).** DSS can be used interchangeably with MIS. A DSS has all of the characteristics of an MIS; however, it emphasizes not just the availability of computer-generated information, but also the act of decision making. In addition, the available information is supportive of other decision processes used by the decision maker.

Of course, at each management level, the specific requirements for information (and hence for the design of the feedback system) vary significantly. The different uses and requirements for information at each management decision level are summarized in figure 4.10. Note that all management systems rely upon exception reporting. The lower level exception reports are closely related to day-to-day operations. Some other significant observations are:

1. The higher the decision level, the greater the reliance on externally generated information and the less the reliance on internally generated information.
2. The higher the decision level, the greater the emphasis upon planning and the use of longer term trend information.
3. The higher the decision level, the greater the necessity to ask "what if" questions as part of the decision process.

Thus, at the higher decision levels in a company, information needs and uses are future-oriented and depend significantly upon external sources of information. This type of information is difficult to quantify and must be coupled with the experience and judgment of the decision maker. It is not surprising to find that decision support systems are at the forefront of the art of information system design. The systems analyst, particularly when working with computer-based business systems, must know what is and what is not practical to attempt. Therefore, it is appropriate to describe briefly current techniques in business information system design.

Information System Design Techniques

Information systems at the operational level can provide immediate payback, usually measurable in dollars. The continuing development of such systems is a meaningful activity not only because they pay off in dollars, but also because they provide the necessary foundation on which to base higher-level systems. Although the payoffs for higher-level systems may be less tangible than dollars, there is increasing management acceptance of such other benefits. Intangible payoffs include (1) improved internal control; (2) better management awareness of problems and opportunities; (3) long-term profitability; and (4) faster response to changes in the business environment.

In seeking to devise solutions that both satisfy information needs and improve the efficiency of the business, the systems analyst must use appropriate tools and techniques. Basic tools and techniques, such as forms design, codes, flowcharts, and the use of tables and graphs are presented in Unit Two of this book. Some advanced techniques representative of current practices in computer-based information system design are discussed briefly here.

Data Management

The successful application of computers in business depends on their ability to store, access, and manipulate large amounts of data. A problem commonly encountered is the need for the same data in many different applications programs. Without an effective method of data management, duplicate data would have to be stored in more than one file or be vulnerable to modification by different applications programmers. Efforts to avoid redundant files began in the Growing Era, were modified in the Refining Era, and are reaching full growth in the Maturing Era. Figure 4.11 shows the successful return to an objective that was not possible with the hardware and software technologies of earlier eras. As background to describing the state of the art in data management, we will examine briefly the stages in the cycle shown in the figure.

During the Growing Era of the 1960s and before the complexities of information system design were well understood, attempts were made to meet a corporation's information needs by trying to develop a total data base that could serve all its information needs. Unfortunately, the software needed to manage massive data bases and create "total information systems" was beyond the technology of that era. Each programmer had to know the location and structure of the data needed by each application. Maintaining the integrity of the data base was extremely difficult since many people had access and could modify this base, affecting others' programs as well as their own. In addition, because of the magnitude of the effort devoted to redefining

Figure 4.11
Progress in data management. Data base management techniques make it possible to organize and access nonredundant data for many different applications of computers to business. This goal was difficult to implement in earlier eras; however, effective data base management systems (DBMS) will be in widespread use during the Maturing Era.

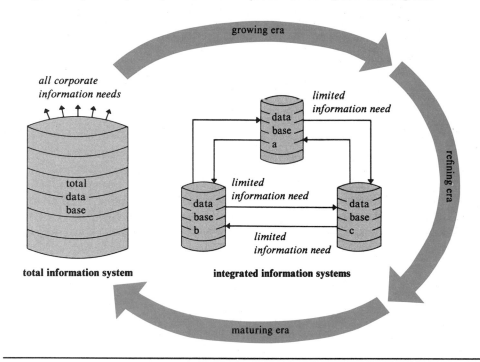

duplicate fields of data, programming projects were large, complex, and costly. Successes in creating a total information system were partial at best, and failures were far more common.

A less ambitious approach to the management of data, using smaller, integrated data bases, was introduced and used with considerable success in the Refining Era of the 1970s. In contrast with the total information system approach of the Growing Era, the integrated data base technique uses multiple linked data bases, each designed to meet a limited corporate information need. The personnel-payroll system, mentioned in chapter 1 as an example of an integrated system, is typical of the use of a common data base for more than one purpose. Integrated data bases can be linked by allowing applications programs to supply data to or access data from more than one data base. The design of integrated data bases reduced the severity of the problem of redundant storage. However, programmers still had to expend much time and effort in locating, accessing, and structuring the data needed by applications that were continuing to grow in variety and scope.

Throughout the latter half of the Refining Era (most of the 1970s), efforts continued in the development of software that would manage large data bases and relieve the programmer of duties other than the structure of data specification for a particular application. These systems, called **data base management systems (DBMS's),** organize and maintain data in a non-redundant structure that can be accessed for processing by more than one application. Data base management systems free the programmer from concern about the location and structure of data. Also, through special query languages, they provide a means for a nonprogrammer to generate needed reports. DBMS's are particularly effective as elements of management information systems because they provide access to large amounts of data that can be structured to meet individual managers' information reporting needs. We will consider data base management systems in greater detail in chapter 19: File Design.

Structured Design and Development

The life-cycle methodology stresses a top-down approach to the design of complex systems. As study phase and design phase techniques have evolved, they have tended to follow this methodology. However, the development phase activities of preparing computer software have traditionally been a bottom-up procedure. The lowest-level modules were completed first, tested, and then combined into higher-level modules. Problems were often encountered in integrating these modules, either in delaying the project or in forcing high-level design changes that necessitated much rework. Top-down procedures, called *structured design and development techniques,* have been introduced to reduce such problems. Among the more significant of these techniques are data flow diagrams, Hierarchy plus Input Process Output (HIPO) charts, structured programming, and pseudocode. Data flow diagrams are useful in describing data flows and transformations in existing systems and in identifying new or improved systems.

HIPO charts resemble organization charts of the type presented in this chapter. They are a graphic design phase tool and are used to describe the functions to be performed by a system by proceeding from the general to the detailed level. Structured programming and pseudocode are compatible development phase techniques. Structured programming is a method for modularizing the logic of computer programs; it is independent of any particular computer language. We will examine these techniques in more detail in later chapters.

Management Science

Management science techniques for the design of business systems are being applied with increasing frequency to the development of systems in order to meet the information needs of high-level decision makers. These techniques include simulation, linear programming, and critical path methods.

1. *Simulation*   It is expensive to create actual systems and to modify them by trial and error. Fortunately, there are techniques, usually involving a computer, for developing mathematical models of systems and for simulating their performance under a series of "what if" conditions. An adequate coverage of simulation techniques is beyond the scope of this text. However, their existence and continuing development are indicative of the potential for the systems analysis profession. Forecasting models, called econometric models, are examples of large-scale simulations. They are based upon mathematical equations that simulate the actual forces affecting our economy.

Figure 4.12
The life cycle of a computer-based business information system. This figure is similar to the life-cycle phases. It also identifies life-cycle management as an important information system design technique, one that makes it possible to refer to the systems analyst as a "life-cycle" manager.

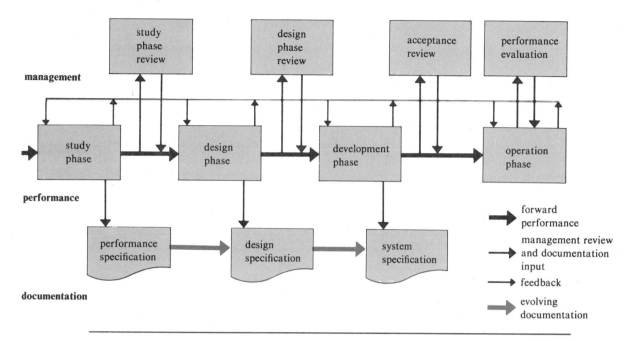

2. *Linear programming* Another advanced technique, which often makes use of the power of the computer, is linear programming. Linear programming involves the use of a mathematical model in order to find the best combination of available resources to achieve a desired result. An example of this technique is the optimum routing of freight-carrying vehicles. Cost savings implicit in the use of this technique are obvious.

3. *Critical path methods* These techniques make possible the determination of the most effective scheduling of time and resources when a complex product, such as a bridge, rocket, or computer system, is being developed. For complicated systems, a computer program is used. One such technique, PERT (Program Evaluation and Review Technique), is described in chapter 22: Preparation for Implementation.

Life-Cycle Management

The *life-cycle management* process, previously described in chapter 2, is a scientific method for the definition, design, development, and operation of effective computer-based business systems. Its methodology provides for the application of all other tools and techniques, as required, throughout the life cycle of a business system. Figure 4.12 is the same as figure 2.2. It emphasizes another extremely important structured design and development technique—structured documentation—without which all other techniques would be of little value. Thus, documentation is cumulative and progresses from the general to the specific. This figure also emphasizes the information system nature of the life-cycle management process. First, the computer-based business system generally involves both horizontal and vertical integration at more than one level. Second, each phase is a systems process. Inputs are transformed into outputs, which, in turn, become the inputs for the next phase. Finally, feedback and control are evident throughout. Management reviews, obstacles internal to each phase, and externally caused changes in one or more baseline specifications are examples of events that may trigger feedback and corrective action.

Thus, the systems analyst is called upon to exercise many of the strategic, tactical, and operational management skills discussed in this chapter. This is why we often refer to the systems analyst as a life-cycle manager.

We have deliberately concluded the unit by returning to the central theme of this book: the life-cycle management process. In the next unit, we will describe some of the basic tools and techniques the analyst uses throughout the entire life-cycle process. The remaining units describe, in time sequence, each of the four phases of the life cycle of computer-based business systems.

Summary

The purposes of a business can be defined in terms of broad goals and specific objectives. Businesses not only use resources to achieve their purposes, they also are governed by constraints, which they cannot alter. Businesses may be considered to be systems of systems, since their major elements are systems that transform inputs into outputs. The relationships between these systems can be displayed by flowcharts that exhibit both product flow and information flow. The information flow network is complex because the objectives of the business and the constraints of its environment act as information generators. Also, information must be prepared in the manner that best meets the operational and management needs of the business. In addition to the operational level, there are three management levels, each with its own information needs: (1) lower management, which is supervisory; (2) middle management, which is tactical; and (3) top management, which is strategic. All management information systems rely on feedback and control to accomplish their purposes. Decision support systems (DSS) are designed to assist managers in their decision making processes.

In order to design effective information systems, the systems analyst must use appropriate techniques. Current design techniques include data base management systems, structured design and development techniques, and management science techniques. Simulation, linear programming, and critical path methods are examples of the latter. The life-cycle methodology, which stresses structured documentation, encompasses all of the tools and techniques for the structured design and development of business information systems. Because they are called upon to exercise management skills, we often refer to systems analysts as life-cycle managers.

For Review

business
goal
objective
production enterprise
service enterprise
system
subsystem
system of systems

functional business
  system
product flow
information flow
information generator
feedback and control
MIS
DSS

DBMS
structured design and
  development
simulation
linear programming
critical path methods
life-cycle management

For Discussion

1. Distinguish between goals and objectives. Name some businesses and identify typical goals and objectives.
2. Why can a business be considered to be a system of systems?
3. How do resources and constraints relate to the systems environment? Give some examples of each.
4. Distinguish between product flow and information flow.
5. Distinguish between external and internal information generators. Give some examples of each.
6. What are the differences between open loop and closed loop systems?
7. Distinguish between feedback and control. How do they relate to the concept of exception reporting?
8. What are the identifying characteristics of a management information system (MIS)?
9. What is a decision support system (DSS)?
10. How does a data base management system (DBMS) compare with a total data base management system of the Growing Era? With an integrated information system of the Growing Era?
11. Distinguish between top-down and bottom-up design.
12. Why does the life-cycle methodology include all of the structured design and development techniques?
13. Why is a systems analyst considered a life-cycle manager?

Tools and Techniques
of Systems Analysis

Unit 2

5
Coding

Preview

In the previous chapter we distinguished between product flow and information flow, noting that information flow takes place through the movement of a large number of information carriers, which generally contain many items of data. A major problem encountered in working with a large amount of data is the retrieval of specific data when it is required. Codes are often used to aid the user in information identification and retrieval. There are several types of common codes used in business. Careful consideration must be given to the selection of the appropriate code for the particular task at hand. This study of code requirements leads to the development of the code plan.

Objectives

1. You will understand how codes are used in business to identify items of data.
2. You will become familiar with the common code types.
3. You will learn how to develop a code plan.

Key Terms

code   A group of characters used to identify an item of data.

code dictionary   A listing of codes and their corresponding data items.

code plan   The identification of particular code characteristics that need to be contained within the code.

Definition of a Code

When large volumes of data are being handled, it is important that the items be identified, sorted, or selected easily and quickly. To accomplish this, each data item must have a unique identification and must be related to other items of data of the same type. Codes can provide brief identifications of data items and replace longer descriptions that would be more awkward to store and to manipulate. We can define a **code** as a group of characters used to identify an item of data.

While identification is the main function of a code, a code may also show relationships between items of data. As an example, a code might identify a particular employee and also indicate the department to which the employee is assigned.

For example, most businesses identify credit customers by numbers or letter-number combinations, a code that is more concise than the customer's name. It also is more precise, since large businesses may have more than one customer with the same name. Also, a customer code may provide more information than just identification. It may contain a digit or letter that indicates such information as geographic area, the store through which credit was obtained, and credit limits. Thus, the code can communicate data about the credit customer and about that customer's relationship to other credit customers.

Other examples of codes are all around us in our daily lives. The postal service uses zip codes to identify delivery areas. We order merchandise through catalog sales orders using inventory codes. We find library books by using their call numbers, a form of code. Social Security numbers, form numbers, and bank account numbers—all are examples of codes.

Coding Considerations

The Need for a Code

A knowledge of codes and coding techniques is essential to the systems analyst. The analyst is called upon frequently to analyze or create documents that contain codes. The ability to interpret codes, evaluate coding schemes, and devise new or improved codes are important skills. The systems analyst must be trained to evaluate the need for a code and to set up a plan if a code is required.

Some codes come to us from external sources. Zip codes are required by the postal service for bulk mailings and are requested for all mail. Federal and state governments require us to use Social Security numbers as an identification code in reporting taxes withheld, earnings, and so forth. The requirement for codes on external reports does not necessarily mean that we must use these codes for internal purposes. Many internal items and data elements that we deal with do not require a code to identify them, while other internal items can be handled more efficiently if they are coded.

To determine whether a code is needed, the systems analyst must find answers to questions related to the items to be coded. Typical questions include these:

1. Who uses the items or data that are to be coded?
2. How often will the coded items be retrieved or used?
3. What are the coded items used for?

4. How much data should the code contain about the coded items?
5. How many items are to be coded?
6. How will the codes be used?
7. How will expansion of the number of coded items be accommodated?
8. What system characteristics must be accommodated by the code?

After the answers to these questions are determined, the analyst can decide whether or not a code is needed and, if so, the type of code best suited for a particular system. If the specific benefits of using a code can be identified, the analyst can proceed to establish a plan for the code.

The Code Plan and the Code Dictionary

The **code plan** identifies the particular characteristics that need to be contained within the code. Only information that makes possible efficient identification and retrieval of coded items should be included. For example, if an analyst were required to prepare a code for a business form, such as an organization chart, a code plan would be developed by selecting the characteristics of the form and its environment that are appropriate to the identification and retrieval of that form. Typical characteristics might include form identification, responsible department identification, revision number, and date of issue.

Characteristics that seem similar may be coded quite differently for unrelated items. For example, consider a tire code and a dress code. The code plan for each would specify an identification of size. In the case of a tire, size is expressed in terms of the ratio between the tire height and width, and of rim size. The size of a dress would, of course, be expressed quite differently.

After preparing a code plan the analyst selects an appropriate coding method. The method selected must be:

1. *Expandable*   The code must provide space for additional entries that may be required.
2. *Precise*   The code must identify the specific item.
3. *Concise*   The code must be brief and yet adequately describe the item.
4. *Convenient*   The code must be easy to encode and decode.
5. *Meaningful*   The code must be useful to the people dealing with it. If possible, it should indicate some of the characteristics of the item.
6. *Operable*   The code should be compatible with present and anticipated methods of data processing—manual or machine.

To allow humans to work easily with code, it is often necessary to develop a **code dictionary.** A code dictionary is a listing of codes and their corresponding data items. The dictionary allows you to translate the code into an identification of the data or to determine the code for a particular data item. The format of the dictionary is usually a column of codes placed against a column of corresponding data descriptions. The format may be in either code or data description sequence; commonly both sequences are used.

To select the coding method that best meets these criteria, the analyst must be familiar with the common types of codes and must know how they are used separately or in combination.

Common Types of Codes

There are five common types of codes: sequence codes, group classification codes, significant digit codes, alphabetic codes, and self-checking codes. In practice, combinations of these codes often are found.

√ Sequence Codes

Simple Sequence Code

A sequence code has no relation to the characteristics of an item. Therefore, a dictionary is required. The assignment of consecutive numbers, for example, 1, 2, 3, . . . , to a list of items as they occur is called a simple sequence code. Figure 5.1 is an example in which employees' names are listed alphabetically and then assigned an employee number in sequence as they appear on the list. The alphabetic order of the items is for the purpose of aiding in the decoding process. However, if a new item were added, it would be assigned the next sequential number. Thus, if "Allen, James" joined the company, he would be assigned the next available number, and the effectiveness of the alphabet sequence would be diminished.

The advantage of the simple sequence is its ability to code an unlimited number of items with the least number of code digits. Its principal disadvantage is the limited amount of information it can convey. However, the simple sequence code often is used as a component of more complex codes.

Block Sequence Code

This code is a modification of the sequence code that makes possible a more homogeneous collection of related items. In a block sequence code, a series of consecutive numbers and/or letters is divided into blocks, each one reserved for identifying a group of items with a common characteristic.

As in the simple sequence code, a list is prepared of items to be coded. The difference is that, in assigning the codes, "blocks" of sequence numbers are set aside for items with some common characteristics. As an example, the simple sequence codes assigned to employees in figure 5.1 could be changed to a block sequence code in which the common characteristic of the items in a block is the first letter of the employee's last name. As shown in the first part of figure 5.2, a "block" of numbers (1–9) is assigned to the A's, another block of numbers (10–19) assigned to the B's, and so on. With this block code we can add new employees and assign them a code from their alphabetic block. Although we cannot maintain the original alphabetic sequence, we can keep the employees grouped by the first letter of their last name.

A common use of the block sequence code is shown in the second part of figure 5.2. In this example of a furniture inventory, blocks of code numbers are assigned by basic characteristics—in this case, the material used in the furniture's construction.

Figure 5.1

Simple sequence code dictionary. This example was developed by listing employee names alphabetically and then assigning employee numbers in sequence.

code	employee name
1	ADDINGTON, HORACE R.
2	ANDERSON, BERTHA A.
3	CONRAD, ROBERT L.
4	CRANE, JAMES M.
5	CUSTER, GEORGE G.
6	DAWSON, PETER R.
7	DUNCAN, HENRY A.
8	ECKEL, GARY T.
*	* * * *
*	* * * *
*	* * * *

Figure 5.2

Block sequence code dictionaries. Block sequence codes assign a block of numbers to a characteristic of the items to be encoded. In these examples, the first letter of the last name and the furniture material are the selected characteristics.

code	employee name	code	data item
1	ADDINGTON, HORACE R.	1	CHAIR, WOOD—TABLE
2	ANDERSON, BERTHA A.	2	CHAIR, WOOD—FOLDING
20	CONRAD, ROBERT L.	3	CHAIR, WOOD—ROCKING
21	CRANE, JAMES M.		
22	CUSTER, GEORGE G.	10	CHAIR, PLASTIC—TABLE
30	DAWSON, PETER R.	11	CHAIR, PLASTIC—FOLDING
31	DUNCAN, HENRY A.	12	CHAIR, PLASTIC—ROCKING
40	ECKEL, GARY T.		
*	* * * *	20	CHAIR, CHROME—TABLE
*	* * * *	21	CHAIR, CHROME—FOLDING
*	* * * *	22	CHAIR, CHROME—ROCKING

Note that, in this example, we could have used the type of chair as the major characteristic just as easily as the construction material. Table chairs could have been assigned numbers 1–10; folding chairs, numbers 11–19; rocking chairs, numbers 20–29; and so on. The characteristics the analyst should select are those that are most meaningful to the users of the code.

Like the simple sequence code, the block sequence code often appears as part of more complex codes.

Group Classification Code

Another common code type, the *group classification code,* designates major, intermediate, and minor data classification by successively lower orders of digits. This code type is useful when the item or information to be coded can be broken down into subclassifications or subdivisions. The zip code is a familiar example of a group classification code. For example, the U.S. Postal Service has recently modified the zip code to one called the zip-plus-4 code. The zip-plus-4 code 91791–0344 relates to major through minor classifications as follows:

```
9  17  91  –  0344
│   │   │       └─the delivery area
│   │   └─the city or a portion of a city
│   └─the major postal center
└─the geographic region
```

The code provides for 10 (0–9) geographic regions, 100 (0–99) postal centers in each region, 100 (0–99) cities or city subdivisions within each postal center, and up to 10,000 (0–9999) delivery areas within each city or city subdivision. The delivery area can refer to one side of the street in a block of homes or to a particular address or post office box for high mail-volume addresses. Each element of the code is a *sequence code.* Code directories are available for the interpretation of each sequence code.

In another example, shown in figure 5.3, the primary objective is to uniquely identify a salesperson within a large department store chain. In addition to the identification of the individual, we want to be able to determine the store and the department where the salesperson works. The highest level of classification is the store, and so the store number becomes the "major" classification. The second subdivision is the department within the identified store; the department number becomes an "intermediate" classification. The number of the employee within an identified department is the "minor" classification. We form the group classification code by combining the numbers of the major, intermediate, and minor classifications. In this example of a code, each salesperson has a unique identification number. Furthermore, we have coded additional valuable data about the work station of the employee.

Note that the store, department, and salesperson numbers are probably simple sequence or block sequence codes. The code is called a group classification code, however, since the dominant part of the code plan is the use of major, intermediate, and minor classifications.

Figure 5.3

Group classification code. Group classification codes designate major, intermediate, and minor data classifications by successively lower orders of digits. In this example, the salesperson group (minor) is within the department group (intermediate), which is within the store group (major).

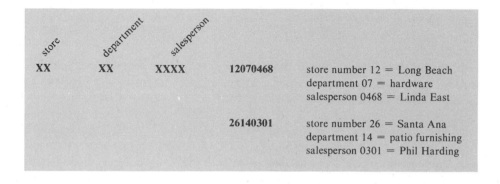

Figure 5.4

Significant digit code. Significant digit codes are numeric codes in which the numbers describe a measurable physical characteristic of the item. In this example, the physically measurable attributes are the ratio of tire height to its width and rim size.

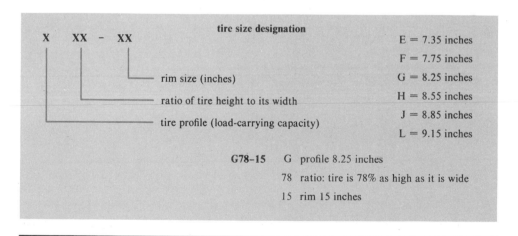

Significant Digit Code

A third common code is the *significant digit code,* a numeric code in which the numbers describe a measurable physical characteristic of the item. The characteristic may be weight, size, length, capacity, time, or any other physically measurable attribute that is part of the code plan. Figure 5.4 gives as an example the code plan for tire size. It has three elements: the tire profile, the ratio of the tire height to its width, and its rim size. This code describes two of the elements as significant digits. These are "ratio of tire height to width" and "rim size." The tire profile is indicated by a letter rather than the actual numerical measurement. A dictionary is required to interpret the profile. As in this example, it is not unusual to see a mix of code types. However, in this case, the tire code is predominantly a significant digit code.

Significant digit codes and group classification codes are often confused with one another. Remember: In a significant digit code, the digits are a measurement of one or more physical characteristics.

Alphabetic Codes

A fourth common type of code is the *alphabetic code,* which describes items by the use of letter and number combinations. There are two categories of alphabetic code: mnemonic codes and alphabetic derivation codes.

Mnemonic Codes

These are letter and number combinations obtained from descriptions of the coded item. A mnemonic, or memory aid, is a reminder of the name or description of an item. Often it is a severe abbreviation of the item's name. Figure 5.5 shows two examples of mnemonic merchandise codes. With a little experience, most people using the code can decode it without reference to a dictionary.

Figure 5.5

Mnemonic codes. Mnemonic codes are letter and number combinations obtained from descriptions of the coded item. A mnemonic is a memory aid, or reminder, of the name or description of an item.

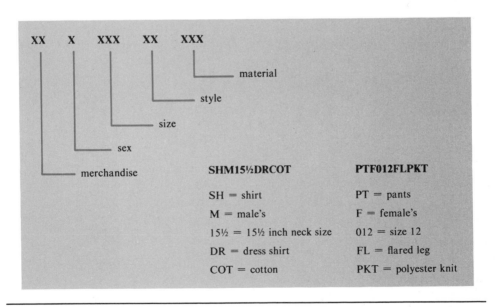

XX X XXX XX XXX

— material
— style
— size
— sex
— merchandise

SHM15½DRCOT

SH = shirt
M = male's
15½ = 15½ inch neck size
DR = dress shirt
COT = cotton

PTF012FLPKT

PT = pants
F = female's
012 = size 12
FL = flared leg
PKT = polyester knit

Many computer language codes used by programmers are symbolic codes. As an example, consider the following assembler language instructions:

Mnemonic	Instruction
MVC	Move Characters
AP	Add Packed Decimal
SP	Subtract Packed Decimal
ST	Store
DC	Define Constant

Alphabetic Derivation Codes

These codes are characters taken or derived from the name or description of the coded item according to a set of rules. Alphabetic derivation codes are used to handle large volume lists that must be maintained and processed in sequence. These codes are used because it is not practical to encode and decode full descriptions or numeric codes. Figure 5.6 is an example of a magazine subscriber's code. The code not only uniquely identifies the subscriber but also provides information about the distribution center, the zip code, the magazine's name, and the expiration date of the subscription. Mailing labels can be printed in zip code sequence within the distribution center for ease of handling for shipment. Renewal notices can be sent within selected time periods based on the subscription's expiration date. The magazine name may identify special interests of the subscriber to which direct mail advertisers can appeal.

There are other forms of alphabetic derivation codes. These are based on relatively elaborate sets of rules for the use of consonants and phonetic characteristics of the item to be coded. An excellent reference for additional information about the codes that we have discussed and about more elaborate codes is provided in an IBM manual on coding methods.[1]

Self-Checking Codes

The fifth code type is the *self-checking code*. It uses a check digit to check the validity of the code. This type of code is an important means of controlling the validity of data that is being processed. For this reason it is important that an analyst have a knowledge of self-checking codes. These codes require the performance of a simple repetitive arithmetic operation. In computer-based systems, the checking operation can be performed rapidly. In on-line systems such as those using point-of-sale terminals, self-checking codes are extremely important. Whenever a clerk is required to enter customer numbers, product numbers, or other coded information, self-checking codes should be used to verify the code.

---

1. *Data Processing Techniques: Coding Methods,* Form GF20–8093 (White Plains, New York: IBM Technical Publications Department).

Figure 5.6
Alphabetic derivation code. Alphabetic derivation codes are made up of characters taken or derived from the name or description of the coded item according to a set of rules. A very common use of this code type is to identify subscribers of magazines.

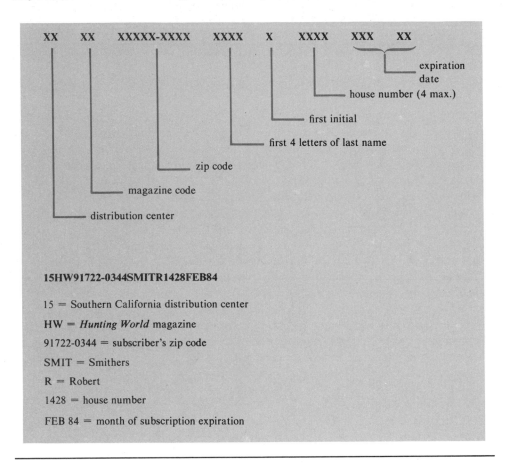

XX   XX   XXXXX-XXXX   XXXX   X   XXXX   XXX   XX

expiration date

house number (4 max.)

first initial

first 4 letters of last name

zip code

magazine code

distribution center

15HW91722-0344SMITR1428FEB84

15 = Southern California distribution center

HW = *Hunting World* magazine

91722-0344 = subscriber's zip code

SMIT = Smithers

R = Robert

1428 = house number

FEB 84 = month of subscription expiration

Figure 5.7 is an example of how such a code can be developed and used. The method for developing the check digit in this example is to multiply the first digit by 1, the second digit by 2, the third by 3, and so forth. After multiplying each code digit by its position, we will then add up the products. The third step is to add the digits of the sum from step two until we have a one digit answer. This final answer is the check digit. If an error has occurred in recording or processing the coded item, the check digit will not "check." As a specific example, figure 5.7 shows a code of 12463 with a check digit of 2. The complete code for the item is 12463–2. If the common error of transposition occurred and the second and third digits were reversed, the code would appear as 14263–2. To check the validity of the code, the computer program would

Figure 5.7
Self-checking code. Self-checking codes are numeric codes to which a "check digit" has been added to provide a validity check. These codes are commonly used in automated systems, especially where numeric codes are entered manually.

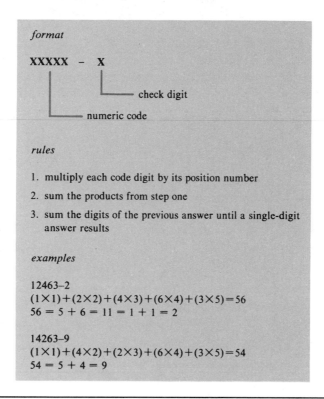

format

XXXXX – X

　　check digit

　　numeric code

rules

1. multiply each code digit by its position number
2. sum the products from step one
3. sum the digits of the previous answer until a single-digit answer results

examples

12463–2
$(1 \times 1) + (2 \times 2) + (4 \times 3) + (6 \times 4) + (3 \times 5) = 56$
$56 = 5 + 6 = 11 = 1 + 1 = 2$

14263–9
$(1 \times 1) + (4 \times 2) + (2 \times 3) + (6 \times 4) + (3 \times 5) = 54$
$54 = 5 + 4 = 9$

multiply each code digit by its position ($1 \times 1 = 1$, $4 \times 2 = 8$, $2 \times 3 = 6$, $6 \times 4 = 24$, $3 \times 5 = 15$), add up the products ($1 + 8 + 6 + 24 + 15 = 54$), and then add the digits of the sum until a single digit answer results ($5 + 4 = 9$), and compare the calculated check digit (9) to the check digit of the code (2). Since they don't match, the program would reject the code as invalid. This method works well for errors due to transposition or the inaccurate recording of the digits.

More sophisticated self-checking codes, often statistical in nature, have been developed for checking the validity of large volumes of data. Often this data has to be interpreted and evaluated by complex processing programs.

The reader should be aware that there is an entire field of mathematics, known as information theory, which deals with information and its validity. However, with respect to codes, our purpose has been to only introduce the analyst to the basic codes and to their uses in implementing effective code plans.

Summary

Identification of data is a major task. The use of codes to identify data has several advantages: (1) codes are relatively brief and can be handled quickly, (2) codes are easier than descriptions to make unique for positive identification, and (3) codes may contain useful information about the coded data.

The first step in generating a coding system is to develop a code plan that identifies the particular characteristics of an item needing to be contained within the code. Careful consideration must be given to the nature of the items to be coded and to how the codes are to be used. The five common types of codes are: sequence, group classification, significant digit, alphabetic, and self-checking codes. They all have different characteristics. The code type selected will depend upon the code plan.

For Review

alphabetic codes	sequence codes	**code plan**
block sequence code	significant digit codes	mnemonic code
code dictionary	alphabetic derivation	self-checking codes
group classification	code	simple sequence codes
code	**code**	

For Discussion

1. Why are codes used in business?
2. What kinds of questions must an analyst consider in order to determine if a code is needed?
3. What is a code plan? Present an example of a code plan.
4. Identify the five common types of codes.
5. Distinguish between simple sequence and block sequence codes.
6. What is the essential characteristic of significant digit codes? Present an example.
7. What are the two types of alphabetic codes discussed in this text? Present an example of each.
8. Why should an analyst be familiar with self-checking codes?
9. Why are alphabetic derivation codes used for large volume mailings?

6
Forms Design

Preview

The cost of collecting raw data and the cost of distributing processed information are two of the major costs of a system. Since most of the data that enters and leaves a system is recorded on forms (source documents and outputs), careful forms design can greatly affect the cost effectiveness of a system.

It is the task of the systems analyst to assist the user in the design of appropriate forms and to coordinate the form production activities. It is also the task of the analyst to control and coordinate the generation of new and modified forms within the organization to prevent costly duplication of forms and form design effort.

Objectives

1. You will learn the basic parts of a form, styles and types of forms, and principles of forms design.
2. You will be introduced to methods of forms control.

Key Terms

forms control   The coordination of the forms needs of multiple departments or users to control the costs forms.

style   The design of the body of a form, that is, boxed or open.

type   The classification of a form by the complexity of its manufacturing process, that is, cut or specialty.

Forms Design Responsibility

The task of drawing a detailed layout for a form is the responsibility of a forms specialist. Forms specialists are experts in types and styles of forms, sizes and styles of print, and details of preparing a form layout for reproduction. These specialists are available to the systems analyst from in-house reproduction centers or from the major forms manufacturers.

The only things the forms specialist does not know are (1) what data the user wants to collect with the form, and (2) how the form is going to be used in the system. These things are known only to the user. The analyst must help users develop their form requirements and guide them in the development of cost-effective forms. The coordination of the forms design effort between the user and the forms specialist is the responsibility of the systems analyst. It is the analyst who must verify that proof copy of the form is exactly as required by the user and the system. The forms manufacturers can provide a wide variety of sample forms for ideas, but the analyst must select and check the final form design.

The systems analyst is also responsible for the input/output screen design on CRT terminals and printed outputs. In this chapter we will concentrate on the design of hard copy source documents. Please refer to chapter 17, Output Design, and to chapter 18, Input Design, for discussions on printed outputs and screen layouts. In all cases, the forms design considerations in this chapter apply.

To be able to evaluate a form intelligently, the analyst must be familiar with the function of each of its parts.

Basic Parts of a Form

Most forms have five basic parts: title, instructions, heading, body, and conclusion.

The *title* identifies the form. It should be as descriptive and as brief as possible. Examples are "Purchase Order," "Memo," and "Sales Order." The title usually is centered at the top of the form.

Instructions tell how to complete or use the form. To be most useful, general instructions should be at the beginning of the form. Instructions should be placed so that they are seen before the form is completed. Forms that have multiple sections or parts may have a new set of instructions at the beginning of each section. Some forms are so commonly used and the instructions are so obvious to any user that general instructions are omitted. Instead, lines or boxes are labeled. Common examples are bank checks and receipts.

The *heading* contains all the general identification data. For example, it might include the date, form sequence number, name, and address. The heading data usually is not the information the form was designed to collect, but that which is necessary for record identification purposes. All data used for reference filing is contained in the heading. The heading often is separated from the remainder of the form by ruled lines or by a box drawn around it.

The specific data the form was designed to collect is called the *body* of the form. An example is a sales invoice that shows quantities, item descriptions, unit prices, and total prices. The body of the form should read from left to right and from top to bottom.

This helps both those who must complete the form and those who must extract data from it. Since the body of many forms is numeric, it is common to see the body set up in columnar form. The appropriate use of columns groups similar data and is an aid in filling out the form.

The last part of a form is the *conclusion*. The conclusion contains the approvals, signatures, and summary data. It appears at the bottom of the form.

The basic parts described here are found on most forms, yet exceptions are common. Forms design is a combination of skill and art, however, you should not allow considerations of art or "pretty forms" to override the principles of good forms design.

Figure 6.1 illustrates the five basic parts of a form.

Styles and Types of Forms

Styles of Forms

Two basic **styles** are used to design the body of forms: the *open style* and the *boxed style*. The open style is the simplest. It consists of headings and open areas in which data can be entered. The boxed style allocates space to each data item. Each box is clearly identified by name or by a brief description.

Forms are seldom pure "open" or pure "boxed." They are usually described as predominantly open, predominantly boxed, or as a combination of boxed and open. Figures 6.2 through 6.5 are examples of form styles.

Figure 6.2 is an example of a predominantly "boxed" form. The only open areas are the spaces for the comments and remarks entries.

Figure 6.3 is predominantly "open." The columns headed Description, Format, and Size divide the form into sections. However, the form is primarily designed for the unrestricted entry of an unspecified number of items.

Figure 6.4 is another example of an almost pure "open" form. The "boxes" on the form serve a function similar to that of ruled lines on a piece of notebook paper. In this case, the characters are flowchart symbols and not letters of the alphabet. The user of the form may place these symbols wherever appropriate.

Figure 6.5 is an example of a combination boxed and open style. This form is divided into well-defined areas. However, the center area, labeled "layout," is completely open.

Types of Forms

The **type** of form is a classification determined by the complexity of its manufacture. The most common forms are printed single sheets of paper, called *cut forms*. Cut forms make up approximately 90 percent of all forms. Cut forms are usually designed by the user and printed in-house or by a local printer. Many ready-made stock forms carried by stationery dealers are also cut forms. Cut forms are often made into pads for convenience, and several copies may be made in one writing by the use of carbon paper.

Specialty forms are more complex. Examples are multiple-copy forms, forms with special binding, and forms designed to be completed with the use of a machine.

Figure 6.1

A typical form. The five basic parts of a typical form are the title, instructions, heading, body, and conclusion. Some forms may not include instructions and/or a conclusion.

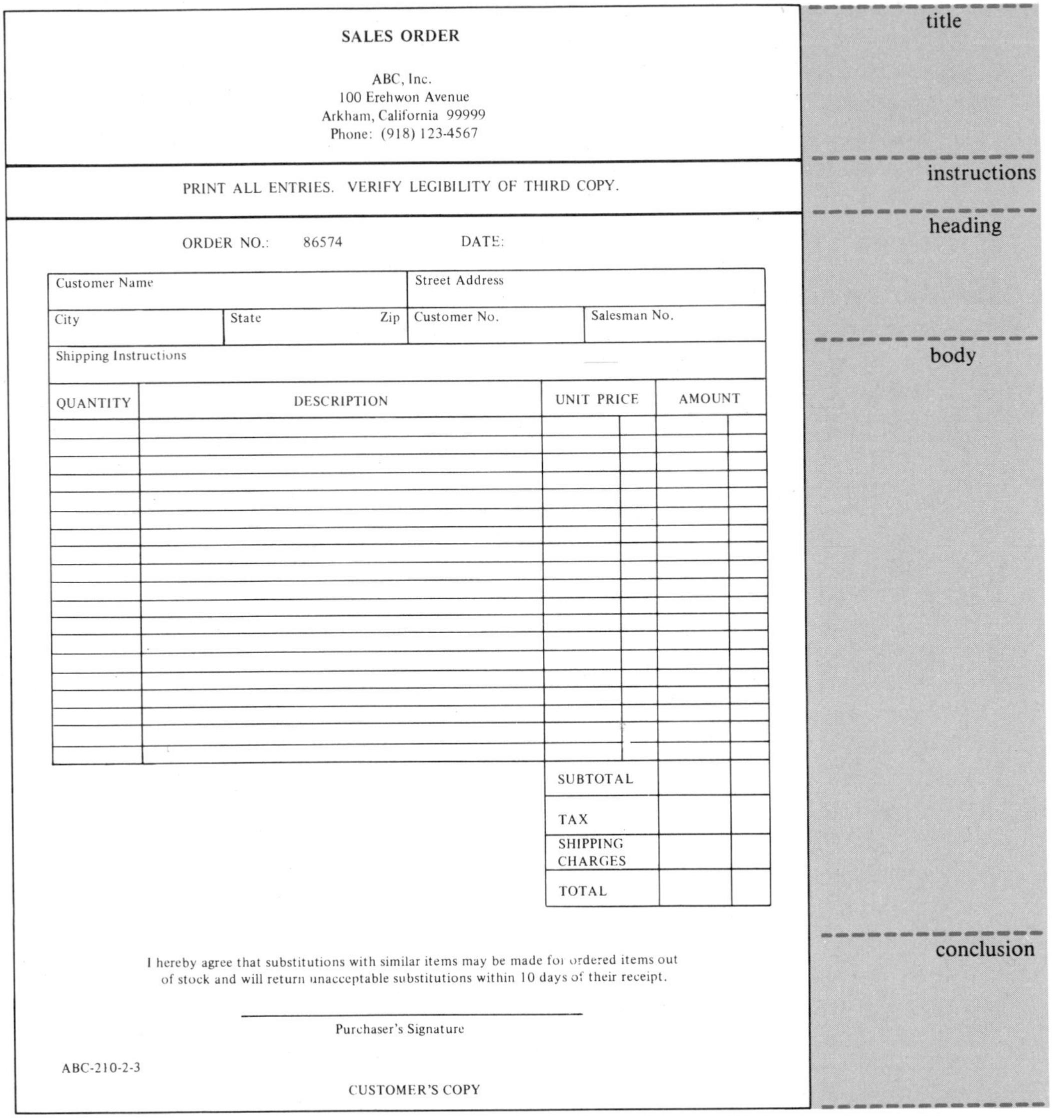

SALES ORDER

ABC, Inc.
100 Erehwon Avenue
Arkham, California 99999
Phone: (918) 123-4567

PRINT ALL ENTRIES. VERIFY LEGIBILITY OF THIRD COPY.

ORDER NO.:   86574   DATE:

Customer Name			Street Address	
City	State	Zip	Customer No.	Salesman No.

Shipping Instructions

QUANTITY	DESCRIPTION	UNIT PRICE	AMOUNT
		SUBTOTAL	
		TAX	
		SHIPPING CHARGES	
		TOTAL	

I hereby agree that substitutions with similar items may be made for ordered items out
of stock and will return unacceptable substitutions within 10 days of their receipt.

Purchaser's Signature

ABC-210-2-3

CUSTOMER'S COPY

- title
- instructions
- heading
- body
- conclusion

Figure 6.2
Information service request. The information service request is an example of a predominantly boxed-style form.

INFORMATION SERVICE REQUEST		Page ___ of ___		

JOB TITLE:

NEW ☐
REV. ☐

REQUESTED DATE: REQUIRED DATE:

AUTHORIZATION

OBJECTIVE:

	LABOR		OTHER	
	HOURS	AMOUNT	HOURS	AMOUNT

ANTICIPATED BENEFITS:

OUTPUT DESCRIPTION	INPUT DESCRIPTION
TITLE:	TITLE:
FREQUENCY: QUANTITY:	FREQUENCY:
PAGES: COPIES:	QUANTITY:
COMMENTS:	COMMENTS:
TITLE:	TITLE:
FREQUENCY: QUANTITY:	FREQUENCY:
PAGES: COPIES:	QUANTITY:
COMMENTS:	COMMENTS:

TO BE FILLED OUT BY REQUESTOR

REQUESTED BY:	DEPARTMENT:	TITLE:	TELEPHONE:
APPROVED BY:	DEPARTMENT:	TITLE:	TELEPHONE:

TO BE FILLED OUT BY INFORMATION SERVICES

FILE NO:	ACCEPTED ☐ NOT ACCEPTED ☐		
SIGNATURE:	DEPARTMENT:	TITLE:	TELEPHONE:

REMARKS:

FORM NO: C–6–1	ADDITIONAL INFORMATION: USE REVERSE SIDE OR EXTRA PAGES

Figure 6.3
Data element list. The data element list is an example of a predominantly open-style form.

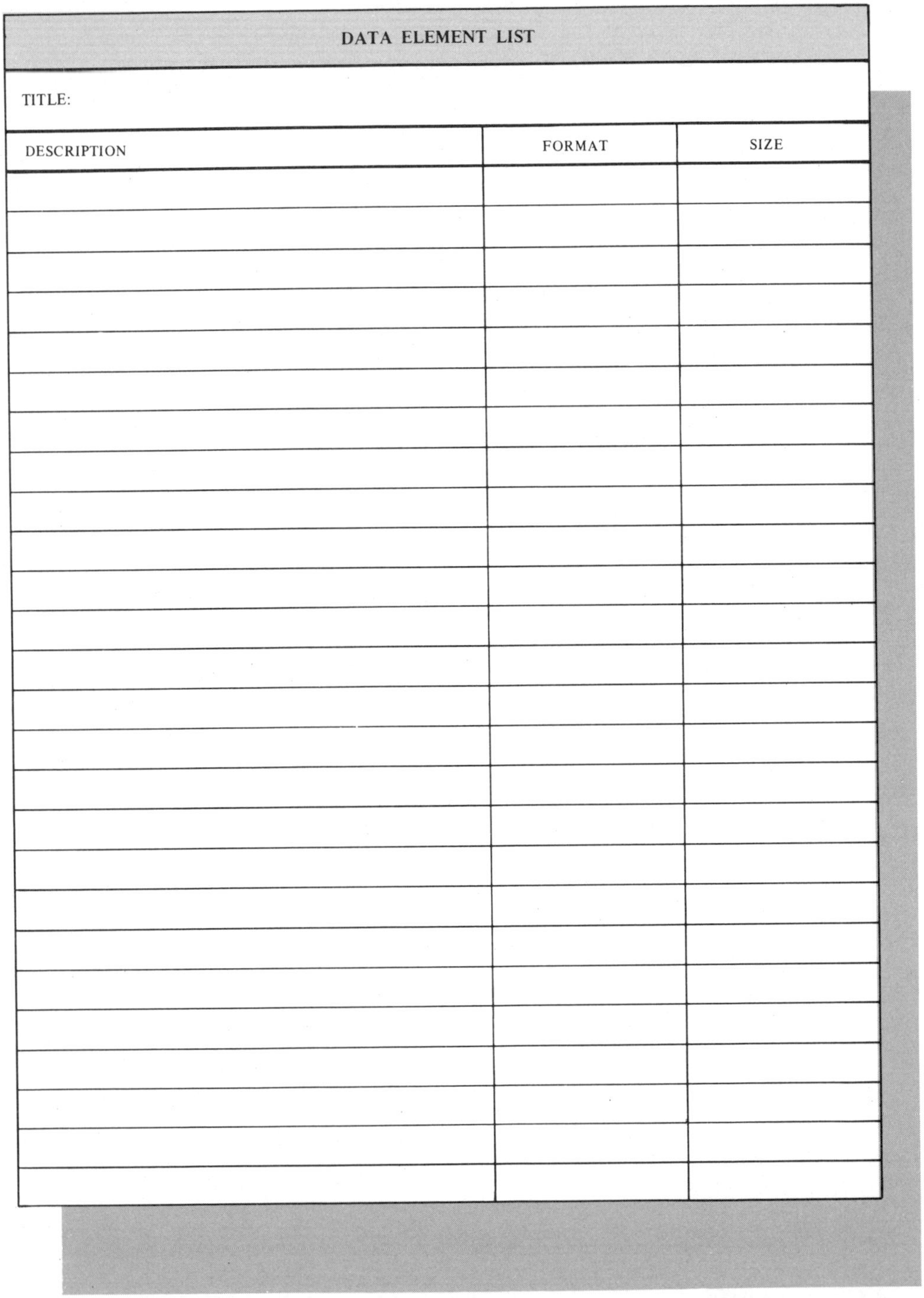

DATA ELEMENT LIST		
TITLE:		
DESCRIPTION	FORMAT	SIZE

Figure 6.4
Flowchart worksheet. The flowchart worksheet is an example of an almost pure open-style form.

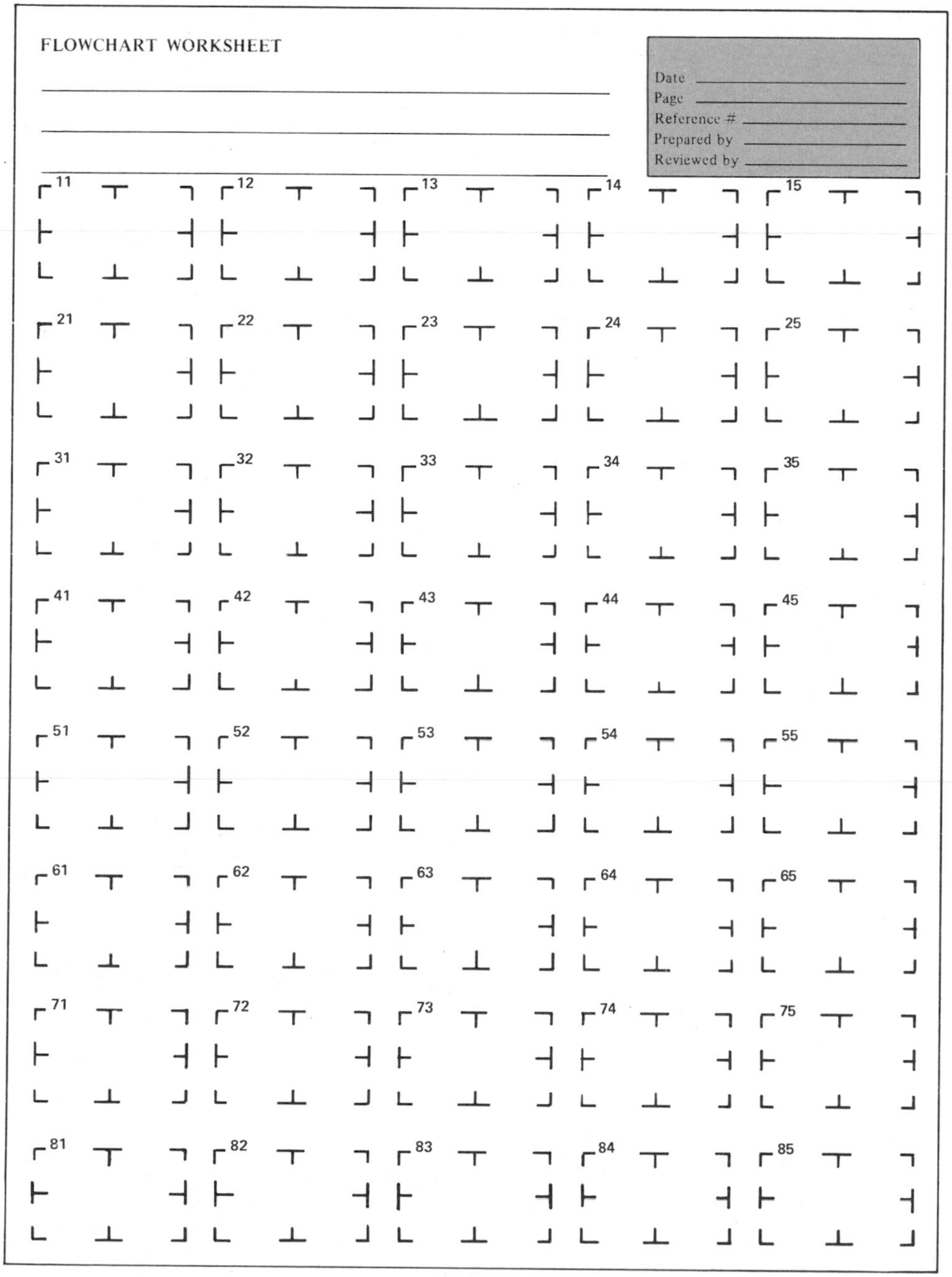

Figure 6.5
Output specification. The output specification is an example of a combination boxed- and open-style form.

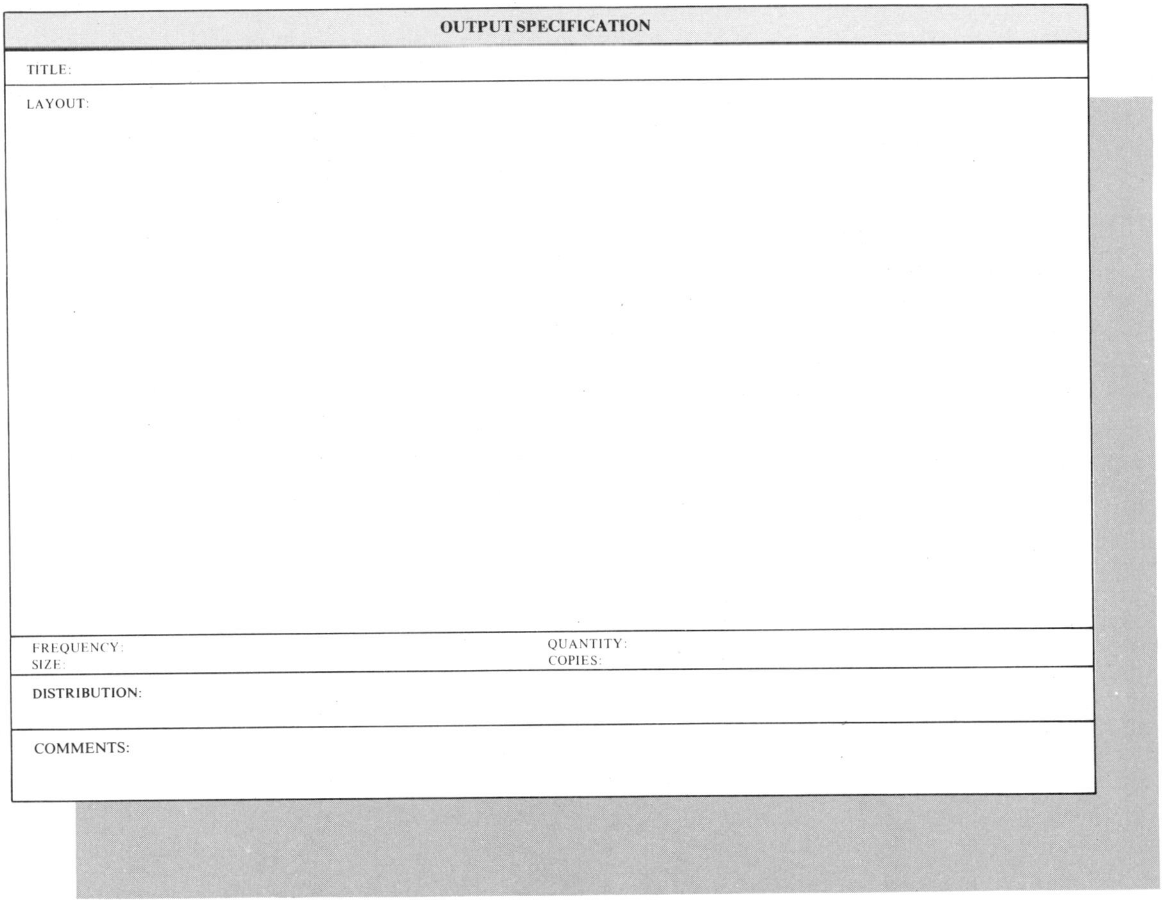

OUTPUT SPECIFICATION
TITLE:
LAYOUT:
FREQUENCY: QUANTITY: SIZE: COPIES:
DISTRIBUTION:
COMMENTS:

Specialty forms are complex enough in their construction to require special equipment for their manufacture or use. Most specialty forms are custom-designed by a forms manufacturer, usually a large printing firm that specializes in forms design and has equipment to produce specialty forms. Some specialty forms are designed to be used only with special equipment. Others are designed to be completed manually. Still others require both manual and machine operations. There are five principal types of specialty forms: (1) forms bound into books; (2) continuous forms—manual; (3) detachable stub set; (4) continuous forms—machine; and (5) mailers.

Examples of these five forms are shown in figure 6.6.

The simplest specialty forms are those that are bound into books; they are very similar to padded cut forms except for a stronger binding. They are designed to be completed manually. Examples are sales books and receipt books.

Figure 6.6
Specialty forms. Specialty forms require special equipment in their manufacture. Common types of specialty forms are forms bound into books, continuous forms—manual, detachable stub sets, continuous forms—machine, and mailers.

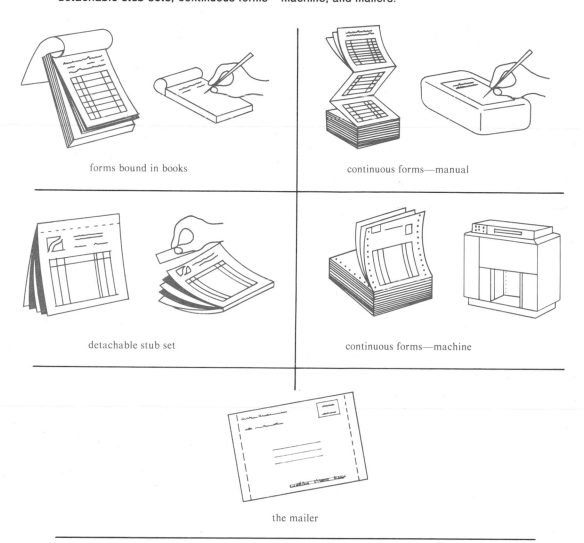

forms bound in books   continuous forms—manual

detachable stub set   continuous forms—machine

the mailer

Continuous forms—manual are attached end-to-end in a long, continuous string. They are designed to be used with a counter-top machine. The machine is used only to feed and hold the forms, which are filled in by hand.

A type of specialty form that can be filled in by hand or by machine is the detachable stub set, which is an original and one or more carbon copies bound together. The set may either use carbonless paper or be bound complete with carbon paper. The binding may be on any of the four edges of the form, but is most often at the top or left edge. The bound portion of the set, or the stub, is perforated. Individual copies may be removed from the set, or the stub may be torn off to separate all the copies. These forms are very popular because of the ease and convenience of use. The form may be completed either manually or by machine. The individual using the form does not have to align the carbon paper and the multiple copies of the form. Also, because of the binding in each sheet, copies do not slip out of alignment as the form is written on or put into a machine. A very common example of the detachable stub set is the form used by credit card companies to record charge sales. Devices to record the information from a customer's Master Charge, Visa, and so forth, are found in many retail stores.

Continuous forms—machine are the forms used by computer printers. They may be single part (no carbons) or multiple part (with carbons) forms. Often forms to be completed by a computer will simply be blank continuous forms. In this case all titles, headings, and so forth, are printed as a function of the computer program. This process is almost as fast as using formatted or printed forms. Using blank paper is much less expensive, and it eliminates the problem of keeping many different types of forms in inventory. Printed continuous forms also are designed for machine printing. The most common example is the paycheck. However, even printing the complete form by the computer does not eliminate the necessity for forms design. The output format must be designed, regardless of how the form is printed. Computer output design and its special considerations are described in chapter 17, Output Design.

The last general type is the mailer. It is a version of continuous forms designed to be filled in with a computer printer; however, its construction is different enough to classify it as a fifth type of specialty form. Mailers are blank forms sealed inside envelopes that are attached end-to-end as continuous forms and covered by a "cover page." Selected areas of the cover page and the front of each envelope are backed by carbon paper. All the data to be printed, including the name and address, is printed on the cover page. Because of the positioning of carbon paper behind the cover page, only the mailing name and address print through on the envelope. Because of the carbon paper on the front inside of the envelope, the form inside the envelope also is printed. The cover page is usually retained as a file copy of the data being mailed; it may be printed with information that does not print on either the envelope or the form. Mailer forms are fairly expensive, but their use saves the cost of stuffing and sealing envelopes. Mailers are usually printed with postage permit information, so no postage needs to be added. The forms may be sent to the post office as soon as they are printed and separated.

Principles of Forms Design

Well-designed business forms can increase clerical efficiency, improve work flow, and lower system costs. To evaluate a form's effectiveness, the analyst should keep four principles in mind:

1. The form must be easy to fill out.
2. The completed form must be easy to use in the system.
3. The form should not collect data that will not be used in the system.
4. The form should not be unnecessarily expensive.

Ease of Data Recording

Business forms should be designed so that they can be filled out quickly and accurately.

It is important that the analyst avoid errors induced by the form's design. Design-induced errors can occur whenever the person completing the form is not sure what data is being requested. Whenever a form is used in more than one department or is used infrequently, it is usually wise to include appropriate instructions. The instructions should be placed on the form just before the section to which they apply. It should be noted, however, that including instructions where they are not required hinders clarity and adds to the expense of producing the form.

Data items should be grouped in a logical pattern. Grouping requires fewer instructions and results in fewer errors. Fewer errors will occur if all the logically related data is collected prior to changing the subject.

All data entry areas must be clearly labeled. Make sure that the user knows whether the label applies to the line above the label or the line below. If the form will be filled in using a typewriter, the data area label should appear above the data entry area so that the label is visible when the form is in the typewriter. Avoid uncommon abbreviations or uncommon words as labels. Although labels should be brief, they must be complete enough to communicate exactly what data is being requested.

Leave adequate space for the response. Common typewriter horizontal spacing is ten or twelve characters per inch. Make sure that the data area is long enough to allow for reasonable responses. Remember also that handwritten entries require more space than typewritten responses. Vertical spacing also is important to forms design. Forms that are expected to be completed with handwritten responses should allow at least one-half inch of vertical spacing. Forms that may be completed using a typewriter should allow at least one-third of an inch and should be a multiple of one-sixth of an inch. (Most standard typewriters print six lines per vertical inch.) Avoid spacing that requires the typist to realign the typewriter for each line. The minimum of one-third inch would be equivalent to double spacing. If the form is always completed using a typewriter, leave out the horizontal lines to make alignment easier. Aligning the form can take as much of the user's time as actually entering the data.

Ease of Use

Sequence the data on the form in the order in which it is to be used. This is especially true when the data is to be entered into a computer system through a key-stroking operation.

The analyst should be aware of the effect of ink and paper color combinations on the legibility and readability of forms. Readability can be improved (or made worse) by the combination of ink color and background color. Figure 6.7 is LeCourier's legibility table, which ranks thirteen ink-background combinations in order of legibility. The common black on white combination ranks sixth, while black on yellow is at the top of the list. Colored paper can be used to distinguish the various copies of a form and to aid in form distribution. However, the use of colored paper should not be allowed to interfere with the legibility of the data. This is particularly problematic for copies at the bottom of a carbon stack.

Figure 6.7
Le Courier's legibility table. Le Courier's legibility table shows the effect of ink color and background color combinations on readability. Note that the traditional black on white ranks sixth in this chart.

order of legibility	color of printing	color of background
1	black	yellow
2	green	white
3	red	white
4	blue	white
5	white	blue
6	black	white
7	yellow	black
8	white	red
9	white	green
10	white	black
11	red	yellow
12	green	red
13	red	green

Many businesses input source data into computerized systems with optical scanners. If optical scanning of forms is to be used, the analyst must consider character size and vertical spacing requirements. It is the analyst's responsibility to verify that the form layout is compatible with the scanner hardware.

Collect Required Data
The analyst should verify that all the data items requested on a form are required and actually used in the system. Many times data is collected on a form simply because it was collected on previous versions of the form. Data items not actually required waste the time of the person completing the form and clutter the form for those who use the data. Wasting clerical time adds unnecessary expense to the system.

Avoid Unnecessary Costs
The costs of using a form are far greater than the costs of producing it. The area with the greatest potential for cost effectiveness is the efficiency of a form in use, rather than the cost of the form itself. Still, forms should not be unnecessarily expensive. If the company has in-house reproduction capability, there has to be a decision on whether to produce the form in-house or out-of-house. This decision must be based on the required quality and complexity of the form relative to in-house capability. For complex or high-quality forms, in-house production is not always the least expensive.

Design forms in a standard size, such as 8½" × 11" or 5½" × 8½". Printers buy their paper stock in standard sizes; if a form of an uncommon size is desired, it may be unnecessarily costly to produce because of the extra cutting or trimming of paper.

In general, *avoid forms larger than 8½" × 11".* Larger forms are often more expensive to store and require larger, more expensive, file storage.

Print forms in reasonable quantities. Bids for printing jobs include a basic set-up charge, which stays the same regardless of the quantity of forms produced. It is advantageous, therefore, to buy forms in as large a quantity as is reasonable, considering the rate of use, the likelihood of modification, and the costs of storage. The larger the quantity, the lower the cost per form.

Forms Control

A form designed for efficient use is a major step toward controlling the cost of collecting data. However, an equally important consideration is the prevention of a flood of new and modified forms from each department of the company. Departments tend to act independently of other departments when it comes to forms. Departments rarely inquire of each other to find out if a form is already in existence. They simply design a "new" one. A system of **forms control** is needed.

The solution to the problem is twofold: (1) establish a central forms authority, and (2) establish control files. The central authority should be an individual or a group with complete control over all company forms. The approval of the central authority must be obtained prior to the design or modification of any form. The advantage of this procedure is that this authority will have knowledge (and samples) of all forms currently in use or being designed. Since many forms may meet the needs of more than one department, the coordination of the design or modification of forms can be handled expediently by one authority. The use of one form rather than several similar forms can be a real cost saving in the production, storage, and distribution of forms.

Two control files are needed to keep track of the forms being used in the company: a numerical file and a functional file. The *numerical file* contains at least one sample of each form being used. The samples are filed by form number, creating a "catalog" of forms being used. Any form can be accessed by means of its coded form number. The *functional file* contains additional copies of each form. They are filed in order of subject, operation, or function. If a form has multiple uses or functions, it will be found more than once in the file. Whenever a new form is required, the functional file can be checked to determine whether a form already exists for that purpose. When the functional file is originally established, many forms usually can be eliminated or consolidated.

The two control files also aid in forms inventory control. Knowledge of which departments use any particular form and their approximate rate of use prevents the build-up of too large an inventory of forms and also is a timely reminder to reorder.

Summary

The user is the final authority on the items of data to be collected, the type and style of form to be used, and the final layout of the form. The forms manufacturer (or in-house reprographics department) is responsible for the final design of the form, according to the requirements specified by the user. It is, however, the responsibility of the systems analyst to coordinate the forms design effort with the user and the forms manufacturer. To assist the user, the analyst must be familiar with the basic parts of a form and with form styles and types. The five parts of a form are: title, instructions, heading, body, and conclusion. The two basic styles are open and boxed, although many forms are a combination of each. The type of form is determined by the complexity of its manufacture. The simplest and most common is the cut form; the most complex, requiring special equipment, is the specialty form.

To assist the user, the analyst must be familiar with the basic parts of a form and with form styles and types. Also, the analyst must always be aware of the following principles of form design:

1. The form must be easy to fill out.
2. The completed form must be easy to use.
3. The form should not collect data that will not be used in the system.
4. The form should not be unnecessarily expensive.

The systems analyst is also responsible for forms control—the coordination of forms design or modification between users and/or departments. This requires the maintenance of two files: a numerical file and a functional file.

For Review

title	open style	functional file
instructions	boxed style	**forms control**
heading	cut forms	**style**
body	specialty forms	**type**
conclusion	numerical file	

For Discussion

1. What are the forms design responsibilities of the user, the systems analyst, and the forms manufacturer?
2. What are the five basic parts of a form?
3. Describe the purposes of the title, heading, and body of a form.
4. What are the two basic styles of forms? Define each.
5. What are the two basic types of forms? Define each.
6. Name and discuss the four principles of forms design.
7. Discuss the "cost" of forms.
8. What are the advantages of formal forms control?
9. What two types of files are required for effective forms control? Why?

7
Charting Techniques

Preview

Designers and users of business information systems must cope with an overwhelming quantity of data in the normal course of their jobs. In addition, a great deal of time can be spent in trying to establish trends or other meaningful relationships among this data. Charts are valuable tools that can assist in these tasks since they are means of presenting large amounts of data, and complex relations between data, in a concise and meaningful manner.

Objectives

1. You will be able to identify and to use the four basic types of charts.
2. You will learn how tables can be used as analytic tools.
3. You will be introduced to decision trees and structured English.
4. You will understand the importance of charts as tools for project management.

Key Terms

chart   Graphical or pictorial expression of relationships or movement.

line chart   A chart constructed by connecting a set of plotted points; also called a graph.

table   A chart made up of intersecting horizontal and vertical lines to form rows and columns; also called a grid chart.

decision table   A table used to describe logical rules.

decision tree   A network-type chart that is the logical equivalent of a decision table.

structured English   A method for displaying a logical process in an outline format.

Gantt-type chart   A horizontal bar chart used to show a project schedule and to report progress on that schedule.

critical path network   A management tool that uses a graphical format to depict the relationships between tasks and schedules.

Basic Charts

The complexity of modern business, coupled with the increasing use of computers, is producing a veritable avalanche of data that must be analyzed and simplified before it can be used. Charts are one effective means of compressing data into a concise, clear format. Charts are graphical or pictorial expressions of relationships or movement. This chapter introduces important basic charts, as well as some specialized charts of particular significance in systems analysis. These include decision tables, charts for project management, and critical path networks.

Types of Basic Charts

Charts inform, compare, emphasize, and, in some cases, predict. Charts inform by displaying relationships or changes in relationships. They compare by relating items of information to an index, or scale. They emphasize significant changes or patterns of movement by accenting them visually. To the degree that past performance is an indicator of future performance, charts can predict by displaying trends.

There are four types of basic charts: bar charts, line charts, pie charts, and step charts. Most of these charts can be used to display relationships and motion. However, some are more suitable than others for each type of display. Figure 7.1 relates each basic chart to the display for which it is best suited. An alternate, less frequent use, is identified as "other use." Each of the four basic types of charts is described in the sections that follow.

Bar Charts

Bar charts depict relationships among elements better than any other type of chart. For this reason, and because they are easily understood in a variety of arrangements, bar charts often are used for management displays. When the bars are separated, the chart displays relationships. When they are spaced closely together, the chart creates an impression of a pattern of movement. Figure 7.2 depicts the appropriate use of bar charts to emphasize a relationship and to emphasize movement. The individual bars may be shaded to enhance the visual impact.

Line Charts

Line charts are the most common type of chart. Line charts often are called graphs because they usually are constructed by connecting a set of previously plotted points. Line charts communicate movement better than any other type of chart. They can display trends, curves, or any relationship where rate of change is important. Two or more lines may be used to compare trends. If more than one "line" is shown on a line chart, different types of lines (solid, broken, dotted, and so forth) or colors may be used to provide contrast. However, care must be taken not to display too many lines. When more than three or four lines are plotted on a single chart, the plots can overlap and obscure the message of the chart. Several lines can be on a single chart as long as they

Figure 7.1

Best use of basic charts. Bar and pie charts are best used to show relationships. Line and step charts best show movement.

display \ type	bar chart	line chart	pie chart	step chart
relationship	✔	0	✔	0
movement	0	✔		✔

✔ = best use
0 = other use

Figure 7.2

Bar charts. Bar charts depict relationships between elements better than any other type of chart. They may also be used to show a pattern of movement.

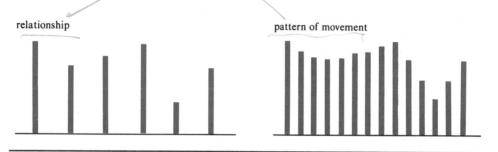

relationship

pattern of movement

do not overlap. If the lines overlap, the chart may confuse the viewer. Figure 7.3 demonstrates both a good and a bad example of the use of line charts. Note that the first part of the figure not only compares the two lines, but also lets the viewer extrapolate a change in relationship on the basis of the pattern of motion shown. The second part is too cluttered to be an effective graph.

Figure 7.3
Line charts. Line charts communicate movement better than any other type of chart. Multiple plots on a single chart can show relationships. Care must be taken to avoid having too many plots and confusing the chart message. Line charts are also called graphs.

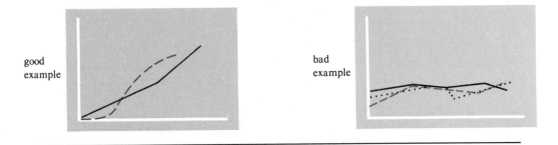

good example

bad example

Pie Charts

Pie charts are excellent charts for presenting relationships as percentages. In the first part of figure 7.4 a single pie is divided up into slices of various sizes. Since the total number of degrees in a circle is 360°, each slice represents a percentage that is the ratio between the angle of its arc and 360°. Thus, slice A of this standard "pie" shown is 50 percent of the total, and slices B and C are each 25 percent of the total.

Two pie charts also may be used to good effect for comparisons. Both charts should be the same size unless the total quantity that is sliced up has changed. For example, if the sources of funds for a fiscal year were presented on one pie chart, the applications of those funds could be presented on another pie chart of the same area. When the total quantity to be presented has changed in size, as might be the case at different points in time, the areas of the charts should be adjusted. When making this type of comparison, it is important to remember that the area increases as the square of the diameter. For example, if ten-year data for annual sales were presented as two pie charts, with sales increasing by a factor of four in the interval, the representation would be as shown in the second part of figure 7.4. Note that a 4:1 increase in surface area is obtained by doubling the diameter.

Step Charts

Step charts often can be used in the place of line charts to convey patterns of motion when relatively few points are plotted and when individual levels are to be emphasized. The first part of figure 7.5 is an example of a common type of step chart. This chart, sometimes called a "staircase chart," is an alternative to the line chart for showing movement. This chart does not convey a "flow" of movement to the same degree as does the line chart. However, it also possesses some of the characteristics of a bar chart. It can display even minor differences in relationships between small increments.

In many cases step charts can be used effectively in lieu of line charts. An interesting example, which brings together the step and line charts, is the histogram.

Figure 7.4

Pie charts. Pie charts are excellent charts for presenting relationships as percentages of the whole. They should not be used to show movement.

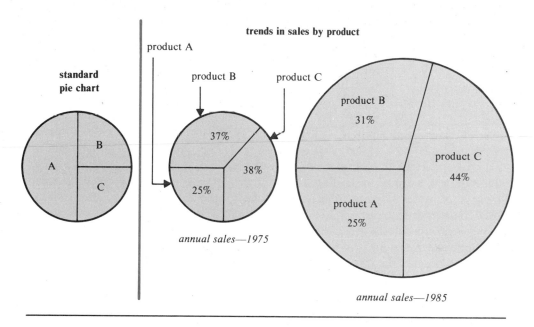

Figure 7.5

Step charts. Step charts often can be used in the place of line charts to convey patterns of motion when relatively few points are plotted and when individual levels are to be emphasized.

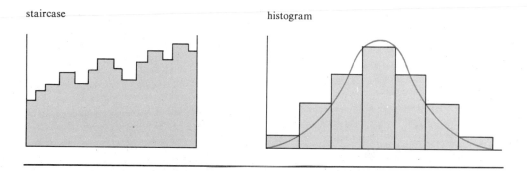

The histogram is used by statisticians to plot the relative frequency of the occurrence of events over definite intervals. The step chart in the second part of figure 7.5 is a typical histogram. The histogram can be used to develop a mathematical approximation to a continuous distribution. This is illustrated by the "equivalent" normal distribution curve that is superimposed on the pictured histogram.

Development of Effective Charts

There are six basic steps in the development of effective charts.

1. Decide upon the message for the chart.
2. Decide upon the best type of chart (for example, line, pie, bar, or step) to communicate the message.
3. Make an initial layout of the chart.
4. Analyze the layout to be sure that the chart stresses its message.
5. Modify the chart to eliminate unnecessary words and distracting detail.
6. Prepare the final chart.

This step-by-step procedure will result in the development of effective and useful charts.

An important thing to remember when preparing a chart is to "keep the chart honest." Charts can be misleading. For example, comparisons between quantities can be distorted by the improper selection of scales for the abscissa (horizontal axis) and the ordinate (vertical axis). Figure 7.6 illustrates how the choice of scale can alter the emphasis of a graph. Both graphs present the same data about corporate computing costs. Each reveals a 10 percent increase in computing costs from 1978 to 1982. The first graph accents the increase in cost; the second graph de-emphasizes it. Whether or not the increase is alarming depends upon factors that are not presented in either graph. However, an alert person would be suspicious of the motive behind such obvious "chartmanship."

Figure 7.6

Effect of scale selection. The choice of vertical and horizontal scales on a chart has a major impact on the chart message. It is important to emphasize the message without distorting it.

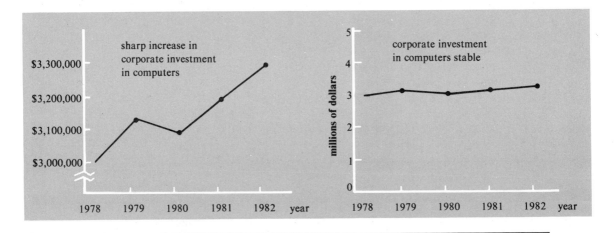

Figure 7.7 presents three examples of charts generated with a computer. As shown, line charts, bar charts, and line-bar chart combinations can be produced. There are several programs available that will take data stored in the computer and generate the required chart in a matter of minutes. These samples were generated and printed with a microcomputer. These charts, of course, may be displayed on a CRT terminal screen, as well as printed.

Figure 7.7
Sample computer-generated charts: (a) line chart, (b) bar chart, (c) combination line-bar chart. Software to generate charts is available for all sizes of computer. Computer-generated charts are easy to create and take only a matter of minutes. These examples were generated using a microcomputer.

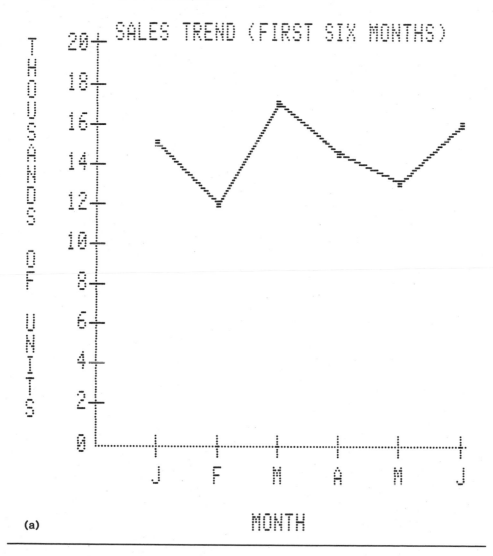

Figure 7.7
continued

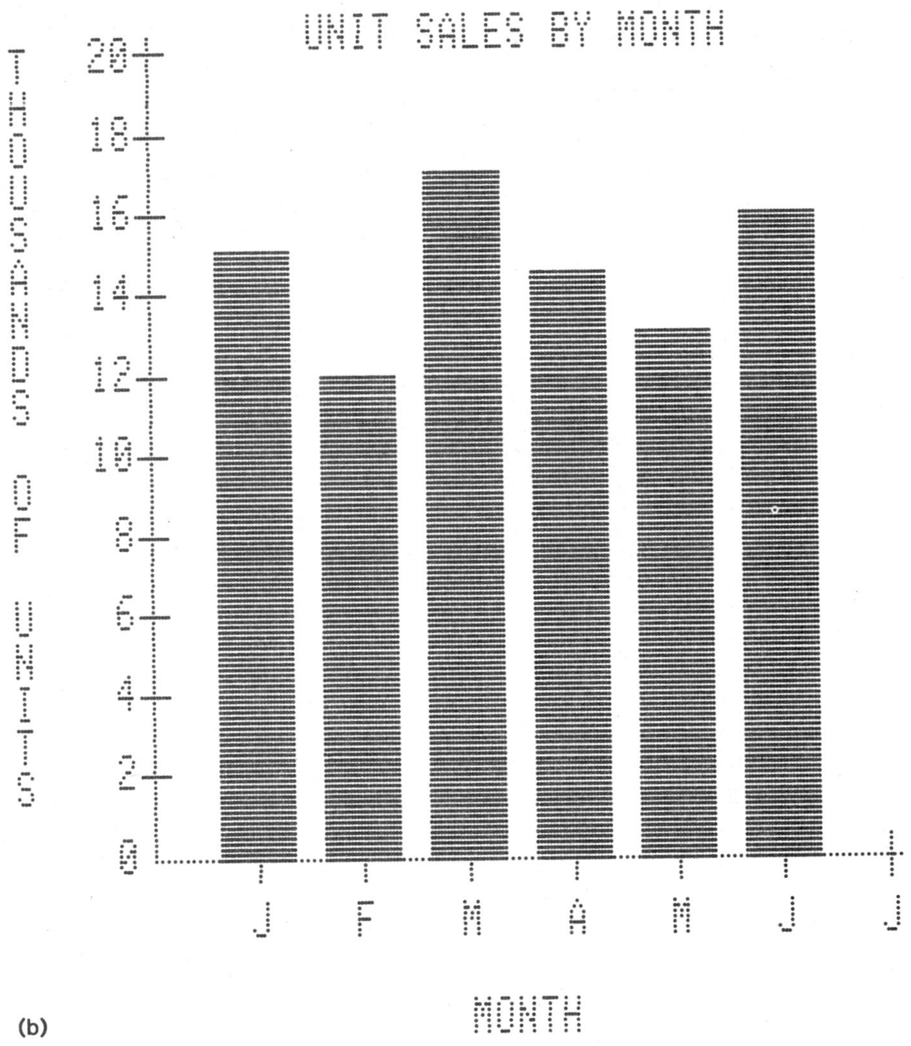

UNIT SALES BY MONTH
THOUSANDS OF UNITS
20
18
16
14
12
10
8
6
4
2
0
J
F
M
A
M
J
J
MONTH

(b)

Tables

Tables as Analytic Tools

Tables are so commonplace that there is a tendency to overlook their usefulness as analytic tools. However, **tables**—sometimes called grid charts because they are made up of intersecting horizontal and vertical lines that form rows and columns—are useful in all systems analysis activities. They provide an organized approach to decision making and are a handy reference for complex data. Setting up a table should be second nature to systems analysts; there is no more convenient method for expressing in summary form the relationships between two or more complex factors.

Figure 7.7
continued

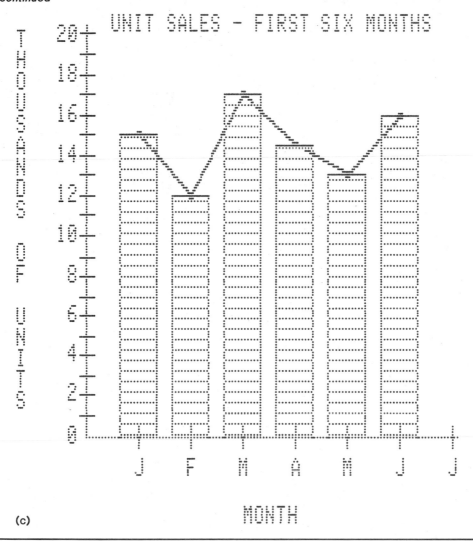

(c)

These relationships may be quantitative, qualitative, or a mixture of both. The first part of figure 7.8 is a segment of the binary multiplication table, and the product of two binary numbers always is a single quantity. The second part presents a qualitative relationship between factors that might be important to the choice of a data processing system in some business application. Although very superficial, this example illustrates one of the principal tasks the systems analyst is called upon to undertake. This task, known as a study of alternatives, involves the identification of

Figure 7.8
Examples of tables. Tables can provide an organized approach to decision making and are a handy reference for complex data. Tables may consist of quantitative or qualitative data.

N_2 / N_1	0	1	10
0	0	0	0
1	0	1	10

**quantitative:
binary
multiplication
table**

factor / processing system	manual	micro-computer	mini-computer	large computer
volume of data	low	low-medium	high	very high
speed of processing	low	low-medium	medium-high	high-very high
accuracy of processing	low	medium-high	high	very high
frequency of processing	low	low-medium	medium-high	high-very high
complexity of processing	low	low-medium	medium	high
total cost	low-medium	low-medium	medium	high

qualitative: selection of a data processing system

potentially suitable systems and the selection of the most appropriate system after evaluation of all factors relevant to a particular application. This important process will be discussed in detail in chapter 13, Feasibility Analysis.

Another particularly useful table for the systems analyst is the decision table, described in the following section.

Decision Tables

Decision tables are a tabular technique for describing logical rules. Most people tend to associate decision tables with the logic of computer programs. However, decision tables also are an important tool of the systems analyst because they are an effective means of expressing the logic of administrative rules and procedures. The first part of figure 7.9 depicts the basic format of a decision table. The table has four parts:

1. Condition stub—lists all conditions to be considered
2. Condition entries—make up the rules to be followed
3. Action entries—point to actions which may be taken
4. Action stub—identifies the action to be followed

Figure 7.9

Decision tables. Decision tables are a tabular technique for describing logical rules. They may be used in both manual and automated systems. This example describes the rules for cashing a customer's check at a retail store.

heading	rule numbers
condition stub	condition entries
action stub	action entries

basic format

check cashing policy		1	2	3	4
conditions	valid store identification card	Y	N	N	N
	purchase > $20.00		N	N	Y
	two other identifications		Y	N	
actions	allow purchase + $25.00	X			
	allow purchase amount		X		
	call store manager			X	X

example: check cashing policy

The decision table is read in the direction indicated by the heavy line and arrowheads. The condition stub is read as an "if" statement, and the action stub is read as a "then" statement. These statements are connected by a rule, which is a combination of the condition entry, usually indicated by a Y(yes) or an N(no), and the action pointed to, usually indicated by X.

We will use a simple example to exhibit the use of a decision table to condense and display systems logic. Consider the following check-cashing policy in a supermarket:

If the customer has a valid store identification card, a check may be cashed for the amount of the purchase plus $25.00. If a customer does not have a valid credit card but can show two other identifications, a check may be cashed for the amount of the purchase, not to exceed $20.00. Otherwise, the store manager must be called to authorize the acceptance of the check.

The "check-cashing policy" decision table appears in the second part of figure 7.9. A typical rule reads: "If a customer does not have a valid credit card, and if the purchase is not greater than $20.00, and if the customer has two identifications, then allow acceptance of the check for the amount of the purchase." Note that, in order to avoid redundancies, it is not necessary to fill in all the spaces in the condition entry section with Ys and Ns. In rule 4, for example, if the amount of the purchase is greater than $20.00, the store policy is that the manager must approve the customer's check if the customer does not have a valid card. Therefore, it would not be necessary to include a Y or N entry corresponding to the condition stub "two identifications."

Systems analysts often use decision tables as a means of communicating the systems logic, embedded in policies and procedures, to the programmer. When this type of logic is made available, the programmer is able to do a more effective job of developing the detailed computer program logic. Decision tables, then, are a technique that is useful in eliminating one of the major pitfalls that has impeded past efforts to develop effective computer-based business systems. This is the pitfall of forcing programmers, through default, to develop logical rules that may not reflect the true procedure that the computer program is intended to implement.

In summary, the decision table has the following major values for the systems analyst:

1. The structure of the table lends itself to a concise and correct statement of decision logic.
2. The table is an effective means of communicating with the computer programmer because it is easily understood.

These examples are only an introduction to the uses of tables. We have encountered and will continue to encounter many kinds of tables in this text. What must be emphasized here is the value of tables as powerful tools for collecting, analyzing, and reporting relationships among data.

Decision Trees and Structured English

Decision Trees

A **decision tree** is a network-type chart that is equivalent to a decision table. Like a decision table, it describes logical rules, showing all of the actions that result from various combinations of conditions. Figure 7.10 shows how network branches are drawn to handle combinations of conditions. This particular network is the logical equivalent of the decision table of figure 7.9.

A decision tree is less compact than a decision table and, in some cases, may be less suitable for presenting complex branching relationships. However, a decision tree provides a very easily understood picture since the branches can be read to show how all of the major and minor logical components go together. Like decision tables, decision trees can represent both system and computer program logical relationships.

Structured English

Decision tables also lend themselves to the use of **structured English,** which is a method for displaying logical processes in an outline format. The "structure" of structured English results from the use of an accepted structured-design terminology for describing the logical operations that computers can perform and from the manner in which the logic is expressed. Figure 7.11 presents the structured English equivalent to the decision tree of figure 7.10. Although structured English, in this instance, describes systems-level logic, this "pseudo language" technique also is used to describe the logic of computer programs. When this is done, the logic is said to be written in *pseudocode,* since it is a "pseudo-programming language." Pseudocode often is used as an alternative to computer program flowcharts, and an example of pseudocode appears in chapter 23, Computer Program Development.

Figure 7.10

A decision tree. The branches of a decision tree form a network that represents the logical rules by which a number of conditions are related. Lower-level relationships are nested within higher-level relationships.

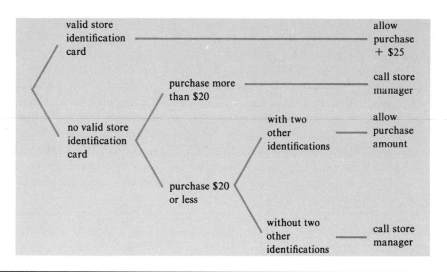

Figure 7.11

An example of structured English. Structured English uses standard logical terms, such as IF, THEN, and ELSE. It also follows rules, such as successive levels of indentation, for expressing logical relationships.

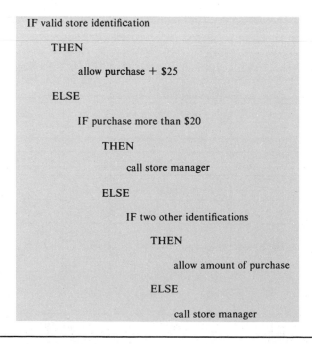

Charts for Project Management

Project Planning and Status Reporting

All projects are made up of significant events, or milestones, that must occur in some time sequence in order for the project to be completed. A project plan is a schedule of milestones over the duration of the project. Charts are an effective means of depicting a project schedule and of reporting progress (or lack of progress) as it occurs. The type of chart most often used for this purpose is a horizontal bar chart, sometimes called a Gantt chart. In figure 7.12 a **Gantt-type chart** illustrates the principles of a project planning and reporting chart. In this chart a project made up of six activities is scheduled over a time period of seven weeks. The length of the horizontal bar corresponds to the duration of the activity. Initially, activities are scheduled by means of an open bar. Then, the bar is "filled in" to show how much of the activity has been completed. In the example of figure 7.12 the reporting date is the end of the fifth week. Activities 1, 2, and 3 have been completed at this point in time. Activity 4 appears to be lagging about one-half week behind schedule. Activity 5 is slightly ahead of schedule, and Activity 6 has not yet been started.

There are many techniques for managing projects. Among the more innovative and effective of these is "management by objectives." This technique depends on the development of a cooperative rather than an authoritative environment, one within which individuals participate in defining their responsibilities and have the opportunity

Figure 7.12

Gantt-type chart. Gantt-type charts are used to schedule the major events of a project and to record actual progress in completing those events. The open bar shows scheduled progress. The filled-in bar shows actual progress. The first "bar" is for the total project.

reporting date: Week 5	reporting period in weeks											
activity	1	2	3	4	5	6	7	8	9	10	11	12
project												
activity 1												
activity 2												
activity 3												
activity 4												
activity 5												
activity 6												

to achieve results and earn recognition for a project (or for a phase of a project). In a joint discussion with a superior, they arrive at mutually understood and accepted objectives, results, and criteria for measuring results.

Figure 7.13 is an example of a Gantt-type chart that can be used to manage projects in this type of an environment. This chart reports:

1. Project title: the name of the project.
2. Programmer/analyst: the name of the responsible individual.
3. Committed date: the date the project is scheduled for completion.
4. Completed date: the date the project actually is completed.
5. Status date: the date of the status report.

Figure 7.13
Project plan and status report. The project plan and status report is an example of a more complex Gantt-type chart. In addition to the major activities, it shows project title, programmer/analyst name, committed date, completed date, and the percent of completion for each of the activities.

PROJECT PLAN AND STATUS REPORT					
PROJECT TITLE	PROJECT STATUS SYMBOLS O Satisfactory □ Caution △ Critical		PROGRAMMER/ANALYST		
	PLANNING/PROGRESS SYMBOLS □ Scheduled Progress V Scheduled Completion ■ Actual Progress ▼ Actual Completion		COMMITTED DATE	COMPLETED DATE	STATUS DATE
ACTIVITY/DOCUMENT	PERCENT COMPLETE	STATUS	PERIOD ENDING (Week)		

6. Activity/document: a line entry for each major activity or document to be completed.
7. Percent complete: the analyst's interpretation of the percentage already completed of a scheduled line entry.
8. Status: the analyst's evaluation of the status of each line entry. Status is reported by means of the following symbols:
   ○ Project status satisfactory
   ☐ Caution: problem encountered but not considered critical
   △ Critical condition: completion of project could be endangered
9. Period ending: the end dates of selected reporting intervals (for example, weeks or months).
10. Project planning progress symbols:
    ☐ Scheduled progress
    ■ Actual progress to date
    ∨ Scheduled or rescheduled completion date
    ▼ Actual completion date

Figure 7.14 is an example of the use of management by objectives. In this figure the principal activities that constitute the study phase for a computer-based business system appear as line entries. The report, which has a status date corresponding to the eighth week of the project, is interpreted as follows:

1. Initial investigation: 100 percent complete; completed one week ahead of schedule (as indicated by the appearance of the actual completion date, ▼ , ahead of the scheduled completion date ∨); status is satisfactory.
2. Project directive: 100 percent complete; completed approximately one-half week ahead of schedule; status is satisfactory.
3. Performance definition: 100 percent complete; completed on schedule; status is satisfactory.
4. Feasibility analysis: 50 percent complete; approximately one-half week behind schedule; rescheduled for completion a week later; status is caution because slippage has occurred.
5. Performance specification, study phase report, and study phase review: not started, but all have slipped a week because of the rescheduled completion date for the feasibility analysis.
6. The study phase: the overall project is 45 percent complete; status is caution because the completion date has been rescheduled. (In this reporting scheme a major line entry, for example, study phase, must not be given a more satisfactory rating than its least satisfactory element, in this case, the feasibility analysis.)

The percentage complete and status ratings shown in figure 7.14 represent the analyst's personal evaluations. For example, even though the shaded-in part of the study phase bar is more than 50 percent of the total area, the analyst feels that only 45 percent has been completed. This could be due to the fact that additional resources have been scheduled for the latter part of the project. The evaluation of caution and

Figure 7.14
Study phase plan and status report. The study phase plan and status report shown is an example of a completed project plan and status report. Note that, as of the eighth week, the project is slightly behind schedule.

PROJECT PLAN AND STATUS REPORT																						

PROJECT TITLE

STUDY PHASE FOR PAYROLL AND PERSONNEL SYSTEM (PAPS)

PROJECT STATUS SYMBOLS
O Satisfactory
☐ Caution
△ Critical

P. H. Eagle

PROGRAMMER/ANALYST

PLANNING/PROGRESS SYMBOLS
☐ Scheduled Progress V Scheduled Completion
■ Actual Progress ▼ Actual Completion

COMMITTED DATE 6/3/xx

COMPLETED DATE

STATUS DATE 4/29/xx

ACTIVITY/DOCUMENT	PERCENT COMPLETE	STATUS	PERIOD ENDING (Week)
			1 2 3 4 5 6 7 8 9 10 11 12 13 14 15 16 17
STUDY PHASE	45	☐	
Initial Investigation	100	O	
Project Directive	100	O	
Performance Definition	100	O	
Feasibility Analysis	50	☐	
Performance Specification	0	O	
Study Phase Report	0	O	
Study Phase Review	0	O	

the analyst's apparent acceptance of a week's lag in completion of the project may not be concurred with by the analyst's supervisor. However, in a project environment in which it is "safe" for an individual to report openly, many possible actions may be taken before temporary difficulties become insurmountable problems. We also should realize that plans and schedules will change, since no one can forecast the future without error. A major advantage of a plan, however, is that it provides a good reference on which to base necessary changes.

Project Cost Reporting

Project status reporting, as described in the previous section of this chapter, does not by itself provide a complete status picture. Cost must be reported as well as progress. All projects operate within the constraints of a budget. A project might be on schedule performance-wise, but at the same time be seriously overexpended. Therefore, a project cost report is required as well as a project plan and status report. Figure 7.15 is an

Figure 7.15
Project cost report. The project cost report is a line chart used to show both estimated and actual system costs. The dashed line represents estimated costs, while the solid line is the actual cost to date.

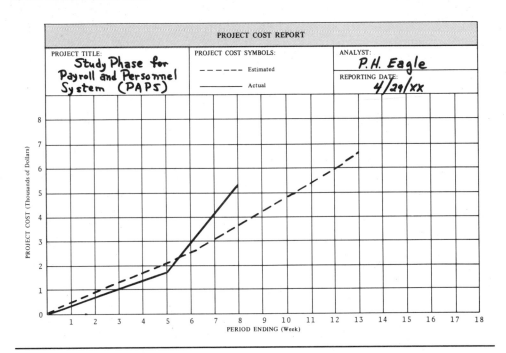

example of a cost report that might accompany the project plan and status report of figure 7.14.

In this example, the project is approximately $1,700 overexpended at the end of the eighth week. This overexpenditure, coupled with the schedule slippage, is reason for a reappraisal meeting between the analyst and supervisor. In fact, this meeting should have taken place at least one week earlier. When appropriate, as in this instance, the analyst should submit an explanatory memorandum along with the project plan and status report and project cost report.

Performance Indices

The project plan and status report and the project cost report always present a current picture of project performance. They do not present historical or trend information. Although this type of information can be obtained by referring to previous reports, it is more desirable to extract the required information from reports as they are presented and to maintain a management control chart. This chart should contain only the data essential to summarizing past performance and to predicting future performance. The following discussion of performance indices is based upon this technique.

We can define three performance indices:

1. *Cost Index* (CI): The ratio of actual costs to planned costs.

$$CI = \frac{\text{actual costs}}{\text{planned costs}}$$

2. *Achievement Index* (AI): The ratio of actual achievement to planned achievement.

$$AI = \frac{\text{actual achievement}}{\text{planned achievement}}$$

3. *Status Index* (SI): The ratio of the achievement index to the cost index.

$$SI = \frac{AI}{CI}$$

The cost index measures the expenditure of money. The information for calculating cost index is obtainable from the project cost reports. A value of CI greater than 1.00 represents overexpenditure.

The achievement index measures results. It can be calculated from the project plan and status report as the ratio between the estimates of actual percentage complete and planned percentage complete. A value of AI less than 1.00 represents under-achievement.

The status index is a single measure of effectiveness that is calculated from the other two indices. It may be viewed as a "rate of return" on expenditure. Note that the status index alone cannot indicate schedule slippage or cost overrun. We still have to refer to either the achievement index or the cost index for this information. For example, a project with an achievement index of 0.80 and cost index of 0.60 would have a status index of 1.33. This means that the rate of return on dollars expended is very high. However, the achievement index indicates that accomplishment is behind schedule.

Let us make some sample calculations showing the value of the three performance indices in reporting trends. Let us assume that the data summarized in figure 7.16 was extracted from the status and cost reports as they were received by management. Note that the cost and achievement entries are cumulative and relate to the overall project. The last entry (week 8) corresponds to the data shown in figures 7.14 and 7.15. Note that we were able to obtain all of the cost data from figure 7.15; however, we would have to rely on previous status reports for the historical achievement information. Figure 7.17 is a graph of the three indices summarized in figure 7.16.

As figure 7.17 dramatically illustrates, costs begin to increase and achievement decreases rapidly by the end of the fifth week. Management action should have been triggered by the declining SI and AI no later than the end of the sixth week. By the end of the eighth week the project is badly out of control.

Figure 7.16
Project trend summary data. This project summary status report worksheet is a convenient form to record data used to calculate performance indices.

	ACHIEVEMENT (%)		COST ($)		PERFORMANCE INDICES		
WEEK	ACTUAL	PLANNED	ACTUAL	PLANNED	AI	CI	SI
1	7	5	450	500	1.40	0.90	1.55
2	13	10	800	900	1.30	0.89	1.46
3	18	15	1,100	1,300	1.20	0.85	1.41
4	23	20	1,500	1,700	1.15	0.88	1.31
5	25	25	1,850	2,050	1.00	0.90	1.11
6	35	35	3,000	2,500	1.00	1.20	0.83
7	40	45	4,200	3,100	.89	1.35	0.66
8	45	50	5,450	3,750	.90	1.45	0.62
9							
10							
11							
12							
13							
14							
15							
16							
17							
18							

PROJECT SUMMARY STATUS REPORT WORKSHEET

Status index graphs for some potential performance situations are shown in figure 7.18. The left part of this figure shows a project that is going out of control; the center part a project that "appears" to be outperforming the plan; and the right part a project that "appears" to be performing consistently as planned. We say "appears" because we must remember to check the achievement index or cost index also.

The following "rules of thumb" apply to the use of the status index:

1. SI between .9 and 1.1: normal range
2. SI between 1.1 and 1.3 or between .7 and .9: management attention, perhaps action
3. SI greater than 1.3 or less than .7: management action usually required.

Figure 7.17
Performance indices for a typical out-of-control project. This performance chart shows that the project has gone badly out of control. Management should have been alerted to the problem when the status index started to decline. Control should have been exercised by no later than the sixth week of the project.

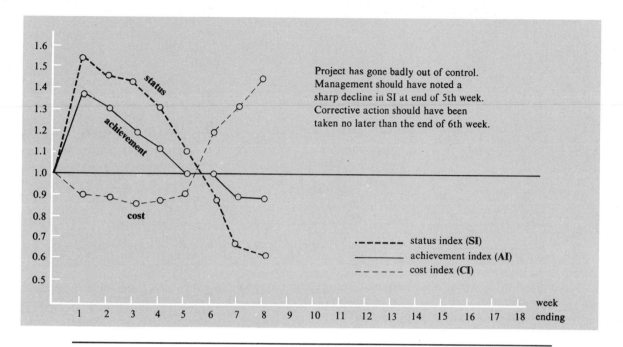

Project has gone badly out of control. Management should have noted a sharp decline in SI at end of 5th week. Corrective action should have been taken no later than the end of 6th week.

- - - - - - status index (SI)
———— achievement index (AI)
– – – – – cost index (CI)

The project management techniques we have discussed thus far usually can be applied without the aid of a computer. More sophisticated techniques are available, but they usually require the assistance of a computer. Among the more powerful and widely used of these techniques are critical path networks.

Critical Path Networks

Critical path networks are planning and management tools that use a graphical format to depict the relationship between tasks and schedules. They differ from conventional scheduling methods in two major respects: (1) they do not use a single time scale; and (2) they facilitate the analysis of many interdependent tasks, some of which must be performed in sequence and some of which should be performed parallel with other tasks.

The two most common critical path network techniques, which are the same in all essential aspects, are CPM (Critical Path Method) and PERT (Program Evaluation Review Technique). Figure 7.19 is a simplified example illustrating the principles

Figure 7.18

Illustrative status index graphs. Projects performing as planned will have a status index that remains near the on-schedule reference line. Projects whose status index continues to move away from the on-schedule reference line require management attention.

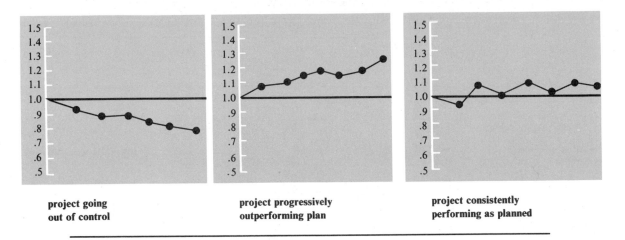

project going
out of control

project progressively
outperforming plan

project consistently
performing as planned

Figure 7.19

Critical path network. Critical path networks are planning and management tools that use a graphical format to depict the relationships between tasks and schedules. The longest network path from start to finish is called the critical path. The critical path determines the minimum time required for the project.

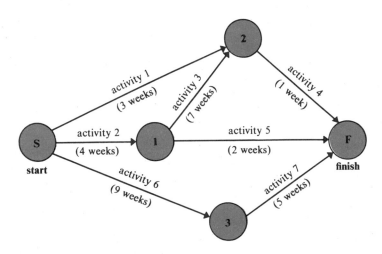

that underlie this powerful management tool. A network such as that shown in figure 7.19 is constructed from two basic elements:

1. Activity: the application of time and resources to achieve an objective.
   Activities are measured in units of time, usually weeks.
2. Event: the point in time at which an activity begins or ends.

Network management techniques are based upon calculation of the time required to proceed from start (S) to finish (F) along each possible path in order to determine the path that requires the longest overall time.

The longest network path from start to finish is called the critical path. In figure 7.19 all the possible paths from start to finish and their respective times are:

Path	"S" to "F" Time
S-1-F	6 weeks
S-1-2-F	12 weeks
S-3-F	14 weeks

Note that S-2-F is not an allowable path because the activities along S-1-2, which take 11 weeks, must be completed before it is possible to proceed from 2 to F. The critical path is S-3-F.

Events along paths other than the critical path can be delayed without causing the time from start to finish to exceed the critical path time. Critical path information is valuable to the project manager because it gives an opportunity to shift resources from activities associated with events not on the critical path to activities that are.

As is apparent from inspection of figure 7.19, a critical path network could become quite complex for even a moderate-sized project. Computer programs that perform critical path network analysis are available. In fact, computer programs, called PERT-COST programs, have been devised to relate network schedules and cost. Nevertheless, the manual construction of critical path networks is of value on small projects or on segments of larger projects. Also, many project planning and monitoring programs are available to use with microcomputers.

Critical path networks have been used most successfully on applications for which all the tasks are well understood and where estimates of times can be based upon valid past experience. For example, the construction industry makes excellent use of critical path techniques. As another example, networks are useful in scheduling the flow of work through computer centers. Critical path networks also are of value in scheduling the many activities in the process of converting from an old system to a new, computer-based system. We will examine the use of a critical path network in considerable detail in chapter 22, Preparation for Implementation.

We shall encounter additional applications of charts throughout this text. Flowcharts, another important class of charts, are powerful pictorial tools that are described in the next chapter.

Summary

Raw data or even tables of numbers can be difficult to analyze for trends or other significant information. Charts can help to solve this problem. They are an excellent method for presenting complex relationships among data in a meaningful manner. There are four basic types of charts: bar charts, line charts, pic charts, and step charts. The best use of bar and pie charts is to show relationships between data items. Pie charts are preferred when the relationships are percentages of a whole. Trends in data are best shown by line charts and step charts. These charts also are effective in making comparisons of trends by plotting multiple lines on a single chart.

Decision table and decision trees are methods for describing logical rules, showing actions that result from combinations of conditions. Structured English is a method for presenting the same logical relationships in an outline format. It is a structured design tool.

Project control is an important use for charts. Horizontal bar charts can communicate planned and actual progress. Line charts can report planned and actual project costs. A status index, which is an overall measure of project performance, can be derived from progress and cost data. This information is very useful to systems analysts, to their managers, and to the users of the system being developed. Critical path networks also are planning and management tools and use a network format to depict time and progress relationships.

For Review

chart	cost index	**decision tree**
pie chart	achievement index	bar chart
line chart	status index	**structured English**
step chart	**critical path network**	**decision table**
decision table	**table**	pseudocode
Gantt-type chart		

For Discussion

1. Name and relate each of the four basic types of charts to its best use.
2. Why are tables useful as systems analysis tools?
3. Why are decision tables used by systems analysts as well as by programmers?
4. What is a decision tree and how does it relate to a decision table?
5. What is structured English and how does it relate to a decision tree? To pseudocode?
6. What is meant by "management by objectives"?
7. What elements of the project plan and status report are most dependent upon the relationship between the analyst and the analyst's supervisor? If this were a poor relationship, what might happen?
8. What correlations should be made between the project status and project cost reports? Explain the use of the performance indices.
9. How do critical path networks differ from Gantt-type charts?
10. What is the critical path, and what is its significance to a project manager?

8
Flowcharting

Preview

The previous chapter introduced charts as a means of compressing large amounts of data or complex relationships into a concise, clear format. Flowcharts also are a method of communicating complex relationships. They are extensively used to describe information flow and processing throughout the analysis, design, and development of computer-based information systems.

Objectives

1. You will learn how to use a standard set of flowcharting symbols to describe information-oriented and process-oriented flowcharts within a system.
2. You will become familiar with the basics of flowcharting using data flow diagrams.
3. You will be introduced to HIPO charts and procedure analysis charts and their uses in information system design and analysis.

Key Terms

flowchart   A pictorial representation that uses predefined symbols to describe data flow and processing in either a business system or the logic of a computer program.

data flow diagram   A network that uses special symbols to describe the flow of data and the processes that change, or transform, data throughout a system.

HIPO charts   A set of charts made up of a hierarchy chart plus input-processing-output charts that emphasize the functions of a system or computer program.

procedure analysis charts   A chart used to record the details of manual procedures in pictorial form for analysis and improvement of those procedures.

System Flowcharts

Definition and Uses of Flowcharts

Definition of a Flowchart

Flowcharting is a graphic technique specifically developed for use in data processing. A **flowchart** is a pictorial representation that uses predefined symbols to describe data flow in a business system or the logic of a computer program.

Figure 8.1 is an example of a simple system flowchart. The symbols shown in this figure are "predefined"; their shapes identify data and communicate what is happening to it. In this example the top and bottom symbols represent input or output data. The middle symbol on the left is the process symbol; it represents the operation performed on input data to convert it into meaningful output. The symbol to the right of the process symbol depicts on-line storage that is available to the computer program. The words within each symbol provide additional information about each step. There are other important symbols that the systems analyst must know in order to make effective use of flowcharts. We will present these after introducing the principal uses of flowcharts.

Uses of Flowcharts

Flowcharts help the analyst to describe and communicate complex sets of data in three principal ways. These are: (1) analyze existing systems; (2) synthesize new systems; and (3) communicate with others.

The flowchart of an existing system enables the analyst to visualize the parts of the system and to record their functions. A *system flowchart* can compress many pages of written description into one informative picture.

Flowcharts help the analyst synthesize new systems. Each "candidate" system can be described quickly and effectively by an appropriate flowchart. Throughout the life cycle the analyst must be able to communicate effectively with users, with programmers (if computer programs are required), and with other analysts. Flowcharts are a powerful method for efficient internal communication since all persons involved can grasp their meaning quickly.

Flowcharts also are an effective method for communication with groups outside the company. External communication is most often accomplished through professional journals, seminars, or meetings with user groups. The user groups may be composed of companies in the same or related industries, or of companies with problems of a similar nature. Knowledge of what other companies have done to solve similar problems often is of great aid to the analyst in solving the company's problems.

Flowcharting Symbols

Standardization of Symbols

For communication to be improved by the use of symbols, the meaning of the symbols must be understood by all their users. Several groups have contributed to the standardization of flowchart symbols. National and international efforts to develop standard flowcharting symbols began in the early 1960s. The effort in the United States resulted in the set of symbols adopted by the American National Standards Institute (ANSI). The ANSI standard was developed through the combined efforts of profes-

Figure 8.1

System flowchart. This system flowchart uses predefined symbol shapes to identify data and communicate what is happening to that data.

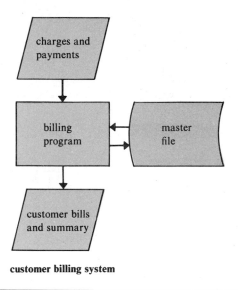

customer billing system

sional associations, including the Association for Computing Machinery (ACM) and the Data Processing Management Association (DPMA). These symbols are a subset of those adopted as a result of the parallel efforts of an international group. This more extensive set of symbols is called the International Organization for Standardization (ISO) flowchart symbols.

The ANSI-ISO standards are conformed to widely by the systems and programming professions. However, they have been augmented in areas in which additional symbols could improve clarity. In particular, the IBM Corporation has developed some additional symbols. Through the production and distribution of a useful plastic "template," IBM has achieved widespread acceptance of its set of symbols.[1] Hence, we use the IBM set of flowcharting symbols in this text.

IBM Flowcharting Symbols

Symbol Groups   Figure 8.2 illustrates the outlines of the flowcharting symbols as they appear on the IBM template. Some of these symbols are used only in systems flowcharts; some are used only in computer program flowcharts; and some are used in both types. We will describe each group, with particular stress on those symbols used to prepare system flowcharts.

Basic Symbols   Figure 8.3 shows the *basic symbols,* which are common to both systems and programming flowcharts. The two symbols in the upper left-hand corner

---

1. *Flowcharting Template,* Form GX20-8020 (White Plains, New York: IBM Corporation).

Figure 8.2
IBM flowchart symbols. The IBM flowcharting template includes symbols used only in systems flowcharts, those used in computer program flowcharts, and those used in both types.

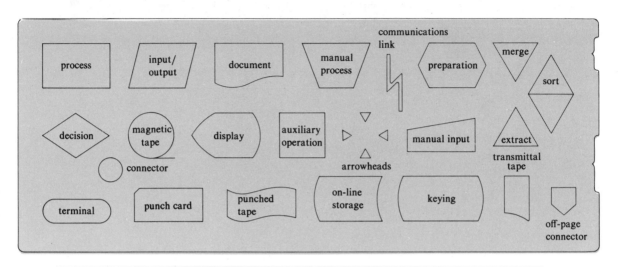

Figure 8.3
Basic flowcharting symbols. Basic flowcharting symbols are used to draw both systems and programming flowcharts. The off-page connector is an IBM extension to the ANSI standard.

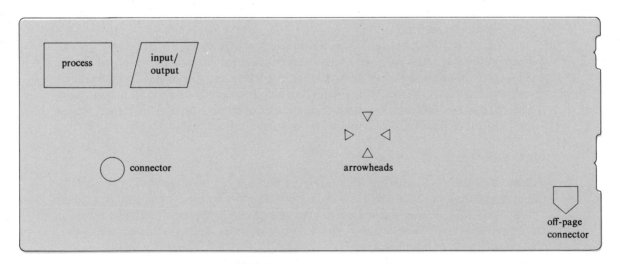

(process and input/output) were introduced in figure 8.1. The process symbol is a rectangle. In a systems flowchart, the "process" depicted can range from a single manual activity to a computer program to the entire computer program component of a system. The amount of "processing" the process symbol represents depends on the intended detail of the flowchart. In program flowcharts the processes are relatively specific, usually representing arithmetic operations or data movement.

The other three basic symbols shown in figure 8.3 are the on-page connector, off-page connector, and arrowheads. All are used to connect, or to link, other flowchart symbols. The use of these symbols, along with flowlines (a line drawn to connect symbols) is illustrated in figure 8.4. The normal flow is downward and by columns from left to right. As shown in this figure, connectors are used in pairs. The connector with the "A" at the bottom of the left column is considered to be "connected" to the "A" at the top of the middle column. The connectors at the bottom of the flowchart columns are called exit connectors. Connectors that "depart" from the normal downward flow also are called exit connectors. Connectors at the top of the flowchart columns and connectors that "join" the downward flow are called entry connectors. If a flowchart is continued on another page, the IBM standard provides for a special off-page connector symbol. The off-page connector appears at the bottom of the right-hand column

Figure 8.4
Flowlines and connectors. The normal logic flow is downward and by columns from left to right. Flowlines and connectors show this logic flow.

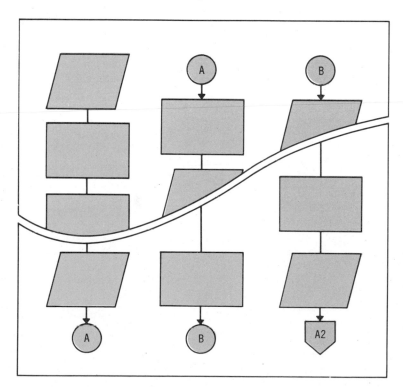

in figure 8.4. Arrowheads are connected to flowlines wherever there might be confusion about the direction of flow. Since the normal direction of flow is down, flowlines that continue down the chart normally do not have arrowheads.

Symbols Related to Systems Fifteen symbol outlines are shown in figure 8.5. Additional symbols can be formed by combining or modifying the symbols shown in this figure. For ease of discussion, we will split the symbols into three groups, shown in separate figures:

1. Symbols for equipment (figure 8.6)
2. Specialized input/output symbols (figure 8.7)
3. Specialized processing symbols (figure 8.8)

Figure 8.6 illustrates three symbols used to show specific types of equipment.

1. The communications link is used in place of a normal flowline to indicate that remotely located equipment is tied into the system through telephone lines, microwave transmitters, or other media. The communications link is commonly used in the description of modern distributed data processing (DDP) systems.
2. Off-line storage is used to show storage that is not accessible to the computer. For example, it is the symbol used to show the manual filing of data in a storage or file cabinet. It can also be used to indicate off-line computer storage, such as magnetic tape. It is a modification of the merge symbol.

Figure 8.5
System flowcharting symbols. Systems flowcharting symbols, plus the basic symbols, are used to construct systems flowcharts. The system symbols consist of symbols for equipment, symbols for specialized input/output, and symbols to show specialized processing.

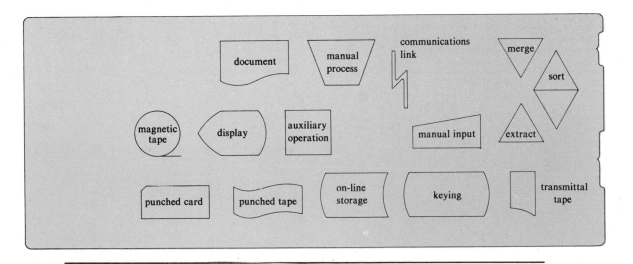

3. On-line storage is storage that is accessible by a computer program as data are being processed. An example of on-line storage is the magnetic disk. This symbol is a generalized symbol and does not identify specific media.

When a specific input or output device or media is known, it usually is helpful to indicate what it is. Figure 8.7 contains nine specialized symbols that may be used in place of the basic input/output symbol (parallelogram).

1. The *document symbol* is used to describe any input or output that is a paper document. It includes source documents and printed reports.
2. *Punched tape* often is produced by attachments to cash registers and bookkeeping machines.

Figure 8.6
Symbols for equipment. Symbols for equipment reflect particular hardware types to be used within the system. These types include communication link, off-line storage, and on-line storage.

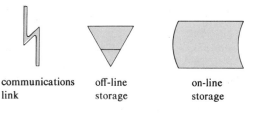

communications
link

off-line
storage

on-line
storage

Figure 8.7
Specialized input/output symbols. Specialized input/output symbols are used to indicate a specific input or output device or media.

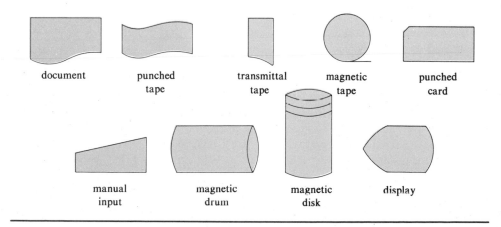

document

punched
tape

transmittal
tape

magnetic
tape

punched
card

manual
input

magnetic
drum

magnetic
disk

display

3. A *transmittal tape* is a proof or other control total produced by a device such as an adding machine.
4. The *magnetic tape* symbol represents the use of magnetic tape as an input, output, or auxiliary storage device.
5. *Punched cards* may be used as input, output, or storage media. Some punched cards also serve as source documents.
6. The *manual input* symbol indicates inputs to computer systems by means of on-line data entry terminals, switches, or buttons. It is not used to indicate keying processes such as keypunching or verifying.
7. *Magnetic drum* and *magnetic disk* are specific replacements for the general on-line storage symbol. Both of these symbols are constructed from the on-line storage outline.
8. The *display* symbol shows visual outputs from on-line devices, such as CRT (cathode ray tube) terminals, console printers, and plotters. As contrasted with hardcopy data carriers represented by the document symbol, the display symbol represents softcopy data carriers.

Symbols used to replace the basic process symbol (rectangle) are illustrated in figure 8.8.

1. *Manual off-line* operations are operations that are performed by humans without the aid of equipment. An example is filling in a source document.
2. The *auxiliary off-line* operations symbol represents off-line operations. These are performed with equipment, but they are not under the control of a computer.
3. The *keying* symbol indicates keypunching, verifying, and other key-driven operations.
4. The *merge, extract, collate,* and *sort* operations are represented by special symbols. The collate and sort symbols are constructed by combining the merge and extract outlines. These symbols are used to represent unit record operations rather than computer operations.

Symbols Related to Programming   Figure 8.9 shows the special symbols most commonly used to prepare flowcharts for computer programs:

1. The *preparation* symbol shows the setting of a program switch or the use of a subroutine within a computer program.
2. The *decision* symbol is used for operations that determine which of two or more alternative paths will be followed in the program.
3. The *terminal* symbol indicates a start, stop, halt, pause, or interrupt in a computer program.

In addition to these three unique *programming symbols,* a fourth symbol called the predefined process is commonly used. The predefined process symbol is a composite symbol drawn by combining the basic process symbol with two vertical lines drawn near the left and right edges. Figure 8.10 illustrates this symbol. It is used to show the

Figure 8.8
Specialized processing symbols. The specialized processing symbols may be used
instead of the process symbol.

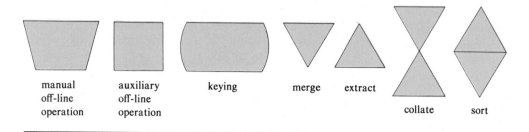

Figure 8.9
Program flowcharting symbols. Program flowcharting symbols consist of the preparation,
decision, and terminal symbols. These three symbols plus the basic symbols make up the
programming symbol set.

Figure 8.10
The predefined process symbol. The predefined process symbol is used to show the
execution of a function that is defined elsewhere in the flowchart or on a separate
flowchart. It is drawn by adding two vertical lines to the process symbol.

execution of a function that is defined elsewhere in the flowchart or on a separate flow-chart. Examples are title-column heading routines, data movement routines, calculation routines, and total routines.

Symbol Sets

The *basic symbols,* the symbols related to systems, and the symbols related to programming are combined in two ways:

1. System flowcharting is performed using both the basic and systems-related symbols.
2. Computer program flowcharting is performed using both the basic and programming-related symbols.

Note: It is not good practice to use programming-related symbols for system flowcharting or to use systems-related symbols for computer program flowcharting.

Because we are primarily concerned with system flowcharting, we next will examine the principal system flowcharts: information-oriented system flowcharts and process-oriented system flowcharts.

Information-Oriented System Flowcharts

Information-oriented flowcharts use a grid structure to trace the flow of data. They identify input data and follow its flow until its subsequent appearance as output information. They do this by identifying specific data carriers. Usually they do not identify processing operations. Hence, the document and display symbols are predominant in information-oriented flowcharts. An example is shown in figure 8.11. This figure shows how data, that is, filled-out forms or an electronic image, flow across organizational boundaries. The storage of hardcopy data is indicated by use of the off-line storage symbol.

It is customary to accompany information-oriented system flowcharts with a narrative description. One technique for doing this is to number the significant information flow steps and to describe them by a narrative on an accompanying sheet of paper. The numbers encircled on the flowchart in figure 8.11 key the symbols in this figure to the narrative shown below the flowchart. Another technique is to describe the inputs and outputs for each labeled column on separate sheets.

Process-Oriented System Flowcharts

Levels of System Flowcharts

Process-oriented flowcharts commonly are referred to as "system flowcharts." They show which data processing operations are converting inputs into outputs. These flow-charts can be drawn to any appropriate level of detail. The highest level and least detailed flowchart is called a high-level system flowchart. Lower levels of detail are represented by expanded flowcharts, which are referred to by names such as "intermediate level" and "detailed level."

Figure 8.12 is an example of the highest level of a process-oriented system flow-chart. It uses the basic symbols to give an overview of the system.

Figure 8.11

Information-oriented flowchart and narrative. Information-oriented flowcharts use a grid structure to trace the flow of data. They identify input data and follow its flow until its subsequent appearance as output information. The narrative is keyed to the flowchart through circled numbers.

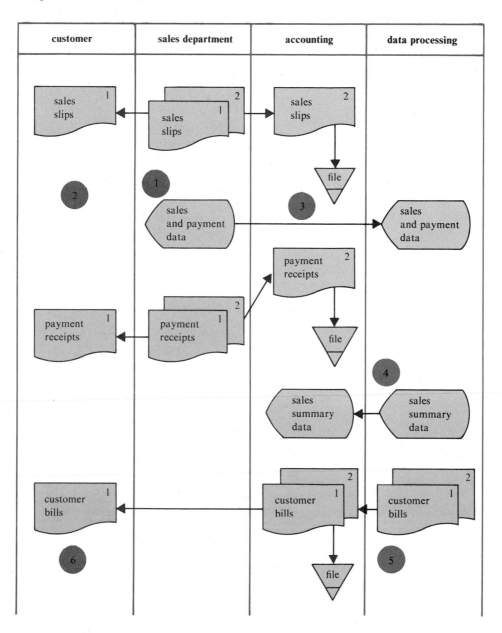

Figure 8.11
continued

CUSTOMER BILLING SYSTEM

1. A point-of-sale terminal is used to capture
    sales and payment data that is transmitted
    to data processing. In addition, sales slips
    and payment receipts are produced in duplicate.

2. The original copies of sales slips and payment
    receipts are given to the customers.

3. The second copies of sales slips and receipts
    are sent to accounting where they are retained
    for future reference in files.

4. Sales summary data is formatted and is
    available for display in the accounting
    department.

5. Two copies of customer bills are produced and
    sent to accounting where the second copies are
    filed.

6. The original copies of the customer bills are
    mailed to the customer.

Figure 8.12
High-level process-oriented system flowchart—basic symbols. High-level system
flowcharts use basic symbols to give an overview of the system.

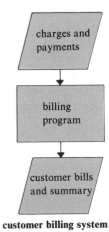

customer billing system

Structure of System Flowcharts

System flowcharts have a characteristic "sandwich" structure, so called because the chart is made up of alternating layers of input/output identifiers and process identifiers. Figure 8.12 exhibits the basic sandwich structure. Charges and payments are the "bread" of the input data; the billing program is the "filling" of the processing operation; and customer bills and summary is the "bread" of the output data. Note that this chart does not explain how processing is performed; it only identifies what processing is done. Also, note that the output of one step may become the input of a successive step.

Process-oriented flowcharts should also be accompanied by a narrative. The flowchart of figure 8.13 is the same as figure 8.1; it also illustrates the use of an accompanying narrative.

Many flowcharts are too complex and too detailed to accommodate the narrative on the same page. Either there is not enough room or the flowchart would appear confusing. Figure 8.14 illustrates the use of a separate page for the narrative.

Figure 8.13
Flowchart and narrative. All flowcharts should be accompanied by a narrative to explain in words what the flowchart has described with symbols. If room permits, the narrative may be on the same page as the flowchart.

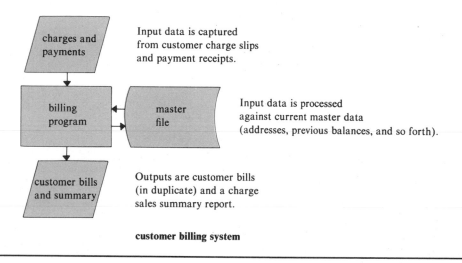

Input data is captured from customer charge slips and payment receipts.

Input data is processed against current master data (addresses, previous balances, and so forth).

Outputs are customer bills (in duplicate) and a charge sales summary report.

customer billing system

Figure 8.14
Flowchart and separate narrative. Many flowcharts are too complex and too detailed to accommodate the narrative on the same page. In these cases the narrative is presented on a separate page but is still keyed to the flowchart.

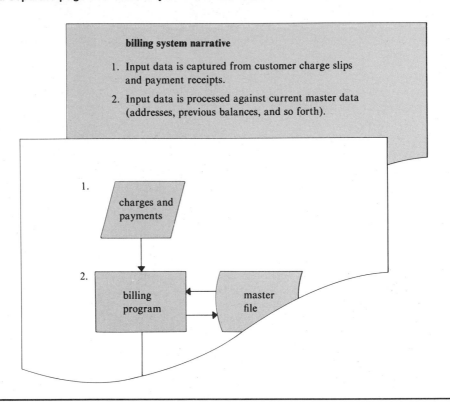

The process-oriented flowcharts we have examined so far have been high-level flowcharts. These flowcharts are expanded in detail as the computer-based business system proceeds through its life cycle. The flowchart levels are associated with certain life-cycle phases:

1. High-level process-oriented flowcharts: study phase. These flowcharts are very valuable at study phase reviews. The participants in these reviews usually are not data processing professionals. Elimination of confusing detail in the flowchart helps to highlight the user-oriented characteristics of the system. If we present management with systems flowcharts that are too detailed, we become "confusers" instead of "simplifiers." Figure 8.15 illustrates a system flowchart suitable for a management presentation.

Figure 8.15
High-level process-oriented flowchart. High-level flowcharts provide an overview of the system by eliminating confusing detail. These flowcharts are very valuable at study phase reviews to highlight the user-oriented characteristics of the system.

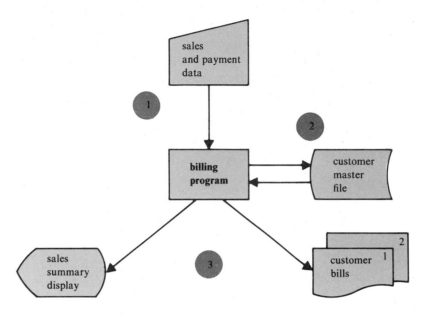

CUSTOMER BILLING SYSTEM

1. Sales and payment data are captured on a point-of-sale terminal and transmitted to data processing.

2. An on-line customer master file supplies data to the program and is updated by the program.

3. Program outputs are a sales summary display and two copies of each customer bill.

2. Intermediate-level process-oriented flowcharts: design phase. These flowcharts identify the specific inputs, outputs, and processes in considerable detail. Figure 8.16 is an example of this type. The accompanying narrative is shown below the flowchart.

   As we will discuss in chapter 16, System Design, major input, processing, and output controls also can be indicated on intermediate-level flowcharts.

3. Detailed-level system flowcharts: development phase. These flowcharts precede the construction of computer program flowcharts by programmers. They identify the inputs, outputs, and processing operations for each of the computer program components. Detailed control operations also are shown. The use of this type of flowchart is described in chapter 23, Computer Program Development.

Figure 8.16
Intermediate-level process-oriented flowchart and narrative. Intermediate-level system
flowcharts identify the specific inputs, outputs, and processes in considerable detail. This
level of flowchart is frequently used during the design phase.

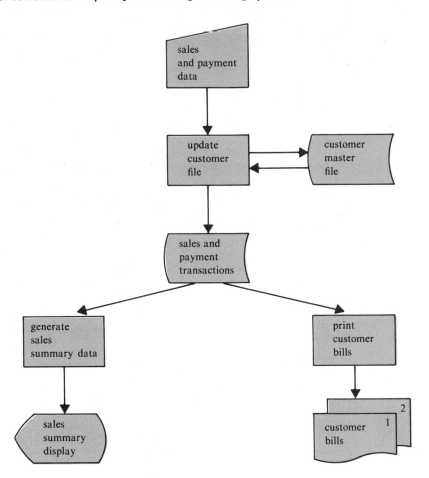

CUSTOMER BILLING SYSTEM

1. Sales and payment data are captured on a point-of-
   sale terminal and transmitted to data processing.

2. An on-line customer master file is updated by the customer
   master file update program.

3. Current sales and payment transactions are the inputs to a
   program that prints two copies of customer bills.

4. Current sales transactions are the inputs to a program
   that provides for a sales summary display.

Data Flow Diagrams

Definition of a Data Flow Diagram

A disadvantage with the use of the IBM flowcharting symbols in preparing system flowcharts is that some of the symbols represent physical media, such as magnetic tape, documents, and CRT displays, and others imply specific physical operations, such as keying or manual input. These symbols may be helpful when the systems analyst is describing an existing system; however, their use may influence the study of alternative solutions for a new or modified system by suggesting a specific implementation before the logical requirements are fully understood.

A **data flow diagram (DFD)** is a structured analysis and design tool that can be used for flowcharting in place of, or in association with, information-oriented and process-oriented systems flowcharts. A DFD is a network that describes the flow of data and the processes that change, or transform, data throughout a system. This network is constructed by using a set of symbols that do not imply a physical implementation. The four basic symbols used to construct data flow diagrams are shown in figure 8.17. They are symbols that represent data sources, data flows, data transformations, and data storage. The points at which data are transformed are represented by enclosed figures, usually circles, which are called nodes.

Figure 8.17
Basic data flow diagram symbols. Data flow diagrams are constructed from four basic symbols. Using these symbols, a systems analyst is able to construct a logic network that traces data streams throughout a system, showing the processes that transform data, the data sources, and the storage of data.

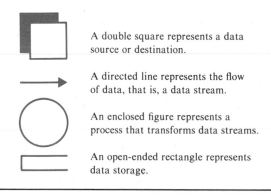

A double square represents a data source or destination.

A directed line represents the flow of data, that is, a data stream.

An enclosed figure represents a process that transforms data streams.

An open-ended rectangle represents data storage.

Figure 8.18

Data flow diagram transformations. A process that transforms data is identified by an enclosed figure, such as a circle. The principal processes that take place at nodes are: combining, splitting, and modifying data streams. DFD networks are constructed from combinations of these processes.

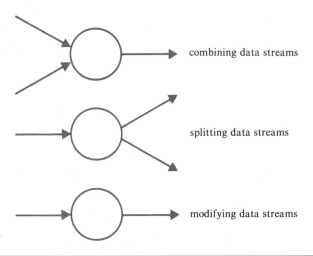

combining data streams

splitting data streams

modifying data streams

Decomposition of Data Flow Diagrams

The three principal types of transformations that take place at the processing nodes are shown in figure 8.18. These transformations include combining, splitting, or modifying data streams. Like other types of system flowcharts, data flow diagrams can be expanded into successively lower levels of detail. Figure 8.19 illustrates the process of expanding a DFD into lower-level networks. This process is called *decomposition,* and it can be continued until modules are arrived at that clarify the origin and use of every data element in the system. The DFD's should be supported by a *system glossary,* which is a document that contains detailed information about data sources, data flows, transformation processes, and data storage.

Examples of Data Flow Diagrams

Figure 8.20 is an example of a high-level DFD for an existing system. The diagram illustrates the data flow, processing, and storage relationships between a customer and a retail store. Notice that the data streams are described by nouns and the processes by action verbs. Because the symbols used to prepare DFD's do not imply a physical implementation, a data flow diagram can be considered to be an abstract of the logic of an information-oriented or a process-oriented system flowchart. For this reason DFD's often are referred to as logical data flow diagrams. In the case of figure 8.20, however, the level of detail is not sufficient to provide a useful description of the present system. Figure 8.21 is an expansion of figure 8.20; it also is a logical abstract of the information-oriented system flowchart of figure 8.11.

Figure 8.19
Hierarchy of data flow diagrams. Data flow diagrams (DFD's) can be decomposed into
lower-level DFD's, each of which is a complete network of more elementary modules.

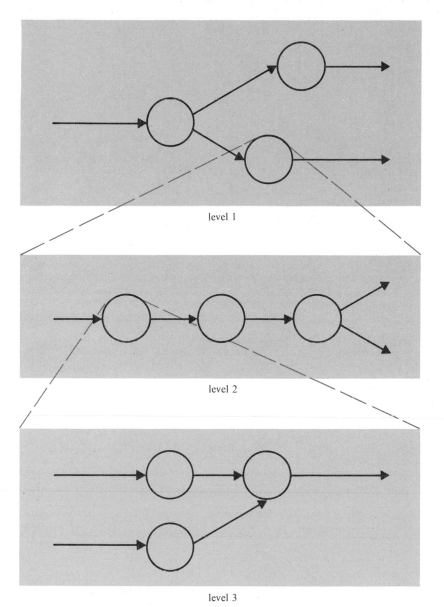

level 1

level 2

level 3

Figure 8.20

A data flow diagram. This high-level data flow diagram uses special symbols to demonstrate the logic relationship between a customer and a retail store.

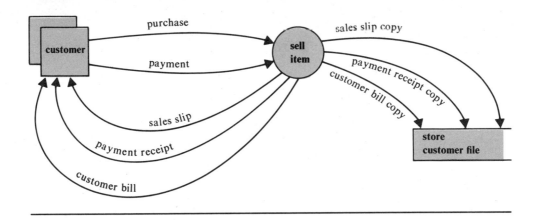

Figure 8.21

An expanded data flow diagram. This data flow diagram is an example of an intermediate-level DFD. At this level of decomposition the network clarifies the logic of the customer-retail store relationship.

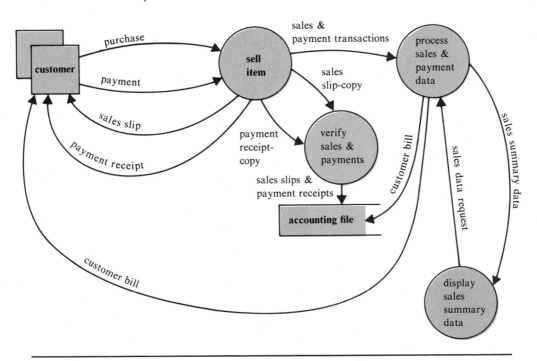

Figure 8.22 provides additional detail; it is a logical abstract of the process-oriented system flowchart of figure 8.16. Decomposition can continue until the objective of understanding the source and use of every data element in the system is achieved.

The power of data flow diagrams in the study phase is not so much in their use in analyzing the present system as it is in working with logical modules in order to synthesize new or improved systems. During analysis it often is difficult to separate the logical DFD from the physical evidence of the existing system. However, during synthesis it is possible to develop alternative solutions for evaluation while avoiding a premature commitment to a specific physical implementation.

A possible problem with data flow diagrams is that, with successive levels of decomposition, the diagrams tend to become quite complex. However, some degree of overdecomposition is preferable to underdecomposition. The former can be remedied; the latter may result in an inadequate understanding of how the system functions. In any event, analysts should not lose sight of the fact that systems do exist in a physical world and that design cannot proceed without a transition from the logical to the physical. In balance, DFD's are important analysis and design tools to be used along with other tools, such as information-oriented and process-oriented flowcharts.

Figure 8.22
Further expansion of a data flow diagram. This data flow diagram illustrates the network created by successive levels of decomposition. It is an expansion of the "process sales and payments" node of figure 8.21.

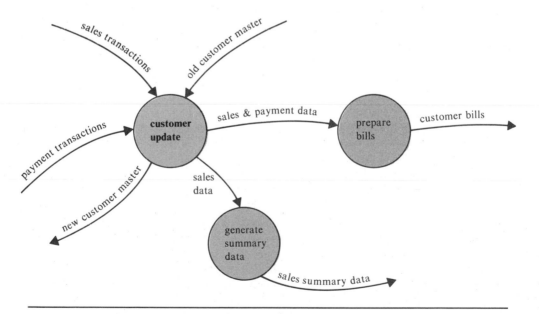

Figure 8.23
Customer billing system hierarchy. Hierarchy charts are a design phase tool with a detail
level about equal to intermediate-level system flowcharts. They describe the hierarchy of
system functions.

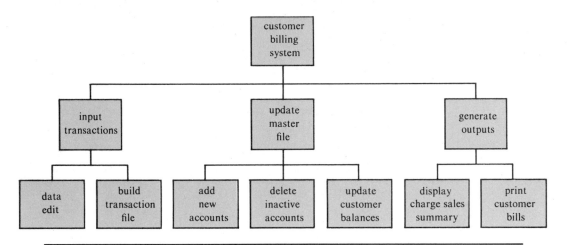

HIPO Charts

The information provided by system flowcharts and by data flow diagrams can be used
to design new systems in conjunction with other charts. Among the more important of
these are **HIPO charts.** HIPO stands for Hierarchy plus Input Processing Output. Actually, two types of charts make up a HIPO package. One shows the hierarchy, the
other depicts the input, processing, and output. The first presents the functional modules of a system (or computer program) as a hierarchy of functions. Levels of hierarchy
were shown on an organization chart in chapter 3. The hierarchy of functions can be
presented in the same way as the hierarchy of positions is shown on organization charts.
Figure 8.23 illustrates the concept of hierarchy of functions. The system is the top
level. The second level is made up of the major functions of the system. For the third
level, each of the major functions is then broken down into its subfunctions. Decom-
posing functions into smaller, more detailed, subfunctions can continue until the chart
shows as much detail as desired or until a basic function is reached. Note that the
hierarchy chart is more detailed than the high-level system flowchart of figure 8.15.
The hierarchy chart is a design phase tool with a detail level about equal to intermediate-level system flowcharts.

The second of the HIPO package charts is a detail-level chart listing the inputs,
processing steps, and outputs of each functional module of the hierarchy chart. These
charts are commonly referred to as IPO charts. One IPO chart is normally prepared
for each function shown on the hierarchy chart. IPO charts do not show hierarchy.
Figure 8.24 is an example.

Figure 8.24

IPO detail chart for data edit function. Each function shown on the hierarchy chart may be described in detail by listing its inputs, processing steps, and outputs. These charts are called IPO charts and are the second part of the HIPO package.

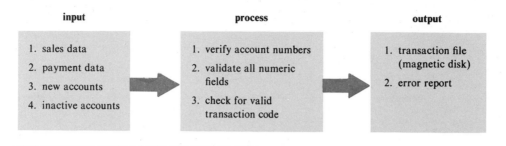

It should be noted that HIPO charts may be used for the top-down planning of programs as well as the planning of systems. The hierarchy of functions charted would be the functions of the program. The program HIPO chart is a common planning tool in a top-down approach to programming called structured programming. The drawing of HIPO charts for a program would be done during the development phase. IPO charts may also be used at a systems level during the study phase for input-output analysis, a fact-finding technique discussed in chapter 11, Initial Investigation.

Computer Program Flowcharts

The definition of a flowchart included the phrase: "uses predefined symbols to describe . . . the logic of a computer program." In this definition, "logic" refers to the types of instructions that business computers normally execute, for example, arithmetic, conditional, and data movement.

This book will not present techniques of computer program logic flowcharting, for these flowcharts are the special tools of programmers and are more properly covered in programming texts.

Computer logic programming takes place during the development phase of the life cycle of a computer-based business system. We will consider a computer program flowchart in chapter 23, Computer Program Development.

Procedure Analysis Flowcharts

Another type of flowchart with which the systems analyst should be familiar is a **procedure analysis chart,** used to record the details of manual procedures in pictorial form. A special set of symbols used with procedure analysis flowcharts is shown in figure 8.25. These are called *procedure analysis flowchart symbols.*

Figure 8.25
Procedure analysis flowchart symbols. Procedure analysis charts record the details of manual procedures in pictorial form using five flowchart symbols: operations, transportation, inspection, delays, and storage.

| operations | transportation | inspection | delays | storage |

Procedure analysis flowcharts are a particularly useful means of making a "step-by-step" analysis of procedures. Figure 8.26 is an example of a procedure analysis flowchart. The details of present (and proposed) procedures can be recorded. Time and distance measurements also can be entered in the analysis sheet. A carefully prepared procedure analysis sheet will point out duplications of effort, time delays, backtracking, and excessive inspection and transportation times. In this example, the procedure analysis sheet reveals several steps that might be modified, such as the walks to and from the A/R ledger files and the searches for customers' cards.

A procedure analysis of an existing system also can stimulate an analyst to conceive of major system changes. Some of these changes might be of an organizational nature. For example, in this case, the analyst might consider a computer-based credit checking system that could lead to the consolidation of the credit and accounts receivable functions into a single department.

Summary

Communication with systems users is of vital importance to the systems analyst. Flowcharts provide the means to effectively describe information flow and processing within a business system; they also provide a description of the logic of a computer program or the procedures within a manual system.

Two major flowcharts for describing systems are information-oriented and process-oriented flowcharts. The information-oriented flowchart is used to describe the flow of data among the various departments of an organization; the process-oriented flowchart describes the data-processing portion of the system. Data flow diagrams also are important tools of structured design. They are a means of describing the logic of data flow in a system, showing data sources, processes that transform data, and data storage. Data flow diagrams are particularly useful in developing possible solutions without premature consideration of physical implementation.

Hierarchy plus Input, Process, and Output (HIPO) charts are important tools in the top-down design of computer-related systems. Procedure analysis flowcharts are useful in a step-by-step analysis of manual procedures.

Systems analysts should be skilled in the use of all the flowcharting techniques. They should be able to select a technique, or combination of techniques, best suited to the solution of particular business information system problems.

Figure 8.26

Procedure analysis sheet. The procedure analysis sheet is an example of a procedure analysis flowchart. It is a useful means of analysis of manual procedure. A carefully prepared procedure analysis sheet will point out duplications of effort, time delays, backtracking, and excessive inspection and transportation times.

PROCEDURE ANALYSIS SHEET

No. _____   Page ____ of ____

JOB ___CREDIT CHECK PROCEDURE___

☐ MAN OR ☐ MATERIAL _____

CHART BEGINS_____ CHART ENDS _____

CHARTED BY_____ DATE _____

SUMMARY

	PRESENT		PROPOSED		DIFFERENCE	
	NO.	TIME	NO.	TIME	NO.	TIME
OPERATIONS	7	6:37				
TRANSPORTATIONS	5	1:17				
INSPECTIONS	2	3:05				
DELAYS	1	:02				
STORAGES	1	:30				
DISTANCE TRAVELLED	140 FT.		FT.		FT.	

#	Details of Present/Proposed Method	Distance in feet	Quantity	Time	Notes
1	Select next sales order			:02	
2	Check salesman's math			3:00	
3	Walk to A/R ledger file	50		:30	
4	Find customer's card			:30	
5	Record customer's balance			:10	
6	Walk to credit memo file	20		:12	
7	Note unprocessed memos			5:00	
8	Return to desk	40		:25	
9	Subtract memos from balance			:30	
10	Add sale amount to adjust balance			:20	
11	Compare new balance to limit			:05	
12	Approve or disapprove credit			:05	
13	Place 3 copies of order in out tray			:02	
14	Take copy 4 to order file	15		:05	
15	File by customer			:30	
16	Return to desk	15		:05	
17					
18					
19					
20					
21					
22					

For Review

flowchart
system flowchart
basic symbols
system symbols
programming symbols
information-oriented
  system flowcharts

process-oriented
  system flowcharts
HIPO charts
hierarchy charts
IPO charts

procedure analysis
  flowchart symbols
**procedure analysis
  flowcharts**
decomposition

For Discussion

1. What is a flowchart?
2. Why is it important to standardize flowchart symbols? Distinguish among the ANSI, ISO, and IBM standards.
3. Relate the IBM basic symbols, the IBM symbols related to programming, and the IBM symbols related to systems to the preparation of system and computer program flowcharts.
4. What is an information-oriented system flowchart?
5. What is a process-oriented system flowchart?
6. What is a data flow diagram?
7. Discuss the similarities and differences between data flow diagrams, information-oriented, and process-oriented flowcharts.
8. What are the two types of charts in a HIPO package?
9. Relate the levels of system flowcharts to the phases of the life cycle of a computer-based business system.
10. What is a procedure analysis flowchart? How is it used?

9
Communications: Technical Writing and Presentations

Preview

A systems analyst must be able to communicate effectively. To do so, the analyst must understand the elements of communication and be able to apply techniques for effective communication. Particularly important are oral communications and technical writing, since the reviews that occur at the end of each phase of the life-cycle process require the preparation of reports and oral presentations to managers and users.

Objectives

1. You will learn the components of effective communication.
2. You will be able to identify the types of technical writing used by systems analysts.
3. You will become familiar with the techniques for making effective presentations.

Key Terms

communication   The process of transferring information from one point to another.

technical writing   A document written for the purpose of communicating facts.

presentation   Oral communication of plans or results made in order to influence people and to obtain decisions.

The Elements of Communication

The Communication Process

Communication is the process of transferring information from one point to another. This transfer may involve both people and machines. In this chapter, however, we are not concerned with the machine aspects of data communication. Our purpose is to present the elements of effective person-to-person communication and to relate them to technical writing and presentations, two of the most essential ways in which systems analysts must be trained to communicate.

Communication consists of sending and receiving messages. Effective communication requires that the sender send the message accurately and that the receiver receive it without distortion. Distortion may occur because of the characteristics of the transmission medium or because of "filtering" by the receiver. Figure 9.1 illustrates the dynamics of communication. Note that feedback is included in this communication model. All the elements of the communication process must function; if not, information will not be transferred without error. Here are some examples of defective communication elements that the analyst should avoid:

1. The sender's message is not clear because the vocabulary used is not understood by the receiver.
2. The transmission medium is incorrect because the situation calls for a face-to-face meeting instead of a memorandum.
3. The receiver has "tuned out" the message because of preoccupation with another matter.
4. There is no feedback because the sender only "gives orders."

The ability to communicate is an essential skill that must be acquired because the systems analyst is sending and receiving information constantly during interaction with managers, programmers, users, and fellow team members. The analyst should, therefore, work continuously toward more effective communication.

Toward More Effective Communication

Communication becomes more effective if the sender and the receiver are sensitive to each other, if an effort is made to seek feedback, and if the appropriate transmission media are selected. Figure 9.2 lists some simple guidelines for the transfer of information.

The sender's thoughts should be organized to stress the purpose of the message. The message should be receiver-oriented, not sender-oriented. The sender should gauge the ability of the receiver to understand the message. The sender should be sensitive to both the status of the user within the organization and the user's attitudes. The sender should use facts and evidence to support the objectives of the message; unsupported opinions must be avoided.

The receiver should try to remain alert and attentive. Attention spans must be adjusted to the requirements of the message. The message must be analyzed and its main points noted as it is presented. The receiver should set aside personal attitudes and be open-minded in order to comprehend the sender's objectives.

Figure 9.1

The communication process. The communication process consists of sending and receiving messages. Effective communication requires that the sender send the message accurately and that the receiver receive it without distortion. The medium carries the message.

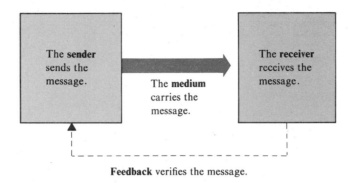

Figure 9.2

Tips for senders and receivers. Communication becomes more effective if the sender and the receiver are sensitive to each other, if an effort is made to seek feedback, and if the appropriate transmission media are selected.

tips for the sender	tips for the receiver
organize your thoughts	be alert and attentive
know your receiver	analyze the message
use facts and evidence	be open-minded

In previous chapters we have emphasized the importance of feedback in information systems. Examples are the life cycle of a computer-based business system, management information systems, and project planning and status reporting. On a person-to-person basis, the analyst not only reports to a supervisor, but also often supervises the work of others. Analysts should be aware that if they encourage and react positively to feedback, they will increase their own effectiveness as managers.

Figure 9.3

Communications and managerial effectiveness. This chart shows four areas, each representing an information-based relationship between a manager and subordinates. By encouraging feedback, a manager is able to reduce the size of the blind area and, by disseminating information, the manager is also able to reduce the size of the hidden area.

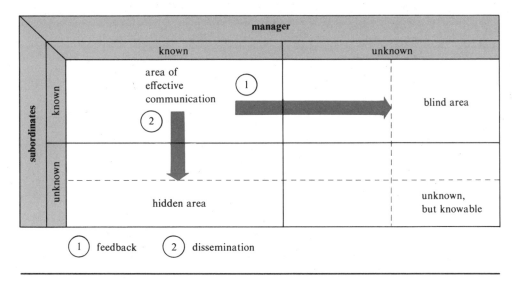

Figure 9.3 depicts the importance of feedback and dissemination (of information) in managerial effectiveness. Four areas are shown, each representing an information-based relationship between a manager and subordinates:

1. Area of effective communication—known both to the manager and to the subordinates
2. Blind area—known to subordinates, but not known to the manager
3. Hidden area—known to the manager, but not known to the subordinates
4. Unknown, but knowable area—not known to the manager or to the subordinates, but potentially knowable.

The manager is able to reduce the size of the blind area by encouraging feedback; by disseminating information the manager is able to reduce the size of the hidden area. As shown by the dashed lines in figure 9.3, the result of these actions is an enlargement of the manager's area of effectiveness and a reduction of the area representing information not known to either the manager or the subordinates. Thus, communication is not an event in time, but a continuous process. The effectiveness of this process depends on both present communication and past actions.

As figure 9.1 shows, the remaining element of the communication process is the medium. It should be selected with both the message and the receiver in mind. The two primary media for transferring information are audio and visual. Effective communicators both "show and tell." All the tools of the system analyst that we have discussed in this text (for example, forms, codes, and charts) can be used to enhance communication. Let us now consider the two formal communication-oriented activities that systems analysts perform most often: technical writing and presentations.

Technical Writing

Definition of Technical Writing

The written word is the most common conveyor of business information. For example, we have identified cumulative documentation as the key to the successful development of computer-based information systems. Throughout the life cycle of a business system, the systems analyst writes many different types of technical documents. **Technical writing,** as contrasted with many forms of nontechnical writing, is direct and to the point. Its purpose is to communicate facts. Therefore, easily understood words and short sentences are used to transmit information from the sender to the receiver. Some of the more important technical documents are policies and procedures, narratives, specifications, manuals, and reports.

As the list implies, the analyst must be qualified to prepare many types of documents. Brief descriptions of these documents illustrate their use as communication media. Examples of the first three types of documents (policies, procedures, and narratives) are included in this chapter because, although we will encounter and use them, we will not describe them in further detail in this book. The remaining three documents (specifications, manuals, and reports) are described in this chapter; however, examples are not provided at this time. In later chapters we will discuss in more detail the specifications, manuals, and reports that are of particular importance to the life-cycle process for developing computer-based business systems.

Types of Technical Writing

Policies and Procedures

Policies are broad written guidelines for conduct or action. They are the result of top level decision making. Policies should be readable; that is, they should use clear sentences, tell who has authority, and state any exceptions. (Occasionally the systems analyst will encounter—and so should be sensitive to—"unwritten" policies.)

Procedures are subordinate to policies. *Procedures* are specific statements that tell how policies are to be carried out. They provide the series of logical steps by which repetitive business operations are performed. They state the necessary action, who is to perform it, and when it is to be performed.[1]

Policies are collected in manuals called policy manuals; procedures are collected in procedure manuals. Since procedures relate to policies, a common practice is to include the policy statement in the procedure. When this is done, the policy appears at the top of the first page and is followed by the procedure, as shown in figure 9.4.

---

1. A network of related procedures for a significant business activity, for example, payroll processing, is called a system. This is within the context of our use of the word "system"; however, it is not the primary meaning of "system" in this text.

Figure 9.4
Policy and procedure statement. This example policy and procedure statement illustrates the format called playscript. Playscript is one of the most effective formats for this type of technical writing.

ABCO CORPORATION	POLICY AND PROCEDURE		
SUBJECT: COMPUTER-BASED BUSINESS SYSTEMS: STUDY PHASE	DATE: 3/17/XX		NUMBER: CS-300-0
	PAGE: __1__ of __1__		SUPERCEDES: NEW
	APPROVED: *G. Worth* PRESIDENT		

POLICY:

       All Study Phase Activities for computer-based business systems shall be completed prior to the authorization of a Design Phase.

RESPONSIBILITY:	ACTION:
User	1. Prepares a Service Request, Form C-6-1.
Information Service	2. Performs an initial investigation.
	3. Prepares a Modified Service Request, Form C-6-1.
User	4. Reviews Modified Service Request.
	5. Issues a Project Directive, Form C-6-1.
Information Service	6. Prepares a formal definition of system performance.
	7. Performs a feasibility analysis.
	8. Prepares a Performance Specification, Form No. C-9-3.
	9. Prepares a Project Plan and Status Schedule, Form C-9-1.
	10. Prepares a Project Cost Estimate, Form C-10-0.
	11. Prepares a Study Phase Report, Form No. C-10-1.
User	12. Reviews Study Phase Report.
	13. Issues Approval to Proceed, Form C-10-1.

The particular format in which the example policy and procedure statement is laid out is called playscript. Although other formats, such as outlines and narratives, are used for policy and procedure statements, playscript is one of the most effective formats. The actors (that is, the doers) are shown clearly on the left side of the page; the steps of the procedure are numbered in sequence; and the actions are expressed in simple sentences using action verbs. The allowance for white space adds to the statement's readability.

Narratives

The *narrative* tells a story and is the most informal type of technical writing. Technical reports frequently contain many passages that are best communicated in a storytelling fashion, for example, introductions, summaries, problem statements, and background discussions. The narrative technique also may be used to describe flowcharts and to define words or concepts in context. Thus, in chapter 4, the description of figure 4.2 is a type of narrative. The paragraph you are reading is an example of a narrative description.

Specifications

Specifications are reference documents that contain basic detailed data. They are the most formal and rigid type of technical document and may even include technical drawings. Specifications may accompany procedures or narratives. For example, if we consider the step-by-step process for rebuilding an automobile engine to be a procedure, the technical description of the engine and its component parts is a specification.

As was illustrated in figure 2.2 of chapter 2, the process of managing the life cycle of computer-based business systems depends upon the creation of three critical specifications: the performance specification, the design specification, and the system specification. We shall describe these specifications in chapters 14, 20, and 24 respectively.

Manuals

Manuals are printed and assembled pages of instructional material. Manuals usually are written for the use of a homogeneous group of people, and so most corporations have many different manuals. However, they are of four basic types:

1. Employee manuals introduce the employee to the company, to its rules, and to company benefits.
2. Policy and procedure manuals are used to collect policies and procedures.
3. Organization manuals contain organization charts and organization function lists.
4. Specialty manuals are prepared in order to meet the needs of different occupational groups.

The manuals with which we will be most concerned are the three types of specialty manuals required before a computer-based business system can be considered operational. These are (1) the programmer's reference manuals; (2) the operator's reference manuals; and (3) the users' reference manuals. We shall discuss the format and content of these reference manuals in chapter 22, Preparing for Implementation.

Reports

A *report* is a formal communication of results and conclusions due to a particular set of actions; it summarizes work that has been performed. The types of reports of most importance to us are the decision-oriented reports prepared at the conclusion of each phase of the computer-based business system life cycle. In particular, these are the study phase report, the design phase report, and the development phase report. These reports are described in chapters 14, 20, and 24 respectively.

Presentations

Preparing for the Presentation

Presentations of plans or results are made in order to influence people and to obtain decisions. Because they are decision-oriented, presentations are a form of selling. Analysts are expected to do more than just present facts; they are expected to have opinions. After all, by the time that an analyst has been immersed at length in a problem, some conclusions and recommendations have to have been developed that the analyst believes to be in the best interest of the company. They are what must be "sold."

All the principles of good communications discussed in this chapter should be applied to the preparation of presentations. The analyst should use both verbal and visual techniques. Some pointers are:

1. *Participate in the selection of people to attend the presentation.* Attempt to have there the individuals who will benefit most from the project and who are most involved in it at present.
2. *Know the names, titles, and attitudes (prejudices?) of all of the attendees.* Prepare to counter anticipated objections. *Above all, know which person is the decision maker.*
3. *Select a title for your subject* that is easy to remember. For example, call it the "Inventory Cleanup Project" instead of "Project 13A."
4. *Keep your presentation simple.* Use words that will be understood. Organize your main points step-by-step so that they lead to your conclusion.
5. *Make the intangible tangible.* "Before" and "after" comparisons are effective. Examples are comparisons of the number of required inventory items; reductions in out-of-inventory items; reductions in cost, time, and personnel.
6. *Use visual aids.* Visual aids that can be used in most conference rooms are flip charts (large sheets of paper that can be clipped together and "flipped" over to accompany a verbal presentation), chalkboards, and overhead projectors.
7. *Keep an eye on the clock.* Do not overstay your welcome. Complete your presentation within the allocated time. Allow approximately 25 percent of your time for discussion.
8. *Rehearse your presentation.* Almost nothing is more disconcerting than not being able to operate equipment. Be particularly aware of unintentional nonverbal communication. You will be communicating to the audience by your dress and manners, by your vocabulary, by your posture, by your sense of humor, and by your enthusiasm.

Scheduling the Presentation

The analyst is frequently faced with the prospect of presenting material to several different groups. These groups may have different interests and represent different levels of management. The analyst must decide whether to start the presentations at the top level or at a lower level. The top level is determined by the scope of the application. A department head is the top level manager for systems affecting only that department; the president of the company may be the top-level manager for a system with corporate-wide impact. It is advantageous if the material to be presented is familiar to top management and if there is genuine top-level interest in the system. A top-level presentation can result in formal management backing, giving the analyst an aura of authority. Also, if the project is rejected at the top management level, there is no need to schedule other meetings.

The advantage of starting at lower levels is the opportunity to inform and to "sell" the system to operational people. Often managers consult with their subordinates after hearing a presentation and before making a decision. Subordinates who feel that they have been left out can "poison" the mind of a supervisor against a good system. Most managers are realists; they know that a poor system may be made to work if accepted and that a good system will not work if not accepted.

A recommended approach is to work with the supervisor who is most directly involved with the system and who has the most to gain from its success. This supervisor can assist the analyst in gaining the support of subordinates and can help to pre-sell management. The supervisor and the analyst, jointly, can decide when and how to present the project to top management.

Sometimes it is desirable to make informal presentations. These provide opportunities to pre-sell and to get valuable feedback without actually seeking a decision. Informal presentations, particularly to senior management, are valuable. However, a word of caution is in order. The analyst should plan for an informal presentation no less carefully than for a formal one. Because an informal presentation is less structured than a formal one, the analyst must be prepared to be responsive to a broad range of topics and questions.

The Presentation Outcome

There are many possible outcomes from a management level presentation. Some typical outcomes are:

1. The analyst's recommendations are accepted.
2. The analyst's recommendations are accepted with modification.
3. Some recommendations are accepted and others are rejected.
4. A decision is deferred on all recommendations.
5. The project is terminated.

In the case of the first two, the analyst is free to move forward. The second two usually mean that additional work must be done and another presentation must be scheduled. These outcomes are not necessarily bad. The analyst can have received valuable feedback and direction; in any event, many system projects have to be "sold" in increments.

One of the most important storm signals that an analyst can sense during a presentation is lack of user identification with the system. If the managers who will be most affected by the system refer to it as "your (their)" system and not as "my (our)" system, the analyst knows that the system will not be accepted or successful until those attitudes are changed. This is the most important reason why it is necessary to work with a user-manager who identifies the system as "ours."

If the project is terminated, the action the analyst should take is clear—update the documentation and file the project—analyze (and rationalize) the failures, smile, and look forward to the next assignment.

Within the context of this book, the outcomes of certain presentations are critical. These are the outcomes of the reviews held at the conclusion of each of the first three major phases of the computer-based business system life cycle: (1) the study phase review; (2) the design phase review; and (3) the development phase review. These reviews are critical because they are a structured interaction with management for the purpose of obtaining a renewed commitment to the system.

Management interest normally is highest when a project is launched and when the system first becomes operational. (Of course, problems encountered as the system is being designed and developed may result in periods of intense management interest.) The interim reviews are a means of reminding management that a significant activity is underway. They also are a means of sustaining interest and support during periods when large expenditures of resources are being made for activities that are not wholly comprehensible to management because of their detailed or technical nature. The importance of these reviews is portrayed in figure 9.5. In this figure the phases of the computer-based business system life cycle are spaced to simulate realistic time spans. Note that management interest is high at the onset of a program and then tends to decay as the project enters the design and development phases. It is high again as the system approaches operational status and drops off after the initial operational problems have been overcome and the system qualifies for normal maintenance. The peaks shown in the graph of management interest are due to the scheduled study, design, and development phase reviews. As the commitment and expenditure graphs indicate, commitment tends to outstrip expenditure. Note that at the end of the study phase, when the cumulative expenditures are only 10 percent, the commitment is 25 percent. Similarly, the commitment is 70 percent at the beginning of the development phase, when expenditures are only 30 percent. Thus, the key management reviews rekindle and peak management interest and generate a new increment of management commitment at the end of each phase.

The communications-oriented concepts of *cumulative documentation* and *incremental commitment* are related in figure 9.6. This figure is our familiar life cycle. In addition, the incremental commitment and cumulative documentation are identified to the ongoing processes of management and documentation.

Having returned to the life cycle of the computer-based business system, we conclude our discussion of communication. We will proceed to study each of the four life-cycle phases in the next four units of this text.

Figure 9.5
Management interest and commitment patterns. Management interest normally is highest when a project is started and when the system first becomes operational. Reviews are a means of reminding management that a significant activity is underway; they are also a means of sustaining interest and support.

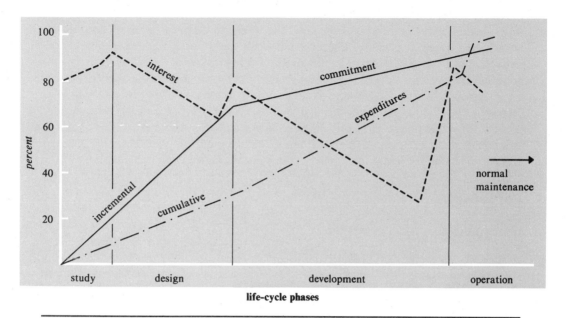

Figure 9.6
Incremental commitment and cumulative documentation in the life cycle of a computer-based business system.

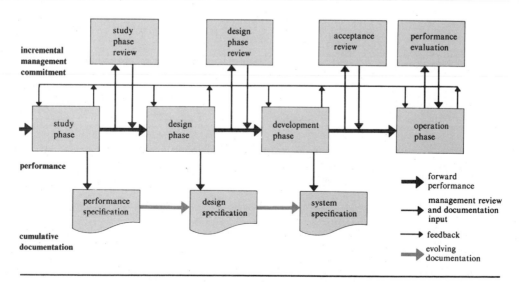

Summary

Communication is the process of transfering information from one point to another. Communication has four components: a sender, receiver, medium, and feedback. A systems analyst must understand the communications process and be skilled in techniques for effective technical writing and oral presentations.

Technical writing involves the creation of several different types of documents, each with a different purpose. These documents include policies and procedures, narratives, specifications, manuals, and reports. All communicate facts as directly and concisely as possible.

Presentations may be either formal or informal. Even informal careful planning. Particularly important is the communication that occurs during the reviews at the end of each life-cycle phase. Oral presentations and technical reports affect management committment to the project. Hence, they should be carefully prepared and directed toward the supervisor who has most to gain from the success of the system.

For Review

communication	specification	**presentation**
policy	manual	**technical writing**
procedure	report	cumulative
narrative	incremental commitment	documentation

For Discussion

1. Give an example of communication and identify each of the four elements (or lack thereof). Why is feedback necessary for effective communication?
2. Why do the authors state that "communication is not an event in time, but a continuous process"?
3. It has been estimated that communication is 80 percent visual. What visual communication processes were described in this chapter? Can you name some others?
4. Distinguish between policies and procedures. What purposes do narratives, specifications, manuals, and reports serve?
5. Discuss different approaches to scheduling presentations. How can an analyst sense a lack of user identification?
6. What are the differences and similarities between formal and informal presentations?
7. What is meant by the terms "cumulative documentation" and "incremental commitment"? How do they relate to effective communication? To the life cycle of computer-based business systems?

The Study Phase

Unit 3

10
Study Phase
Overview

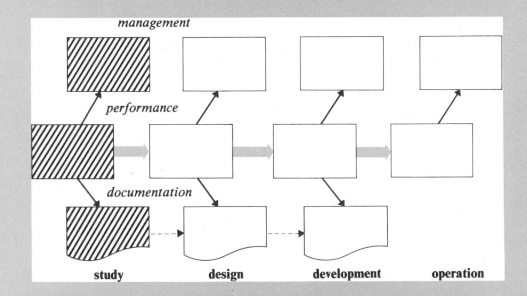

Preview

The study phase is the first of the four life-cycle phases. It is the phase in which a problem is identified, alternate solutions evaluated, and the most feasible solution recommended. An initial investigation is performed to establish a written "contract," called a project directive, between the user and the information services organization. A feasibility analysis is performed in order to evaluate possible solutions and to recommend the most feasible one. At the end of the study phase a report is completed and recommendations are made to the user.

Objectives

1. You will become familiar with the major study phase activities.
2. You will acquire a perspective of the study phase that will serve as a reference and guide to the specific chapters in this unit.

Key Terms

study phase   The life-cycle phase in which a problem is defined and a system is recommended as a solution.

information service request   A written request for information services support.

initial investigation   An investigation performed to clarify the business information system problem and to develop a project directive.

project directive   The final version of an information service request; the written contract between the user and the information systems organization.

system performance definition   A process involving the statement of general constraints, identification of specific objectives, and description of outputs.

feasibility analysis   A procedure for identifying candidate systems and selecting the most feasible.

study phase report   A comprehensive report prepared at the conclusion of the study phase activities.

Study Phase Activities

The **study phase** is the first life-cycle phase used in the creation of a computer-based business information system, either a new system or a modification of an existing system. During the study phase a preliminary analysis is carried out in sufficient depth to permit a technical and economic evaluation of the proposed system. At the conclusion of the study phase, a decision is made whether or not to proceed with a design phase. A formal project may not be established until a design phase is initiated. However, the study phase is conducted in an organized, project-like manner. The principal study phase activities, as depicted in figure 10.1, are listed here.

User Need

The creation of a computer-based business information system begins with a stated *user need*. This need may be a requirement for new information or for the solution of a problem. The statement of need is a written request for information systems service, which we shall refer to as an **information service request.** The information service request may define the user's needs completely and may be sufficient for an analyst to proceed with the system design. In this case, it would be accepted as a "contract" between the sponsor and the information services organization. However, normally an **initial investigation** must be completed before a fully informed response can be made. When this is the case, the request for service is identified as a limited information service request, and a systems analyst is assigned to conduct an initial investigation.

Initial Investigation

The first steps in the initial investigation are directed toward clarifying the problem and strengthening the analyst's background in the problem area. If there is an existing system that is performing some or all of the functions the new system is to perform, the analyst must study this system. After becoming familiar with the system, the analyst can investigate specific operations, particularly problem areas, in detail.

The analyst begins the initial investigation by studying the organization responsible for the current system and identifying product flow and information flow. The study of the existing organization provides a background knowledge of the problem environment. The identification of product flow familiarizes the analyst with the physical processes involved in making the product. The identification of information flow deals with the documents and other data carriers that control the operation of the current system. The systems professional studies information flow by finding and analyzing facts. Fact-finding activities include reviewing existing manuals and procedures, preparing questionnaires, and conducting personal interviews. Fact analysis is accomplished by techniques such as data element analysis, input-output analysis, recurring data analysis, and report use analysis.

After completing the initial investigation, the systems analyst organizes and summarizes the results of the fact-finding and fact-analysis activities. The analyst now has a current information file and a comprehensive knowledge of the existing system. This knowledge includes the relation of the cost and performance of the present system to the stated objectives of the study, the identification of the performance requirements

Figure 10.1
Study phase activity flowchart. The principal activity sequences of the study phase are the identification of user need, documented as a project directive; system performance definition; and feasibility analysis. A study phase report is prepared and reviewed prior to proceeding with the design phase.

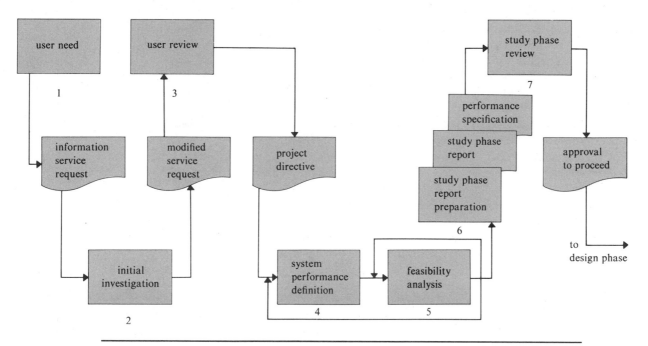

for the new or modified system, and the identification of problem constraints. The analyst has reached some conclusions about discarding, modifying, or continuing the present system or some of its elements. The analyst then summarizes and presents the results of the initial investigation to the user in the form of a modified information service request. The initial investigation activities are discussed in detail in chapter 11, Initial Investigation.

User Review
The modified information service request reflects the analyst's understanding of the problem and states his understanding of the system's objectives. The modified request is discussed with the user-sponsor, and additional revisions are made if necessary. With the concurrence of the user, the modified information service request becomes the formal contract between the user-sponsor and the analyst. This contract is called a **project directive.** The project directive authorizes the analyst to proceed to define the formal, user-oriented performance requirements for the new system and to complete the study phase.

The study phase proceeds in two major sequences:

1. **System performance definition:** (a) statement of general constraints; (b) identification and ranking of specific objectives; and (c) description of outputs.
2. **Feasibility analysis:** (a) selection and description of candidate systems, including flowcharts, specific constraints, inputs, processing, and storage; and (b) selection (from the candidates) of the most feasible system.

System Performance Definition

The performance of the new system is defined by listing its specific objectives in order of priority and by describing in detail the outputs to be produced. The outputs must be described in terms meaningful to potential users, not in "computerese." Typically, there are external and internal constraints that will limit the number of possible problem solutions the analyst can consider. The systems analyst must take these into account while relating outputs to objectives. Performance definition is the subject of chapter 12, System Performance Definition.

Feasibility Analysis

The feasibility of designing the system is determined by evaluating alternate methods of converting available input data into the required outputs to fulfill the system objectives. Each of these alternate methods is termed a candidate system. For each candidate a data flow diagram and/or information-oriented flowchart is drawn. These flowcharts trace information from input to output by identifying all data carriers (reports, forms, CRT displays and so forth) required by the system. Then, a high-level system flowchart is prepared. It identifies the system data base and the major processing operations that must be performed on the input data to produce the desired outputs.

The constraints unique to each candidate are stated. Both the unique and general constraints are taken into account as each candidate system is evaluated.

Candidate systems are evaluated by identifying factors that significantly affect system cost and performance and by ranking each candidate in terms of these factors. Typical factors are development costs, operating costs, response time, development time, accuracy, and reliability. The feasibility study is concluded by the selection of the most suitable candidate.

As shown in figure 10.1, two major *feedback* loops can occur during the study phase. The loop around the feasibility analysis block indicates the consideration of more than one system variation, or candidate. The loop around the system performance definition block indicates that the process of selecting a feasible system could include modification of the initially desired outputs. The feasibility analysis processes are described in chapter 13, Feasibility Analysis.

Study Phase Report Preparation

After the feasibility study has been completed, a **study phase report** is prepared for the user-sponsor of the system. It contains a summary of the feasibility study and presents recommendations related to proceeding with the design phase. An essential part of the study phase report is a user-oriented *performance specification*. This specification is the first of the three major baseline specifications. If the recommendation is to proceed with a design phase, a project plan and a cost schedule are prepared, and they are included as part of the study phase report. These schedules provide detailed estimates for the design phase and gross estimates for the development phase of the system project. They also serve as bases for continuing and expanding the project control functions.

Study Phase Review

The study phase report is reviewed with the user-sponsor and other affected management. If the recommendations of the report are accepted, the user issues a written approval to proceed. This approval includes an authorization for manpower and other resource expenditures required for the design phase.

Project Control

The diversity of the Maturing Era information resource applications and the complexity of the information-handling technologies are major challenges of the future. The decades of the 1980s and 1990s will witness increased emphasis upon methods for organizing and controlling projects that have the creation of cost-effective, usable business information systems as their objective. As stated in the preceding discussion of study phase report preparation, a project plan and cost schedule are prepared before the design phase is initiated. It is good professional practice to introduce project control at the onset of the study phase. The minimal elements of project control are (1) a project file, in which pertinent documents and working papers are organized and stored; (2) a *progress plan and status report* chart; and (3) a *project cost report* chart. Projects increase in complexity as they progress from the study phase into the design and development phases. Techniques for project management, such as the critical path methods introduced in chapter 7, are expanded upon in chapter 22, Preparing for Implementation. Also, in chapter 26, Changeover and Routine Operation, there is a discussion of organizing for the management of information resource projects.

Figure 10.2 displays a typical project plan. In this example the study phase is of twenty-four weeks duration. Figure 10.3 is a typical estimate of accompanying costs. These charts conform to the format presented in chapter 7, Charting Techniques. The activities projected in figure 10.2—the principal study phase activities—will be discussed in additional detail in the next chapters of this unit. As we proceed, we shall make use of an example accounts receivable system to provide a coherent and continuous thread between related life-cycle activities.

Summary

The study phase is the first of the four life-cycle phases. It is the phase in which the business information system problem is identified, the system performance defined, alternate solutions evaluated, and the most feasible solution recommended for system

Figure 10.2
Study phase project plan and status report. A project plan, with milestones against which progress can be reported, is an important element of project control. This plan is necessary because of the complexity of computer-related business information systems.

	PROJECT PLAN AND STATUS REPORT																	

PROJECT TITLE

PROJECT X

PROJECT STATUS SYMBOLS
O Satisfactory
□ Caution
△ Critical

S. A. Nallist

PROGRAMMER/ANALYST

PLANNING/PROGRESS SYMBOLS
□ Scheduled Progress ∨ Scheduled Completion
■ Actual Progress ▼ Actual Completion

COMMITTED DATE
8/15/xx

COMPLETED DATE

STATUS DATE
3/1/xx

ACTIVITY/DOCUMENT	PERCENT COMPLETE	STATUS	PERIOD ENDING (Week) 2	4	6	8	10	12	14	16	18	20	22	24	26			
STUDY PHASE	0	0												▼				
Initial Investigation	0	0	▼															
User Review	0	0	∨															
Project Directive	0	0	∨															
Sys. Perf. Definition	0	0		▼														
Feasibility Analysis	0	0				▼												
Performance Spec.	0	0				▼												
Study Phase Report	0	0				▼												
Study Phase Review	0	0					▼											

Figure 10.3
Study phase project cost report. A project cost report accompanies a project plan and status report. It is a means of providing an estimate of cost to which actual cost can be compared as the information system project proceeds.

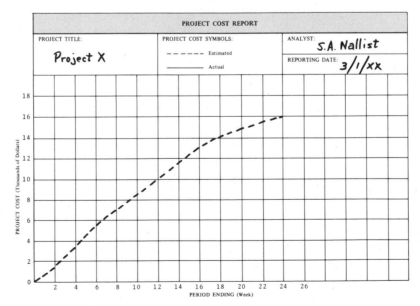

design. The study phase begins with a written statement of user's need, called an information service request. Usually this is a limited information service request since the problem is not completely defined and since the systems analyst requires a familiarization period. An initial investigation is performed for these purposes. At the conclusion of the initial investigation, the analyst prepares a modified information service request for review with the user. The result of this review is a written contract, called a project directive, between the user and the information services organization.

Subsequent to the initial investigation, two major activities occur. First, the system performance is defined more fully by listing the specific objectives and describing in detail the outputs to be produced. Second, alternate system solutions, called candidates, are identified and evaluated by means of a feasibility analysis. The most feasible system is recommended. The recommendation is part of a study phase report, which is submitted to the user. The study phase report contains a user-oriented performance specification, which is the first of the three major life-cycle baseline specifications. The study phase report also contains project plan and cost schedules, provided that the recommendation is to proceed. If that recommendation is approved, the project enters the design phase.

For Review

life-cycle phases	**project directive**	project plan and status
study phase	**system performance**	report
information service	**definition**	project cost report
request	**feasibility analysis**	feedback
initial investigation	**study phase report**	study phase review
user-need	performance	
user-review	specification	

For Discussion

1. Define and explain the purpose of:
   a. information service request
   b. initial investigation
   c. project directive
   d. feasibility analysis
   e. system performance definition
2. What is a candidate system and how does it relate to the feasibility analysis?
3. What is the purpose of the study phase report?
4. Why should a user's need be stated in written form?
5. How does the initial investigation benefit the user? The systems analyst?
6. Why is it important to introduce the elements of project control in the study phase?
7. With reference to figure 10.1, identify some instances when feedback might occur during the study phase. What might cause this feedback?

11
Initial Investigation

Preview

After the need for a new information system has been identified, the systems analyst performs an initial investigation to define the problem in detail. An information service request (ISR) generates the initial investigation, which includes: background analysis, fact-finding, and fact-analysis activities. When the initial investigation is completed, the user and the systems analyst prepare a final version of the ISR, called a project directive. The project directive becomes the contract between the user and the information services organization and enables the study phase activities to continue.

Objectives

1. You will learn the purposes of an initial investigation.
2. You will understand the importance of the ISR as a means of providing feedback between user and systems analyst.
3. You will be able to describe the differences between an initial ISR, a modified ISR, and a project directive.
4. You will become familiar with several techniques for fact-finding and fact analysis.
5. You will understand why it is important to identify the principal user.

Key Terms

information service request (ISR)   A written request for information services support.

limited ISR   An initial ISR, often incomplete, which usually initiates an initial investigation.

modified ISR   An ISR prepared by the systems analyst after completion of the initial investigation.

project directive   The final version of the ISR, which is the contract between the user and the information services organization.

initial investigation   An investigation conducted by the systems analyst to clarify the problem and define it in a detail that is mutually understood by the user and the analyst.

principal user   The person who, in practice, will accept or reject the computer-based business information system.

Problem Identification

Need Identification

Either a user or a systems analyst may identify the need for a new or improved system. Users must react to external information requirements, such as government regulations; they must respond to their own management's request for additional information; they may become aware of the unsatisfactory performance of systems for which they are responsible. For instance, the manager of an accounts receivable department may become concerned about the repeated late billing of customers or about an increase in the percentage of delinquent accounts.

Similarly, a systems analyst who is familiar with an operational or administrative area may suggest improvements. Frequently an analyst is able to view systems and their interactions with a perspective that individuals involved in daily operations lack. Often problems come into focus after joint discussions between a user and an analyst, each of whom provides an individual expertise and viewpoint.

The Information Service Request

For illustrative purposes, we will assume a typical industrial systems environment, one in which the systems analysis department is part of a larger information organization, which we will call information services. Information services also is responsible for programming support and for information resource management, including communications, data processing, and office automation equipment and operations. We will identify the formal request for information services support as an **information service request** (ISR). Figure 11.1 is our ISR form. As a typical document of this type, it provides for:

1. *Job title:* Name assigned by user to the work requested.
2. *New or rev:* Identifies the request as a new job or a revised job.
3. *Requested date:* Date the request is submitted.
4. *Required date:* Date the job should be completed.
5. *Objective:* Briefly states the principal purposes of the job.
6. *Labor:* Authorization to expend labor hours and dollars (amount).
7. *Other:* Authorization to expend nonlabor (for example, computer time) hours and dollars (amount).
8. *Anticipated benefits:* Lists the principal benefits (for example, cost savings, faster response) the company will derive from the system.
9. *Output descriptions:* For each principal output:
   frequency—how often the output is required (for example, daily, weekly, on demand).
   quantity—number of distinct outputs, hardcopy (for example, printed document) and softcopy (for example, screen display).
   pages—number of pages per hardcopy output.
   copies—number of sets of hardcopy output.
   comments—pertinent output information (for example, distribution, special paper, controls, type of screen display).

Figure 11.1
Information service request. An information service request is a formal request from a user group for support from the information services organization. It provides for statements of objectives and anticipated benefits, and for the description of outputs and inputs.

INFORMATION SERVICE REQUEST — Page ____ of ____

JOB TITLE:	NEW ☐ REV. ☐	REQUESTED DATE:	REQUIRED DATE:

OBJECTIVE:

AUTHORIZATION			
LABOR		OTHER	
HOURS	AMOUNT	HOURS	AMOUNT

ANTICIPATED BENEFITS:

OUTPUT DESCRIPTION		INPUT DESCRIPTION
TITLE:		TITLE:
FREQUENCY:	QUANTITY:	FREQUENCY:
PAGES:	COPIES:	QUANTITY:
COMMENTS:		COMMENTS:
TITLE:		TITLE:
FREQUENCY:	QUANTITY:	FREQUENCY:
PAGES:	COPIES:	QUANTITY:
COMMENTS:		COMMENTS:

TO BE FILLED OUT BY REQUESTOR

REQUESTED BY:	DEPARTMENT:	TITLE:	TELEPHONE:
APPROVED BY:	DEPARTMENT:	TITLE:	TELEPHONE:

TO BE FILLED OUT BY INFORMATION SERVICES

FILE NO:	ACCEPTED ☐ NOT ACCEPTED ☐		
SIGNATURE:	DEPARTMENT:	TITLE:	TELEPHONE:

REMARKS:

FORM NO: C–6–1 — ADDITIONAL INFORMATION: USE REVERSE SIDE OR EXTRA PAGES

10. *Input description:* For each principal input:
    frequency—how often the input will be received (for example, daily, weekly, continuously).
    quantity—number of inputs (for example, source documents, on-line transactions, user queries).
    comments—pertinent input information (for example, location, availability, controls, type of data carrier).
11. *To be filled out by requestor:*
    Requested by—name, department, title, telephone.
    Approved by—name, department, title, telephone.
12. *To be filled out by information services:*
    File number—identifier assigned to request by information services.
    Accepted or not accepted—explained, as necessary, in remarks.
    Signature—name, department, title, telephone.
    Remarks—filled in by information services as appropriate, for example, explanation of nonacceptance, indication of limits on the ISR, request for additional information, or identification of an analyst assigned to the job.
13. *Additional information:* The requestor may use the reverse side of the form, additional pages of the form, or other supplemental pages, as appropriate, to describe more fully any part of the information service request.

It is important that the ISR identify all of the desired outputs and input sources as completely as possible, even though all of the information may not be firm at this time. *Hardcopy* and *softcopy* outputs should be included; some outputs will be combinations of each. Similarly, input sources, such as data entry devices, documents, and CRT terminals, should be described. The information services organization can be helpful at this time. Normally, a substantive ISR is not submitted until after discussions have taken place between representatives of the user organization and information services.

There are two types of affirmative responses to an ISR. The first response is a "can do" response. If all the data and other resources required to perform the task within the authorized expenditure and time limits are available, the ISR can be accepted without modification as a project directive. This means that work can commence on the entire job.

However, if the job is new, if the system is large, or if many factors are unknown, another type of response usually is made. The ISR is identified as "limited" in the remarks section of the information service request. Typically, a **limited ISR** authorizes an **initial investigation** so that the analyst can study the problem and develop a more definitive ISR before major expenditures are authorized for the remainder of the study phase. Often, the limited ISR cannot identify or define all the outputs and inputs of the system under study. In this case preliminary best estimates are made or a "to be determined" (T.B.D.) entry is made. After the initial investigation has been completed, a **modified ISR** is prepared by the analyst and reviewed with the user-sponsor. The modified ISR is the project directive suggested by the analyst. During the review of the modified ISR with the user, additional changes may be made. If the result of the review is to proceed with the job, a final ISR is drafted. When approved by the appropriate user and information service managers, the final ISR becomes a contract between the user and the analyst.

This contract, the **project directive,** as distinguished from less comprehensive or intermediate information service requests, is the formal, mutual commitment that binds the user and the analyst throughout an information system project. Thus, the iterative revision of the ISR is a feedback process that structures a goal-oriented and documented interaction between the user and systems analyst.

As we indicated, the limited ISR usually authorizes an initial investigation. We will proceed to identify and describe the principal elements of an initial investigation. We will then conclude this chapter with a discussion of the project directive.

The Initial Investigation

[handwritten: INFORMATION SERVICE REQUEST]

Project Initiation

The analyst commences an initial investigation armed with a limited ISR. Figure 11.2 is an example of one page of a limited ISR. This ISR authorizes an initial investigation for an accounts receivable system. Unknown items (for example, number of copies) are identified as "to be determined" (T.B.D.). When starting the initial investigation, the systems analyst must contact individuals in the user's organization and in other organizations that may be affected by the system. These individuals will be concerned (and often with cause) about the analyst's activities. Therefore, it is a good practice for a senior user-manager to issue an information memorandum stating the general purpose of the investigation and establishing the identity and responsibilities of the systems analyst. This memorandum should originate at the managerial level, where responsibility lies for all activities the system may affect. Figure 11.3 shows an information memorandum related to an initial investigation of an on-line accounts receivable system. This system has been given the acronym of OARS.

The scope of the initial investigation may vary from a brief one-person effort to an extensive series of activities requiring the participation of many individuals. Regardless of the size of the initial investigation, the analyst should perform the investigation within a project management framework. This framework should include (1) a project file; (2) a project plan and status report chart; and (3) a project cost report chart. *[handwritten: PROJECT MANAGEMENT FRAMEWORK]*

A project file is essential to the management of systems projects because of the volume of data that must be collected, organized, digested, and summarized. The major elements of a project file are:

[handwritten: PROJECT FILE]

1. The information service request and other directives and memoranda received by the project.
2. Plans and schedules.
3. Collected documentation and working papers.
4. Memoranda and reports produced by the project.

Figure 11.2
Limited information service request—partial. A limited information service request enables the analyst to study the business problem and to fully define an information systems project.

INFORMATION SERVICE REQUEST					Page 1 of 3	

JOB TITLE: Initial Investigation of an On-line Accounts Receivable System	NEW ☑ REV. ☐	REQUESTED DATE: 9/1/XX		REQUIRED DATE: 9/22/XX	

		AUTHORIZATION			
OBJECTIVE: To improve the efficiency of customer billing and account collection		LABOR		OTHER	
		HOURS	AMOUNT	HOURS	AMOUNT
		100	$2,000	0	0

ANTICIPATED BENEFITS: 1. Faster customer billing and collection
2. Reduction of cash flow problems

OUTPUT DESCRIPTION	INPUT DESCRIPTION
TITLE: Customer Monthly Statement	TITLE: Sales Order
FREQUENCY: monthly QUANTITY: 6,000/cycle	FREQUENCY: daily
PAGES: 1 - 5 COPIES: T.B.D.	QUANTITY: 600
COMMENTS: To be printed on a multi-part form.	COMMENTS: Sales transaction **generated**
TITLE: A/R Transaction Register	TITLE: Payment/Credit Data
FREQUENCY: daily QUANTITY: 1	FREQUENCY: daily
PAGES: T.B.D. COPIES: T.B.D.	QUANTITY: 2,000 min.
COMMENTS: For use of A/R and Credit Departments	COMMENTS: Entered by Accounts Receivable Department

TO BE FILLED OUT BY REQUESTOR			
REQUESTED BY: G. Davis	DEPARTMENT: 310	TITLE: Head, A/R Dept.	TELEPHONE: X3250
APPROVED BY: Ben Franklin	DEPARTMENT: 300	TITLE: Manager, Accounting Div.	TELEPHONE: X3208

TO BE FILLED OUT BY INFORMATION SERVICES			
FILE NO: ISR-310-1	ACCEPTED ☑ NOT ACCEPTED ☐		
SIGNATURE: C. Hampton	DEPARTMENT: 200	TITLE: Manager, Info. Ser. Div.	TELEPHONE: X2670

REMARKS:

This is a limited ISR. All output and input descriptions are tentative.
J. Herring, Senior Systems Analyst, is assigned to conduct an
Initial Investigation.

FORM NO: C-6-1	ADDITIONAL INFORMATION: USE REVERSE SIDE OR EXTRA PAGES

Figure 11.3
Information memorandum. An information memorandum gives the purpose of the system, defines the role of the systems analyst, and demonstrates user-management support.

<u>MEMORANDUM</u>

TO: All Department Heads and Supervisors, Accounting Division.

COPIES TO: Vice President, Finance; Vice President, Sales; Division
             Managers; Head, Systems Analysis Department, J. Herring; File.

FROM: Manager, Accounting Division

SUBJECT: Study of an On-line Accounts Receivable System (OARS)

DATE: September 1, 19XX

I have requested that the Systems Analysis department of our Information
Services Division initiate a study of the feasibility of modifying our
present accounts receivable system. As you are aware, we are currently
experiencing delays in collecting account payments. One reason is the
overload and obsolescence of the batch-oriented computer system installed
five years ago. An additional reason, stemming from our business success,
is an anticipated accelerated growth in the number of new accounts and
in the daily volume of invoices. Another is the corporate plan to
establish regional cost centers.

Ms. J. Herring has been assigned the responsibility for conducting an
initial investigation. She will be working most closely with Mr. G.
Davis, Head of the Accounts Receivable Department. However, I have
asked that Ms. Herring visit with each Accounting Division department
head preparatory to beginning her investigation in order to explain her
approach to this assignment. I will appreciate your cooperation
in aiding her to familiarize herself with all of the current accounting
operations and documentation related to accounts receivable.

Please inform your personnel of Ms. Herring's assignment and solicit
their participation in an area which can contribute significantly to
the profitability of our corporation.

Ben Franklin

Ben Franklin
Manager, Accounting Division

Approved:

Alex Hamilton

Alex Hamilton
Vice President, Finance

Of course, the scope and depth of the initial investigation and of the project management framework must be scaled to the size of the assignment. Regardless of project size, effective project management is required in order to provide documentation of completed work and a sound basis for continuing the study phase.

The principal activities managed and performed during an initial investigation are background analysis, fact finding, fact analysis, and the organization and presentation of results.

Background Analysis

The analyst makes background analyses related to the proposed application in order to become familiar with the organization environment and the physical processes related to the new or revised system. The analyst must understand the structure of the organization within which the current system is operating and within which (often after considerable alteration) the new system will be expected to operate. It is necessary to determine the interactions between procedures and organization. Often, complex procedures are the result of inefficient organization. The analyst may have occasion to recommend organizational changes. Therefore, the systems analyst should (1) obtain or prepare organization charts; (2) obtain or prepare organization function lists; and (3) learn the names and duties of the people shown in the organization charts.

Since product flow deals with the movement of material and with the physical operations performed upon that material, the analyst observes these physical processes to acquire a "feel" for them. This feel is important if a person expects to conceive and implement systems that will perform in an actual working environment. As an example, the manufacturing processes for producing a large volume of small components, such as integrated circuits, are quite different from those for producing a relatively low volume of large items, such as computers, although each computer contains large quantities of integrated circuits. The systems for controlling each of these types of operations are different. The former may be highly repetitive and component-oriented, while the latter may be nonrepetitive and system-oriented.

After acquiring the necessary background knowledge, the systems analyst investigates the information environment in which the proposed system is to operate. To do this, the analyst finds and analyzes facts and then organizes and summarizes them.

Fact-Finding Techniques

The analyst collects data from two principal sources: written documents and personnel who are knowledgeable about or involved in the operation of the system under study. The analyst selects the fact-finding techniques judged to be most appropriate to the situation. Some systems are well documented; others are not. In some instances, interviewing all operating personnel may be effective; in others, interviews should be conducted on a very selective basis.

Fact-finding techniques that analysts often employ include: (1) *data collection,* (2) *correspondence and questionnaires,* (3) *personal interviews,* (4) *observation,* and (5) *research.*

1. **Data Collection**   *GATHERING   EXISTING   DOCUMENTATION*

In this first fact-finding step, the analyst gathers and organizes all documentation related to data carriers for the system under investigation. Examples of *data carriers* are forms, records, reports, manuals, procedures, and CRT display layouts. Data carriers themselves may not be readable, as in the case of electronic record storage media; however, descriptive documentation should exist. The analyst must be cautious in relying upon the validity of collected documents. Procedures, for example, may not have been updated to include recent changes to the system. Day-to-day problems may have introduced changes that are not reflected in the system documentation. And, of course, some people have a tendency to ignore procedures. Therefore, unless recently familiarized with the system and with its operating personnel, the analyst must have current information. This information can be obtained through correspondence, including questionnaires; through personal interviews; and by direct observation.

PROBLEM: EXISTING DOCUMENTATION MAY BE INCOMPLETE

pg. 2 ORANGE BOOK

2. **Correspondence and Questionnaires**

One method by which the systems analyst can determine if a particular procedure is current and being followed is to request that the individuals responsible for specific activities verify the procedure. The analyst may accomplish this by marking or reproducing appropriate sections of manuals or procedures and sending them to the responsible persons along with an explanatory letter.

Correspondence enables the analyst to explain the purpose of the investigation activities and to inform people of what is expected from them. It is particularly important that interviews be preceded by correspondence defining the subject area and the specific topics to be reviewed.

PROS

The questionnaire is an important and often effective type of correspondence. For example, it may be the only efficient method of obtaining responses from a large number of people, particularly if they were widely scattered or in remote locations. Questionnaires should be brief in order to increase the promptness and probability of response. The questionnaire also can be used to solicit responses to specific questions from individuals. However, because of the possibility of misinterpretation, questionnaires should be followed up by personal interviews whenever possible. Figure 11.4 is an example of question-oriented correspondence between an analyst and an individual, in this case, an accounts receivable department manager. Note that an effort has been made to make the questions straightforward and unambiguous.

3. **Personal Interview**

The personal interview is one of the most fruitful methods of obtaining information. An interview is a person-to-person communication. Hence, the guidelines for effective communications described in chapter 9, Communications, should be observed. The analyst is more of a receiver than a sender when conducting an interview. Although it is valid for analysts to use interviews to explain their projects and to "sell" themselves, they are primarily seeking information. Therefore, they must remember to be good listeners.

Interviews are critical because people are the most important ingredient of any system. The success or failure of a system often depends upon the acceptance of the

Figure 11.4
A questionnaire. A questionnaire helps the systems analyst to perform a background investigation. It supplements personal interviews.

```
TO:       George Davis, Head, Accounts Receivable Department

FROM:     Judy Herring

SUBJECT:  On-line Accounts Receivable System (OARS)

DATE:     October 1, 19XX


I have used the manuals and procedures which you sent me to
prepare a grid flowchart and an input-output analysis sheet.
These reflect my understanding of the flow of documents be-
tween the customer, the Shipping department, and the Accounts
Receivable department.  A copy of my flowchart and accom-
panying input-output analysis sheets are attached to this
memorandum.  I would like to discuss the chart with you and
will call you for an appointment in a few days.  I also
would appreciate it if, at the same meeting, you could pro-
vide me with answers to the following questions:

     1. Have you observed an increasing delay in receipt of
        customer payments?  If so, to what do you attribute
        the delay?

     2. Do the customer statements contain all of the infor-
        mation you need?  If not, what changes would you
        suggest?

     3. How will the proposed regional cost center concept
        affect your operations?

     4. What is the delay between date of sale and date of
        customer billing?

     5. Why are...
```

analyst by the personnel who are affected by the system. These personnel determine its usability. The following are some interview guidelines:

1. Plan the interview just as carefully as you would plan a presentation.
2. Adhere to your plan by keeping the interview pertinent. However, be flexible. Do not force the interview to follow a preconceived pattern.
3. Be informed, but do not attempt to present yourself as "the expert."
4. Arrange for a meeting time and place free from interruptions and other distractions.
5. Be punctual.
6. Know the name and position of the person you are interviewing.
7. Be courteous at all times.
8. Avoid the use of potentially "threatening" devices, such as tape recorders and cameras.

These guidelines are intended to help the analyst to create an atmosphere of cooperation, confidence, and understanding. This type of atmosphere is conducive to effective communication; however, it is difficult to create because the factors by which individuals are motivated are complex. Abraham Maslow defined a widely accepted ascending hierarchy of the needs of individuals in his classic book, *Motivation and Personality*.[1] These needs are:

1. Physiological needs
2. Safety needs
3. Belonging and love needs
4. Need for self-esteem and the esteem of others
5. Self-actualization needs
6. Cognitive needs
7. Aesthetic needs

Except for self-actualization and cognitive needs, the list is self-explanatory. Self-actualization refers to a self-started growth that encourages people to be what they are best suited to be. Cognitive needs refer to one's impulse to understand and to explain. Higher order needs usually emerge only after the lower order needs are satisfied. Thus, it is unlikely that cognitive needs could be gratified by an individual who perceived himself to be deprived of physiological or safety needs.

When it is likely that the solution to a problem will involve the use of a computer, many individuals become fearful. They sense a threat to the fulfillment of basic needs, such as physiological needs and safety needs. Unfortunately, their fears often are justified because computers can introduce major changes. However, most companies do not want to lose the services of skilled and loyal employees. Very often these employees are or can become qualified to perform important functions in the new system. Also, it usually is less expensive (and more humanitarian) to retrain employees of proven worth to the company than it is to recruit and indoctrinate new employees.

As an analyst, you should attempt to motivate individuals to work toward the success of the new system. Frederick Herzberg,[2] a psychologist who devoted many years to the study of motivation, developed foundational insights that should be of value to the analyst. Professor Herzberg distinguishes between *motivating factors* and *hygiene factors*. The motivators are the primary cause of job satisfaction; they relate to job content. The hygiene factors do not motivate, but cause dissatisfaction if they are absent. They relate to job environment. Motivators include achievement of something useful, recognition of achievement, meaningful work, responsibility for decisions, advancement, and growth.

Hygiene factors include relationship with supervisors, salary, status, security, and working conditions. Unhappiness results if these factors are not present. However, their presence does not contribute nearly as much to job satisfaction as does the presence of motivators. Hence, a systems analyst should describe the roles of individuals in the new system in terms of motivators whenever possible.

---

1. Abraham Maslow, *Motivation and Personality* (New York: Harper and Brothers, 1954).

2. Frederick Herzberg, "One more time: How do you motivate employees?" *Harvard Business Review* (Jan.–Feb. 1968).

It is good professional practice for the analyst to schedule the first interviews with management personnel. The analyst should solicit their aid in scheduling interviews with employees under their supervision. The analyst should attempt to enlist user-managers as allies in quieting the concerns of their subordinates and in encouraging support for the new system.

If successful in the conduct of interviews, the systems analyst will have not only obtained information, but also gained the support and confidence of the people who can make the project succeed or fail. This support is essential throughout all the phases of the life cycle of the business system.

Observation

In the course of data collection, interviewing, and other fact-finding activities, an experienced analyst observes the operation of the ongoing system and begins to formulate questions and draw conclusions on the basis of what is observed. Skilled analysts are able to discipline their powers of observation and recall. By "walking through" operations and seeing for themselves, they are able to correlate work flow and data flow and identify anomalies.

Observation is a continuous process. It usually is informal. However, there also are formal observation techniques that analysts may employ. For example, they may sample operations at predetermined or random times. They may perform statistical analyses. One observation technique that often is effective with manual activities is the construction of data flow diagrams and procedure analysis charts of the type described in chapter 8, Flowcharting. For example, a procedure analysis flowchart might show that a clerk in the credit department makes frequent, lengthy trips to the accounts receivable department to locate specific items for which payment may have been made. From this observation the analyst might conclude that one possible reason for delays in account collections had been identified.

Research

The final fact-finding technique we will mention is research. Research is of particular importance when a new application is being considered because it is a means of stimulating creative approaches to problem solving. All the fact-finding methods we have discussed are forms of in-house research. However, there are many out-of-house sources of information. These include trade and professional publications, such as the *Journal for Systems Management,* published by the Association for Systems Management (ASM), and *Data Management,* published by the Data Processing Management Association (DPMA). Other organizations such as the American Management Association, also publish books and reports that provide detailed information in specific applications areas.

Computer-oriented news publications, such as *Datamation* and *Computer* (published by the Institute of Electrical and Electronic Engineers Computer Society), *Computer World,* provide current articles to help an analyst to keep informed about hardware, software, and application developments.

Government publications often are pertinent, particularly as a means of obtaining background information. And, of course, libraries not only are sources of information, but also contain indexes to a large volume of periodical literature.

USE OF EXTERNAL LITERATURE

A major problem with much of the literature available to an analyst is that it may be out of date by the time it is in print. Two relatively time-current research resources are vendors and personal contacts. Vendors, such as the IBM Corporation, have found that by providing "applications" assistance to their customers, they can increase the effectiveness of and enlarge the market for their products. An analyst who can distinguish between real system needs and the possible overenthusiasm of a vendor can tap this rich research resource. Also, there are industry and applications-oriented publications that are updated monthly and frequently supply special reports. An example is *Datapro* publications, which prints many reports, including data about minicomputers, data processing, and office automation. Also, information resource-oriented features regularly appear in magazines such as *Time, Business Week, Newsweek,* and *Forbes.*

Analysts should establish and maintain contacts with their counterparts in other companies. One highly recommended method for making such contacts is membership and active participation in a professional society, such as the Association for Systems Management or the Data Processing Management Association. These organizations conduct many professional seminars related to current topics. Also, visits to companies with similar problems and the exchange of ideas with their analysts can be rewarding.

Fact-Analysis Techniques

Fact finding and *fact analysis* are related activities. As they collect information, efficient analysts organize, analyze and use it to identify additional information needs. There are many useful techniques for the organization and analysis of collected documents. These techniques provide the analyst with insight into the interaction among organizational elements, personnel, and information flow. Four techniques we will discuss are:

1. Data element analysis
2. Input-output analysis   *IDENTIFY MAJOR INPUT/OUTPUTS*
3. Recurring data analysis
4. Report use analysis

Data Element Analysis

Through this technique systems analysts assure themselves that they understand the meaning of the data names and the codes that appear in the manuals, procedures, charts, and other forms of documentation they have collected. One method of *data element analysis* has two steps:

1. Assign a number to each data element or code that appears upon a data carrier, such as a document or a CRT screen display.
2. Head a separate piece of paper with the title or other identification of the data carrier and write the meaning of each numbered data element or code.

Figure 11.5
Document numbered for data element analysis. The first of two steps in data element analysis is to assign a number to each data element on a document.

I A·

Figure 11.5 is an example of a document, in this case a customer monthly statement, taken through the first step. Figure 11.6 is a partial analysis of the same document. Analysts employ their knowledge of code planning and construction in data element analysis. They must be able not only to understand the codes and their elements, but also to recommend meaningful improvements.

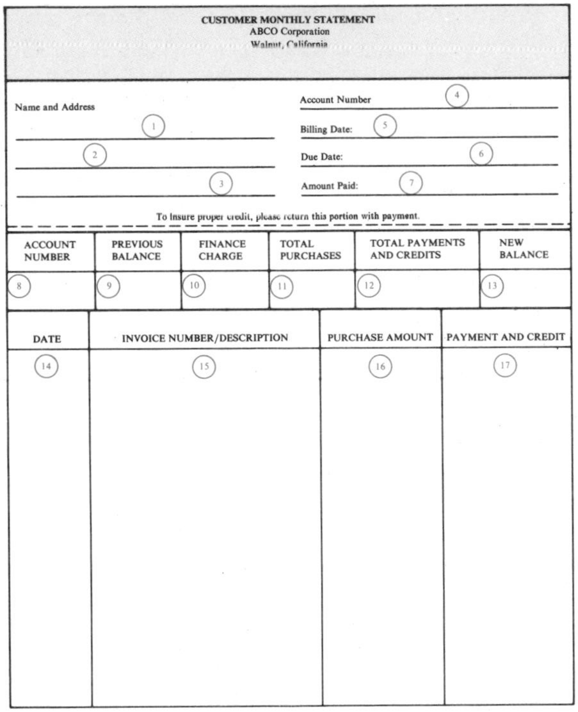

Similarly, analysts use their knowledge of forms analysis and design to determine whether a form is adequate. Very often a systems analyst will recommend the redesign of forms as a means of reducing error and improving information flow.

Input-Output Analysis

Input-output analysis is a general term for analysis techniques based upon the perception of a system as a process that converts inputs into outputs. Information-oriented system flowcharts, process-oriented system flowcharts, and data flow diagrams are excellent tools for input-output analysis. As an aid to the preparation of an information-oriented system flowchart, it often is useful for the analyst to construct a grid that identifies sources, processors, receivers, and storers of information. Figure 11.7 is an

Figure 11.6
Data element analysis. The second and final step in data element analysis is to record the meaning of each data element, in order to be certain of a common understanding by the user and the analyst.

1 ß.

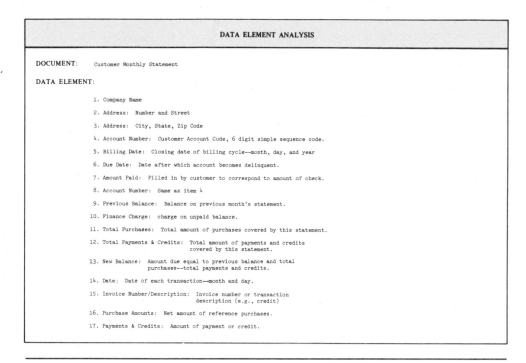

DATA ELEMENT ANALYSIS

DOCUMENT: Customer Monthly Statement

DATA ELEMENT:

1. Company Name

2. Address: Number and Street

3. Address: City, State, Zip Code

4. Account Number: Customer Account Code, 6 digit simple sequence code.

5. Billing Date: Closing date of billing cycle--month, day, and year

6. Due Date: Date after which account becomes delinquent.

7. Amount Paid: Filled in by customer to correspond to amount of check.

8. Account Number: Same as item 4

9. Previous Balance: Balance on previous month's statement.

10. Finance Charge: charge on unpaid balance.

11. Total Purchases: Total amount of purchases covered by this statement.

12. Total Payments & Credits: Total amount of payments and credits covered by this statement.

13. New Balance: Amount due equal to previous balance and total purchases--total payments and credits.

14. Date: Date of each transaction--month and day.

15. Invoice Number/Description: Invoice number or transaction description (e.g., credit)

16. Purchase Amounts: Net amount of reference purchases.

17. Payments & Credits: Amount of payment or credit.

example of the grid technique. This flowchart accomplishes two things. First, it provides the analyst with meaningful "pictures" of the connections between the information elements of the system. Second, the effort to draw the flowchart serves to pinpoint areas that are not completely understood and that may require additional study. Documentation that accompanies the information-oriented system flowchart may be a narrative description or be in the form of an input-output analysis sheet. Figure 11.8 is an example of an input-output analysis sheet prepared for an accounts receivable department. In dealing with complex information systems the systems analyst should prepare similar sheets for each organization affected by the system. Input-output analysis sheets are examples of high-level IPO (input-processing-output) charts.

A dramatic input-output analysis technique that analysts sometimes employ is to mount actual forms and reports on a wall of a room. Information flow can be displayed by colored tape or string. The values of this technique are that it provides the analyst with a "life-size" model; it keeps all of the data carriers in view; and it provides "impact" for presentations and group discussions.

The information-oriented system flowchart does not provide a picture of the operations performed by a data processing center. This detail is provided, at the systems level, by a process-oriented system flowchart. Figure 11.9 displays the principal data processing operations for an accounts receivable system.

Figure 11.7
Information-oriented flowchart for an accounts receivable system. Information-oriented flowcharts are systems flowcharts that display the relationships among data carriers for systems or subsystems. The files shown are external to data processing.

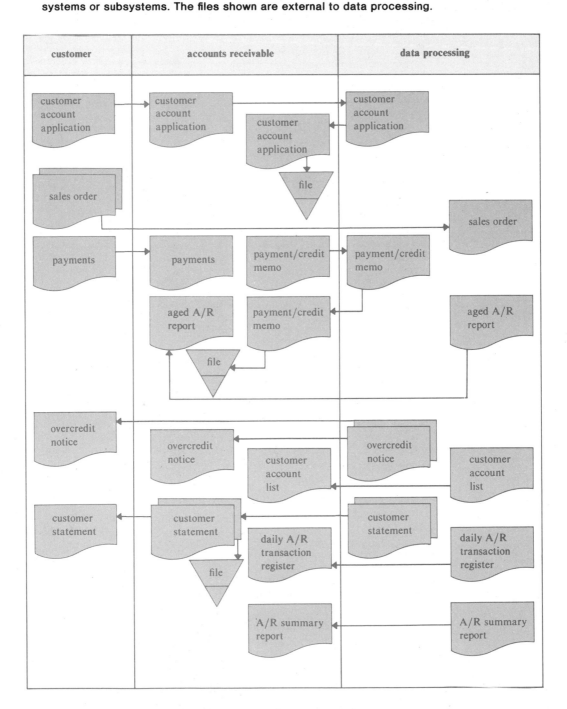

Figure 11.8

Input-output analysis sheet for an accounts receivable system. Input-output analysis sheets are system-level IPO (input-processing-output) charts that describe the relationships between inputs, processing functions, and outputs for an organizational entity. The processes and files described are external to data processing.

INPUT-OUTPUT ANALYSIS SHEET		
ORGANIZATION: Accounts Receivable Dept. SYSTEM: Accounts Receivable DATE: 9/3/XX		
INPUT	PROCESSING FUNCTIONS/FILES	OUTPUT
Customer Account Application	• Customers submit account applications, which are processed by the Accounts Receivable/Credit department. • If an application is accepted, it is sent to Data Processing for entry into the system. Data Processing returns the application for filing.	Customer Account Application
Sales Order	• Sales orders are sent directly to Data Processing for order processing. The customer retains the carbon copy.	Sales Order
Customer Payment	• Payments are sent to the Accounts Receivable department.	Payment/Credit Memo
	• A payment/credit memo is generated and sent to Data Processing. Data Processing returns the memo for filing.	
	• An aged A/R report is sent to the Accounts Receivable department from Data Processing each month.	Aged A/R Report
	• Data Processing sends overcredit notices to a credit clerk whenever a new order would exceed the customer's credit limit. If the additional credit is approved, the notice is returned to Data Processing with an authorization to process the order. If the additional credit is disapproved, the order and the notice are returned to the customer.	Overcredit Notice
	• Customer account lists are produced on demand and sent to the Accounts Receivable department for distribution.	Customer Account List
	• Customer statements are sent to Accounts Receivable in duplicate. The original copy is sent to the customer; the duplicate is filed. One-third of the statements are produced each ten days of the month, that is, on the 1st, 10th, and 20th.	Customer Statement
	• A daily A/R transaction register is sent to the Accounts Receivable department.	Daily A/R Register
	• The accounts receivable summary report is prepared weekly and sent to the Accounts Receivable department for distribution.	A/R Summary Report

Figure 11.9

Process-oriented flowchart for an accounts receivable system. Process-oriented flowcharts are systems flowcharts that display the principal information processing operations and the sequence in which they occur. The narrative that accompanies a process-oriented flowchart describes these operations. The processes and files described are internal to data processing.

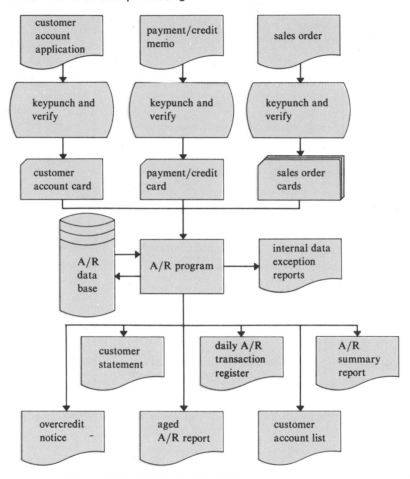

PROCESS-ORIENTED FLOWCHART NARRATIVE

1. Customer account applications, payment/credit memos, and sales orders are the three major system inputs.

2. Each source document is keypunched and verified.

3. The A/R program inputs the transaction data cards and reads/updates the master file on magnetic disk. Program edit routines produce exception reports whenever invalid input data is detected.

4. The six major outputs of the system are the customer statements, A/R transaction register, A/R summary report, overcredit notice, aged A/R report, and a customer account list.

Data flow diagrams also can be used to analyze existing systems; they may be used in addition to or in place of other types of flowcharts. In situations where there are few constraints upon the development of new or improved systems, data flow diagrams can be very effective. Figure 11.10 is a data flow diagram that is equivalent in logic content to the information-oriented and process-oriented system flowcharts of figures 11.7 and 11.9. This data flow diagram has the following advantages:

1. The symbols do not imply specific physical media; thus, the analyst is able to deal with only the logic of the system in developing alternatives for improving it. For example, these alternatives might be based upon various combinations of processing nodes or they might result from the concept of using CRT screen displays as much as possible.
2. The diagram may be used to identify logical errors or omissions "skimmed over" in other types of flowcharts. The networks could be developed, through further decomposition, to the level of detail required to identify and clarify each data element and processing transformation.

In balance, the systems analyst should use the combination of input-output analysis techniques that is most suitable for analyzing the present system and for identifying alternative solutions, bearing in mind that these solutions require both a logical and a physical implementation.

Recurring Data Analysis

After becoming familiar with the content and meaning of the principal system documents, the systems analyst usually analyzes recurring data. For this purpose a form is prepared like the one in figure 11.11. Document names and identifying numbers are entered across the top of the sheet. All the data elements associated with the first form are listed in the column headed Data Element. This process is continued for each form, moving from left to right across the sheet. Only previously unlisted data elements are added to the Data Element column. A check mark is entered at the intersection of corresponding forms and data elements. The analyst must be familiar with the forms being analyzed to avoid being deceived by the same name appearing with different meanings. An example might be the term "quantity," which could mean quantity ordered on a sales order or quantity shipped on an invoice. Similarly, the analyst should be able to distinguish between different names with the same meaning. For example, "employee number" and "badge number" might have the same meaning.

The significance of *recurring data analysis* is twofold: (1) Unnecessary input and output data duplication can be detected. This leads to form simplification, consolidation, and elimination. (2) Redundant files can be located. This leads to more efficient use of file media and may suggest the use of shared data bases.

Figure 11.10
Data flow diagram for an accounts receivable system. A data flow diagram is an important input-output analysis technique. It enables the analyst to both focus upon the logic of a system and develop alternatives for improving the system without concern about premature physical implementation.

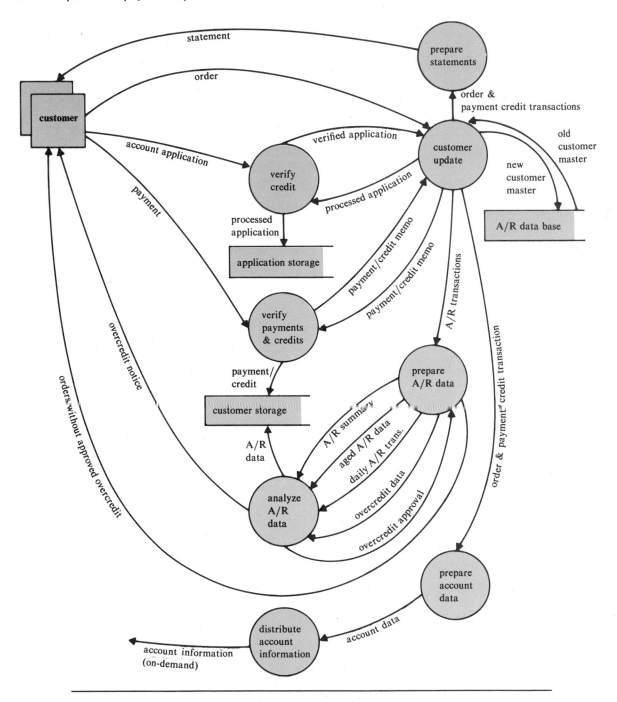

Figure 11.11
Recurring data analysis sheet. Recurring data analysis sheets are used to identify data duplication in order to eliminate, or consolidate, forms and files.

RECURRING DATA ANALYSIS SHEET	Sales Order	Shipping Order	Shipping Invoice									
Customer Name	√	√	√									
Customer Address	√	√	√									
Invoice No.	√	√	√									
Account No.	√	√	√									
Date	√	√	√									
Quantity Ordered	√	√	√									
Description	√	√	√									
Unit Price	√		√									
Amount	√		√									
Customer Signature	√											
Sales Person	√											
Total Amount	√		√									
Ship to		√										
How Shipped		√										
Quantity Shipped			√									

Report Use Analysis

Reports and copies of reports tend to proliferate. Many individuals are collectors of reports, more because of fear of being left out than because of any legitimate need for information. A useful technique for dealing with reports that are suspect because of a lengthy distribution list is a *report use analysis*. A form of the type shown in figure 11.12 is prepared and the data elements are associated with identified users of the report. A completed report use analysis sheet can be correlated with information obtained from other sources, such as user interviews. It may disclose data elements (and possibly entire reports) not required by many of the individuals or groups on the distribution list. It is not unusual to find reports that no one uses. The report use analysis sheet can provide insight into the true information needs of an organization and can help the analyst to develop more meaningful reports for the new or revised system.

Having concluded the fact finding and fact analysis, the analyst is at this point prepared to organize and summarize the results of these activities.

Figure 11.12
Report use analysis sheet. Report use analysis sheets assist the systems analyst in identifying data elements not used by individuals on the report distribution list. This leads to reduced distribution, consolidation, or elimination of reports.

REPORT USE ANALYSIS SHEET						
REPORT DESCRIPTION: Customer Monthly Statement						
USER NAME/FUNCTION / DATA ELEMENT	Customer	Customer Account Credit Clerk	Accounts Receivable Department (File copy)			
Customer Name	√					
Customer Address	√					
Account No.	√	√				
Billing Date	√	√				
Due Date	√	√				
Amount Paid	√	√				
Previous Balance	√	√				
Total Purchase	√	√				
Total Payments & Credits	√	√				
New Balance	√	√				

Results of Analysis

The systems analyst usually collects and analyzes large amounts of data. In the course of the initial investigation, the analyst has to discard data that is irrelevant and organize and summarize that which is relevant. After completing the process of organizing and summarizing data, the analyst should have a file of current information and a thorough knowledge of the current system. The information file should include:

1. Updated system documentation, including copies of all pertinent forms and reports
2. Correspondence and completed questionnaires
3. Interview records
4. Results of fact analysis, including flowcharts prepared during the investigation

The file should document the knowledge that the analyst has acquired in the course of the initial investigation. This knowledge should include:

1. A comprehensive understanding of how the current system operates, including its cost
2. Familiarity with the names, positions, and personalities of personnel operating in or affected by the system
3. Identification of the good and bad features of the current system
4. Correlation of actual system problems with the problems listed on the ISR that initiated the investigation

With respect to correlation between actual system problems and those listed in the ISR, the analyst should be able to distinguish between those problems that can be solved by a new or revised system and those that cannot. For example, a new system might be completely ineffective if the problem is an "uncontrollable" executive personality.

The single most important result of the initial investigation is the identification of the principal user. We will discuss the significance of this in the following section.

Identification of the Principal User

Identification of the principal user is critical to the success of the computer-based business information system. While organizing the results of the initial investigation and considering the anticipated benefits of the system, the analyst should remember that these benefits are not measured by the value system of the analyst, but by the value systems of users. In particular, the analyst should identify the principal user and be sure that the anticipated benefits are "meaningful" in the value system of that user. The *principal user* is the person who, in practice, will accept or reject the computer-based business system. The principal user may be the individual who issues the project directive; usually this is the case. However, the principal user also may be a superior or a subordinate of that individual. In some cases this individual may be a member of a different organization. For example, the head of the purchasing department may request a report that identifies vendors by amount of money subcontracted and by geographic location. This report may not be used by the purchasing department, but by the director of contracts to demonstrate to the federal government that the company subcontracts a certain percentage of its work with local small businesses.

Systems analysts must be sure that their own concepts of meaningful benefits are compatible with those of the principal user. As responsible professionals, analysts should seek to modify the principal user's concept of "meaningful" when it is important to the cost or performance of the computer-based business system. For example, the principal user may have in mind instantaneous access, via a desk-top terminal, to a data base that is always time-current. The analyst may know that the volatility of the data base is so low that a relatively inexpensive weekly report would be adequate. Further, the analyst may know that the activity on the terminal would not be high and that its use would be no more convenient than consulting a computer print-out. Under such circumstances, the analyst should try to persuade the user to forego the "glamour" of a terminal (which may be operated only by the user's secretary after a few weeks) in favor of a well-organized weekly report and a brief daily change listing.

[handwritten: MOTTO: GIVE THEM WHAT THEY WANT, PROVISO: ETHICAL - PRE-SELLING ALLOWS FOR THE DIPLOMATIC UNION OF WANTS AND NEEDS.]

The analyst may or may not succeed in influencing the principal user's expectations of system performance. However, a systems analyst should not knowingly undertake the development of a system that will produce outputs contrary to the principal user's expectations. Many of the failures of computer-based systems can be traced to analyst attitudes such as "He doesn't know what he needs," and "When she sees how useful it is, she will buy it."

Before spending corporate resources to produce a doubtful system, a systems professional invests the personal time and effort required to persuade the principal user to follow a reasonable course of action. We call this type of persuasion "ethical" pre-selling. Some methods of "unethical" pre-selling, which a true professional avoids, are taking advantage of the user's credulity about computers to sell glamorous but unneeded features; and climbing on the bandwagon and suggesting that the user request even more of something that isn't needed. Unethical pre-selling can, in the short term, lead to fancy systems and large computer centers. In the long run, however, such practices seldom are successful. These systems do not pay their own way and often collapse from their own weight. In addition, they leave behind a stigma that affects the development of other, legitimate computer-based information systems.

Identifying the principal user aids the analyst in another important way. It helps to establish the scope of the project, which should not extend beyond the level of responsibility of the principal user. Thus, if the individual who is to accept or reject the system is the manager of the accounts receivable department, the scope of the system is different than if it were to be used by the manager of accounting. In the one case, the scope would be limited to the accounts receivable department and its immediate interfaces. In the other case, the scope might include several accounting departments and might interface with many other elements of the company.

Finally, armed with the results of the initial investigation, the analyst prepares for a formal user review.

User Review

Modified Information Service Request

The analyst presents both the results of the initial investigation and recommendations to the user-sponsor. (We will assume that the user-sponsor is the principal user. If not, it is vital that the principal user also attend the review.) If the analyst has concluded that the project should be continued, a modified information service request is included in the presentation. The modified ISR may suggest modifications to the objectives, benefits, output descriptions, and input descriptions from those put forth in the original ISR. Figure 11.13 is an example of a modified ISR for an accounts receivable system. It is identified as a modified ISR in the remarks section of the bottom of the form. Some of the differences from the limited information service request of figure 11.2 are:

1. Expanded description of anticipated benefits
2. Resolution of to be determined (TBD) information
3. Labor authorized to complete the study phase

Figure 11.13
Modified information service request—partial. At the conclusion of the initial investigation the systems analyst may present the user with a modified information service request, based on findings that occur during the initial investigation.

INFORMATION SERVICE REQUEST		Page _1_ of _3_		

JOB TITLE: Study Phase for the On-line Accounts Receivable System (OARS)

NEW ☑ REV. ☐

REQUESTED DATE: 9/1/XX	REQUIRED DATE: 12/19/XX

OBJECTIVE: To improve the efficiency of customer billing and account collection

AUTHORIZATION

LABOR		OTHER	
HOURS	AMOUNT	HOURS	AMOUNT
600	$12,000	0	0

ANTICIPATED BENEFITS:
1. Faster billing and collection 2. Reduction of cash flow problems 3. Improved controls

OUTPUT DESCRIPTION	INPUT DESCRIPTION
TITLE: Customer Monthly Statement	TITLE: Sales Order
FREQUENCY: monthly QUANTITY: 6,000/cycle	FREQUENCY: daily
PAGES: 1 - 5 COPIES: 2	QUANTITY: 600 max.
COMMENTS: Three billing cycles. Multi-part form.	COMMENTS: Sales transaction generated
TITLE: A/R Transaction Register	TITLE: Payment/Credit Memorandum
FREQUENCY: daily QUANTITY: 1	FREQUENCY: daily
PAGES: 40 max. COPIES: 1	QUANTITY: 2,000 max.
COMMENTS: Listing for A/R Department file. Also, CRT display.	COMMENTS: Entered by Accounts Receivable Dept.

TO BE FILLED OUT BY REQUESTOR

REQUESTED BY:	DEPARTMENT:	TITLE:	TELEPHONE:
APPROVED BY:	DEPARTMENT:	TITLE:	TELEPHONE:

TO BE FILLED OUT BY INFORMATION SERVICES

FILE NO: ISR-310-1A	ACCEPTED ☑ NOT ACCEPTED ☐		

SIGNATURE: *C. Hampton*

DEPARTMENT: 200	TITLE: Manager	TELEPHONE: X2870

REMARKS:
This is a Modified Information Service Request prepared for Requestor Acceptance as a Project Directive. J. Herring is designated as Project Leader.

FORM NO: C-6-1   ADDITIONAL INFORMATION: USE REVERSE SIDE OR EXTRA PAGES

However, the initial investigation produced no essential changes in the anticipated benefits of the system.

The analyst should provide all those invited to the presentation with copies of the modified ISR and other pertinent written material beforehand. This affords them a chance to familiarize themselves with the material and to prepare questions. The user review typically includes the user-sponsor, the analyst, the analyst's supervisor, and other appropriate management and operational personnel.

The analyst should discuss the key elements of his initial investigation. The analyst should be able to support the recommendations, whatever their nature. If the recommendation is to proceed with the study phase, the resources required should be identified and a project plan and a cost schedule should be presented for the remainder of the study phase.

As a result of the user review, the project may be terminated, modified, or continued. If the decision is to proceed, it is documented by the issuance of a project directive.

Project Directive

The modified ISR really is a draft of a proposed project directive. The project directive is an authorization document issued by the user after the review of the initial investigation has been completed; it reflects the results of discussions and decisions made during that review. It may or may not be identical to the modified ISR prepared by the analyst.

When the project directive is signed by the user and accepted by information services, it becomes a contract under which both organizations are accountable for performance. The format and content of the project directive are similar to those of the ISR. Often the same form is used for both. This is the practice that we will follow in this book. Figure 11.14 is based on the modified ISR of figure 11.13. Note the following changes: The requested date has been advanced, and the labor allocation has been increased slightly. In the remarks section, this ISR is designated a project directive. Also, the requestor is to be informed when 90 percent of the authorized funds have been spent. This is a safeguard against an unauthorized cost overrun.

The project directive is the first of many incremental commitments made by management in the course of the life cycle of a computer-based business system. At this time, the project directive may authorize all the resources required to develop the new system. Usually it authorizes only the resources required to complete the study phase. Additional resources are authorized after successful reviews of the study phase and of subsequent phases. In figure 11.14, as the job title indicates, the analyst is authorized to complete the study phase.

The project directive initiates a comprehensive study of the feasibility of the proposed system. The feasibility analysis is preceded by the development of a detailed user-oriented definition of expected new system performance. The process by which the expected system performance is arrived at is described in chapter 12, System Performance Definition. The feasibility analysis itself involves the evaluation of alternative systems and the selection of the one that best meets the detailed system performance requirements. This process is discussed in chapter 13, Feasibility Analysis.

Figure 11.14

Project directive—partial. The final version of the information service request (ISR), based upon discussions between the user and the analyst of the modified ISR, is called a project directive. It is the "contract" between the user and the systems analyst.

INFORMATION SERVICE REQUEST		Page 1 of 3

JOB TITLE: Study Phase for the On-line A/R System (OARS)

NEW ☑ REV. ☐

REQUESTED DATE: 11/7/XX **REQUIRED DATE:** 12/19/XX

AUTHORIZATION

	LABOR		OTHER	
	HOURS	AMOUNT	HOURS	AMOUNT
	600	$12,500		

OBJECTIVE: To improve the efficiency of customer billing and collection

ANTICIPATED BENEFITS: 1. faster billing   2. reduce A/R payment "float"
3. expand to 20,000 accounts   4. allow additional billing cycles

OUTPUT DESCRIPTION	INPUT DESCRIPTION
TITLE: Customer Monthly Statement	**TITLE:** Sales Order
FREQUENCY: monthly **QUANTITY:** 20,000	**FREQUENCY:** on-line
PAGES: 1 to 3 **COPIES:** 2	**QUANTITY:** 1,000 per day
COMMENTS: 2-part form; multiple billing cycles per month	**COMMENTS:** Sales transaction generated
TITLE: A/R Transaction Register	**TITLE:** Customer Account Application
FREQUENCY: daily **QUANTITY:** 1	**FREQUENCY:** daily
PAGES: 40 max. **COPIES:** 1	**QUANTITY:** 20 max.
COMMENTS: printed copy for A/R file; data available for on-line screen display	**COMMENTS:** application form submitted by customer batched and entered each day after approval

TO BE FILLED OUT BY REQUESTOR

REQUESTED BY: *G. Davis*	DEPARTMENT: 310	TITLE: head, A/R dept.	TELEPHONE: X3250
APPROVED BY: *Ben Franklin*	DEPARTMENT: 300	TITLE: manager, accounting div.	TELEPHONE: X3208

TO BE FILLED OUT BY INFORMATION SERVICES

FILE NO: ISR-310-24	ACCEPTED ☑ NOT ACCEPTED ☐		
SIGNATURE: *C. Hampton*	DEPARTMENT: 200	TITLE: manager, Info. services	TELEPHONE: X2670

REMARKS:
This is a project directive.
J. Herring is appointed project leader. Advise requestor when funds are 90% expended.

FORM NO: C-6-1	ADDITIONAL INFORMATION: USE REVERSE SIDE OR EXTRA PAGES

Summary

After the need for a new system has been identified, an initial investigation is performed in order to define the problem in detail and to insure that the systems analyst and the user have a common understanding of the information to be provided by the system.

The request for information services support is a written request, called an information service request (ISR). The ISR may undergo several modifications before it becomes a written contract, called a project directive, between the user and the information services organization.

The first version of the ISR usually is a limited ISR, and it generates an initial investigation. Four major activities that are performed during the initial investigation are: (1) background analysis; (2) fact-finding; (3) fact-analysis; and (4) organization and presentation of results. Some useful fact-finding techniques are: data collection, correspondence and questionnaires, personal interviews, observation, and research. Fact-analysis techniques include: data element analysis, input-output analysis, recurring data analysis, and report use analysis.

After completing the fact-analysis activities, the systems analyst will have identified the principal user of the computer-based business information system. This is the person who, in practice, will accept or reject the system. Also, after fact-finding, the analyst will be able to prepare a modified ISR for review with the user. The result of this review is the final version of the ISR—the project directive. It initiates a detailed, user-oriented definition of expected system performance and a comprehensive analysis of the feasibility of the proposed information system.

ACTIVITIES

For Review

information service request
hardcopy
softcopy
limited information service request
modified information service request
project directive
fact-finding

data collection
data carrier
correspondence and questionnaires
personal interviews
motivating factor
hygiene factor
observation
research

fact-analysis
data element analysis
input-output analysis
recurring data analysis
report use analysis
principal user
IPO chart
initial investigation
data flow diagram

ISR → INITIAL INVESTIGATION → PROJECT DIRECTIVE

For Discussion

1. Describe several ways in which a business system information problem might be identified. *IDENTIFY MAJOR INPUTS/OUTPUTS*

2. Distinguish between the terms information service request, modified information service request, and project directive.

3. Why should an information memorandum be issued at the time an analyst begins an initial investigation?

4. What are the elements of project control that aid in the management of an initial investigation?

5. Why does an analyst perform a background analysis? Under what conditions might such an analysis not be necessary?

6. What tentative conclusions can you draw from the recurring data analysis and report use analysis sheets of figure 11.10 and figure 11.11?

7. How does an analyst apply skills in coding and forms design during an initial investigation?

8. Discuss "motivators" and "hygiene factors" as they relate to personal interviews.

9. What should an analyst's information file contain and what knowledge should the analyst have at the conclusion of an initial investigation?

10. Discuss the importance of identifying the principal user.

12
System Performance Definition

Preview

In the previous chapter we discussed the activities of the initial investigation. At this point in the life cycle the systems analyst has studied the existing system and should have learned all that is possible about it. The analyst has presented the principal user with conclusions and recommendations and has received approval to proceed with the project. The next step is to determine exactly what the new or modified system is to do. In defining the required performance of a system, the analyst must do three things: (1) state the general constraints—the limitations placed on the system; (2) rank and list the specific objectives of the system; and (3) describe the outputs through sketches and lists of the data elements appearing on the sketches.

Objectives

1. You will learn how to state general constraints.
2. You will understand how to identify and rank specific objectives.
3. You will be able to describe outputs using output specification sheets and data element lists.

Key Terms

constraint   A condition, such as time or money, that limits the solutions a systems analyst may consider.

specific objective   A measurable performance outcome for a system.

Performance Definition

Overview

The performance required of a system is defined by the description of its outputs in a user-oriented format. Chapter 10, Study Phase Overview, listed five steps in which the study phase proceeds after the completion of the initial investigation. The first three of these steps result in the definition of the performance of the new system. They are:

1. Statement of general constraints
2. Identification and ranking of specific objectives
3. Description of outputs

These steps are described in this chapter. The final two steps:

4. Selection and description of candidate systems
5. Selection of the most feasible system

relate to the analysis of feasibility, and they are discussed in chapter 13, Feasibility Analysis.

Throughout the study phase, the description of system performance is user-oriented. System performance definition follows the initial investigation, a user review, and the issuance of a project directive.

In general, the initial investigation will have eliminated extraneous factors and identified the major system problems. Similarly, the project directive will have meaningful (to both the user and analyst) statements about constraints, objectives, and outputs. The purpose of defining system performance is to describe the system outputs at an additional level of detail. At this level the analyst is able to make sketches of all system outputs and to prepare accompanying data element lists. Sometimes all the required information is available to the analyst from the initial investigation and the project directive. Other times, it may be necessary to conduct additional investigative activities.

In this chapter, we will use the on-line accounts receivable system (OARS) of the ABCO corporation as an illustrative example of the procedure for defining system performance. We will continue to use the OARS example in subsequent chapters. Therefore, we will begin this chapter by presenting additional detail about the current accounts receivable system.

Example System: Additional Information

Let us again consider the ABCO corporation's current accounts receivable system (A/R system) presented in chapter 11, Initial Investigation. In that chapter we described a variety of techniques for fact finding and fact analysis. Elements of the A/R system were used for illustrative purposes throughout that chapter. Now, we will present some additional information about the ABCO corporation and its accounts receivable system. This will enable us to use this system as a coherent example.

The ABCO corporation is a small business that designs and manufactures specialty household items. It does not sell to the public, but is a nationwide wholesale and retail supplier. It currently has three regional sales divisions through which merchandise is supplied to its customers.

Currently, the ABCO corporation has 10,000 customer accounts distributed over the three regional divisions. This level of accounts appears to be the maximum capacity of the present accounts receivable system. The number of accounts is expected to double to 20,000 over the next five years. Sales are expected to increase by approximately 2,000 accounts each year for the next five years. The current system cannot meet this projected growth and satisfy the corporate goal of distributing information processing resources to regional profit centers. In addition, the number of regions is expected to grow to five from the current three regions to better serve ABCO customers. Specific problems that have been identified are:

1. Saturation of the capacity of the present computer system causing difficulties in adding new accounts and in obtaining information about the status of existing accounts.
2. Processing delays in preparing customer billing statements because of the batch-oriented design of the current accounts receivable system.
3. Excessive elapsed time between mailing of customer statements and receipt of payments, which creates a high-cost, four-day float (the time that the money is not in use, even though the customer has sent a check).
4. Inadequate control of credit limits.
5. Inability to provide the regional centers with timely customer-related information.

The ABCO corporation has a medium-sized computer that is being used to its capacity. In addition, its design is not oriented toward on-line systems. Mr. B. Franklin, manager of the accounting division, felt that the best way to reduce the time required to process orders and customer statements, and provide a faster response to requests for customer information to the regions, is through a distributed data processing system. By placing computer terminals within each region, ABCO regions would have fast access to the data base and customer orders could be entered locally, rather than mailed to a central site. Entering sales data through terminals would also eliminate the use of punched cards, a source of delays and errors.

Accordingly, Mr. Franklin instructed Mr. G. Davis, head of the A/R department, to ask the information services division to study the feasibility of an on-line accounts receivable system. The initial request for service (presented in chapter 11 as figure 11.1) was accepted as a limited information service request. Ms. J. Herring was assigned the responsibility for conducting the initial investigation.

We shall review and elaborate on the results of that investigation as an "example system" as we proceed to discuss the steps involved in system performance definition.

General Constraints

Statement of General Constraints

General **constraints** are those limiting all problem solutions that the systems analyst may consider. Typically, these are constraints that have been imposed from the outset of the project or that have been identified in the course of the initial investigation and discussed with the user-sponsor. There are many possible general constraints. Among the most common are management policy, legal requirements, equipment and facilities, for example, an existing computer center, audit and internal control requirements, fixed organizational responsibilities, cost, and time.

With respect to the cost and time constraints, it is important for the analyst to realize that management often is willing to accept something less than an optimum system provided that it meets basic needs, that its development cost is not excessive, and that it will be available when needed. The analyst who is not able to appreciate the realities of the corporate environment will not enjoy a long and rewarding career. Incidentally, this is one reason why it is often a good strategy, when possible, to segment large systems so that some elements can be installed and can begin to "pay their own way" while the development of other elements continues.

Example System: General Constraints

For the example system, the ABCO corporation's OARS, some general constraints might be as follows:

1. Development of the on-line accounts receivable system (OARS) is to be completed within fourteen months.
2. OARS is to have a growth potential to handle 20,000 customer accounts.
3. OARS is to interface with the existing perpetual inventory system.
4. OARS is to be designed as an on-line system operating in a distributed data processing environment.
5. The design must be compatible with corporate plans to proceed toward a regional profit center concept.

After the system constraints are established, the analyst proceeds to identify specific performance objectives.

Specific Objectives

Identification and Ranking of Specific Objectives

The objective stated in the project directive is the major system objective. However, this objective is usually a general statement of purpose. The systems analyst must derive from this general statement of specific objectives to which each system output can be related. The first step is to analyze the anticipated benefits stated in the project directive. These benefits, in association with the general system objective, help to formulate **specific objectives,** which are measurable performance outcomes. The experienced analyst does not lose sight of the fact that the benefits of a system must be meaningful and measurable in the value system of the principal user.

Figure 12.1
System benefits. Whenever possible, benefits must be translated into specific objectives that can be stated in measurable terms. This, however, does not mean that intangible benefits should not be considered.

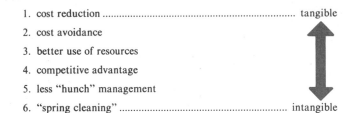

1. cost reduction .... tangible
2. cost avoidance
3. better use of resources
4. competitive advantage
5. less "hunch" management
6. "spring cleaning" .... intangible

The anticipated benefits may be tangible or intangible. Figure 12.1 lists six categories of system benefits. They range from the very tangible, for example, cost reduction, to the very intangible, for example, "spring cleaning" (which means change for the sake of change). Historically, management has tended to emphasize tangible benefits. More recently, many companies have progressed to the use of computer-based systems to achieve competitive advantages and to provide information that aids in policy and planning decisions. These benefits may be visible only in the corporation's profit and loss statement.

Whenever possible, benefits must be translated into specific objectives that can be stated in measurable terms. For example, "To improve customer service," a general objective and statement of purpose, might be supported by the following specific objectives:

1. To increase the percentage of goods shipped on schedule from 40 percent to 80 percent within six months and to 95 percent within a year.
2. To reduce the number of order cancellations from 35 percent to 5 percent within six months.
3. To reduce the number of back orders from 25 percent to 5 percent at a rate of 5 percent per month.

As illustrated above, percentage improvements are an effective means of comparing the performance of a new system with that of the existing system. However, the analyst should be careful not to use the present system's performance as the only reference for establishing the new system's objectives. The analyst should focus also on new needs, some of which may not be related to elements of the current system. Frequently, in the course of developing specific objectives—both tangible and intangible—the analyst has to extend fact finding and analysis to areas or depths not covered in the initial investigation.

Example System: Specific Objectives

Let us again consider the ABCO corporation's OARS. As the first step in identifying and ranking specific objectives, we shall review the salient results of the initial investigation. Figures 12.2 through 12.9 summarize and supplement the information presented in chapter 11. The material is presented as follows: existing system flowcharts, a partial organization chart, existing system outputs and inputs, existing system costs, new system project directive, and new system specific objectives.

Existing system flowcharts   Figure 12.2 depicts the reporting relationships between the major organizations affected by the A/R system. This is a working chart used by the analyst to record background information pertinent to the organizational environment. Other useful information, such as department numbers and telephone numbers, also can be indicated.

Figure 12.2
ABCO Corporation—partial organization chart. Organization charts are frequently used by the analyst to record background information pertinent to the organizational environment. These charts should identify by name and function the individuals with whom the analyst expects to come into contact.

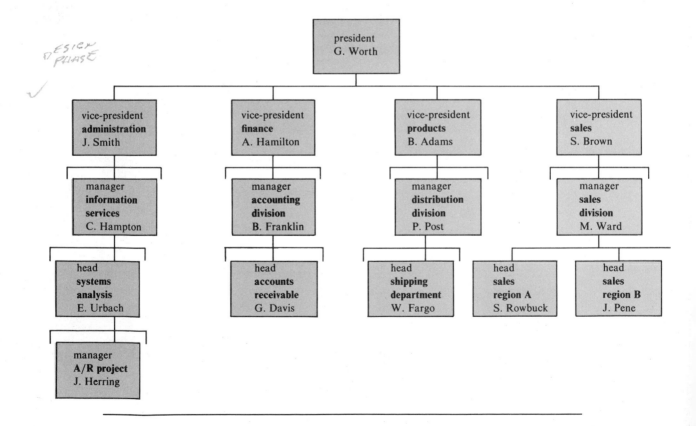

To show additional detail, such as the organizational breakdown within the accounts receivable department, the analyst can prepare additional charts.

These charts should identify by name and function the individuals with whom the analyst expects to come into contact. Organizational function lists for each department shown on the organization chart should also be developed. This additional detail is not included as part of the example; however, it is implicit to the development of the detail presented in the subsequent figures that describe the example company's existing A/R system.

Figure 12.3 is the same as figure 11.7 of the preceding chapter. It identifies the flow of data from the customer through the shipping, accounts receivable, and data processing departments and back to the customer. Figure 12.4 is an input-output analysis for the accounts receivable department. The narrative content was prepared from organizational background information (for example, flowcharts and organization function lists) and from information gathered during fact finding and analysis. It is the same as figure 11.8.

Figure 12.5 displays the accounts receivable processing operations. The system is a punched card, batch-oriented system with six printed outputs. The narrative description for this system is shown on the bottom of the flowchart.

Current system outputs and inputs Figure 12.6 shows the six outputs of the current system. These are the customer statement, overcredit notice, aged A/R report, daily A/R transaction report, A/R summary report, and the customer account list. Figure 12.7 illustrates the inputs to the current system. These are the customer account application, payment/credit memo, and the sales order.

Current system costs The direct operating costs of the current A/R system were found to be a combination of computer, key punch, and clerical costs. These costs are summarized in figure 12.8. The total current operating cost is $18,500 a month ($1.85 per account for 10,000 accounts).

New system project directive The most significant conclusion reached by J. Herring as a result of the initial investigation was that most of the problems of the current system were related to its batch-processing nature. It appeared that a terminal-oriented, on-line system would be the solution. Her conclusion is reflected in the project directive, which appears as figure 12.9. This figure completes the example project directive. The six outputs for the new system are:

1. Customer monthly statement
2. A/R transaction display and report
3. A/R summary display
4. Aged accounts receivable report
5. Customer account data display
6. Overcredit notification

Tentatively, the inputs for the new system are the same as for the existing system. Data will be entered directly from the regions through local terminals.

Figure 12.3
Information-oriented flowchart for an accounts receivable system. Information-oriented flowcharts are systems flowcharts that display the relationships between data carriers for systems or subsystems. The files shown here are external to data processing.

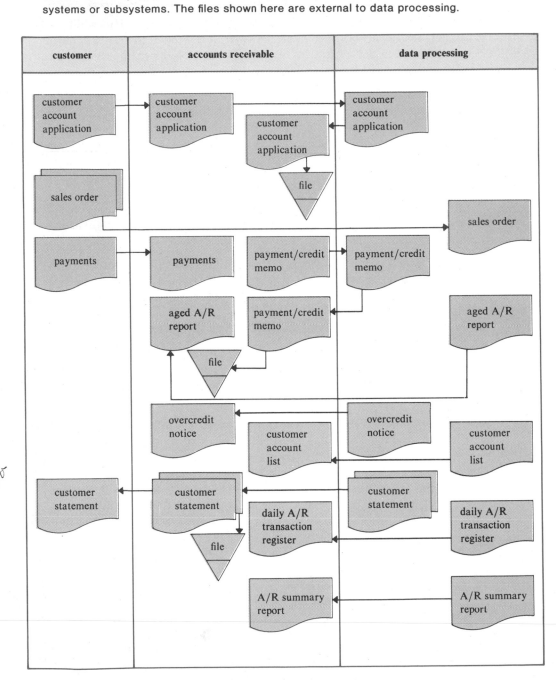

Figure 12.4

Input-output analysis sheet for an accounts receivable system. Input-output analysis sheets are system-level IPO (input-processing-output) charts that describe the relationships between inputs, processing functions, and outputs for an organizational entity. The processes and files described are external to data processing.

INPUT-OUTPUT ANALYSIS SHEET		
ORGANIZATION: Accounts Receivable Dept. SYSTEM: Accounts Receivable DATE: 9/3/XX		
INPUT	**PROCESSING FUNCTIONS/FILES**	**OUTPUT**
Customer Account Application	• Customers submit account applications, which are processed by the Accounts Receivable/Credit department. • If an application is accepted, it is sent to Data Processing for entry into the system. Data Processing returns the application for filing.	Customer Account Application
Sales Order	• Sales orders are sent directly to Data Processing for order processing. The customer retains the carbon copy.	Sales Order
Customer Payment	• Payments are sent to the Accounts Receivable department. • A payment/credit memo is generated and sent to Data Processing. Data Processing returns the memo for filing.	Payment/Credit Memo
	• An aged A/R report is sent to the Accounts Receivable department from Data Processing each month.	Aged A/R Report
	• Data Processing sends overcredit notices to a credit clerk whenever a new order would exceed the customer's credit limit. If the additional credit is approved, the notice is returned to Data Processing with an authorization to process the order. If the additional credit is disapproved, the order and the notice are returned to the customer.	Overcredit Notice
	• Customer account lists are produced on demand and sent to the Accounts Receivable department for distribution.	Customer Account List
	• Customer statements are sent to Accounts Receivable in duplicate. The original copy is sent to the customer; the duplicate is filed. One-third of the statements are produced each ten days of the month, that is, on the 1st, 10th, and 20th.	Customer Statement
	• A daily A/R transaction register is sent to the Accounts Receivable department.	Daily A/R Register
	• The accounts receivable summary report is prepared weekly and sent to the Accounts Receivable department for distribution.	A/R Summary Report

218 **Figure 12.5**

Process-oriented flowchart for an accounts receivable system. Process-oriented flowcharts are systems flowcharts that display the principal information processing operations and the sequence in which they occur. The narrative that accompanies a process-oriented flowchart describes these operations. The processes and files described are internal to data processing.

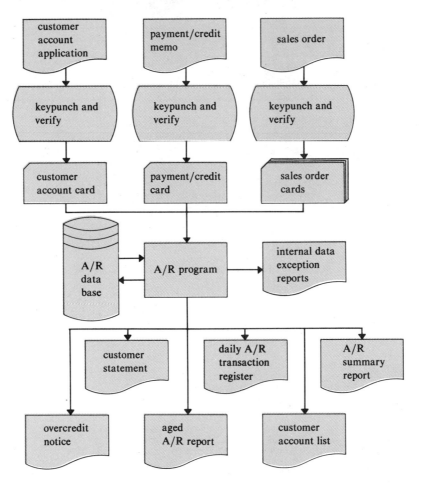

DESIGN
PHASE ✓

PROCESS-ORIENTED FLOWCHART NARRATIVE

1. Customer account applications, payment/credit memos, and sales orders are the three major system inputs.

2. Each source document is keypunched and verified.

3. The A/R program inputs the transaction data cards and reads/updates the master file on magnetic disk. Program edit routines produce exception reports whenever invalid input data is detected.

4. The six major outputs of the system are the customer statements, A/R transaction register, A/R summary report, overcredit notice, aged A/R report, and a customer account list.

Figure 12.6
Outputs from existing A/R system. The systems analyst reviews the outputs of the existing accounts receivable system and related information collected during the initial investigation.

INITIAL
INVESTIGATION
FEES.

✓

Figure 12.6
continued

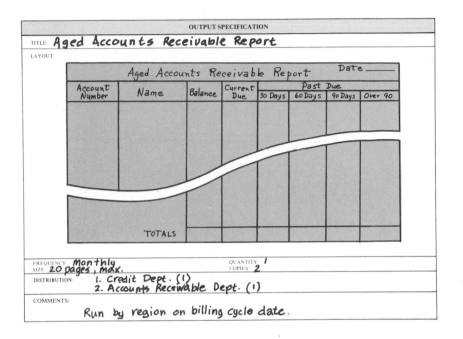

OUTPUT SPECIFICATION

TITLE: **Aged Accounts Receivable Report**

LAYOUT:

Aged Accounts Receivable Report   Date_____

Account Number	Name	Balance	Current Due	Past Due			
				30 Days	60 Days	90 Days	Over 90

TOTALS

FREQUENCY: **Monthly**
SIZE: **20 pages, max.**
QUANTITY COPIES: **2**
DISTRIBUTION: 1. Credit Dept. (1)
2. Accounts Receivable Dept. (1)

COMMENTS: Run by region on billing cycle date.

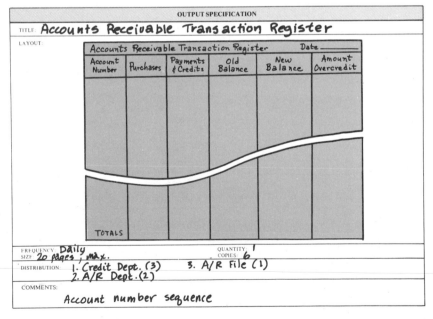

OUTPUT SPECIFICATION

TITLE: **Accounts Receivable Transaction Register**

LAYOUT:

Accounts Receivable Transaction Register   Date_____

Account Number	Purchases	Payments & Credits	Old Balance	New Balance	Amount Overcredit

TOTALS

FREQUENCY: **Daily**
SIZE: **20 pages, max.**
QUANTITY COPIES: **6**
DISTRIBUTION: 1. Credit Dept. (3)   3. A/R File (1)
2. A/R Dept. (2)

COMMENTS: Account number sequence

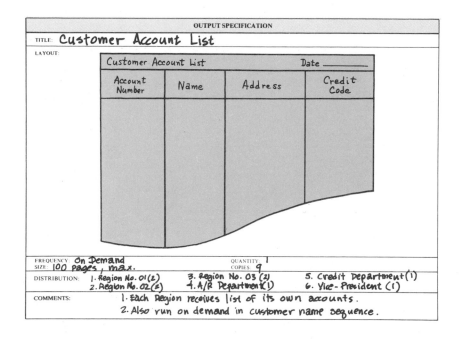

OUTPUT SPECIFICATION

TITLE: **Customer Account List**

LAYOUT:

Customer Account List			Date _____
Account Number	Name	Address	Credit Code

FREQUENCY: On Demand
SIZE: 100 pages, max.
QUANTITY COPIES: 9

DISTRIBUTION:
1. Region No. 01(2) 3. Region No. 03 (2) 5. Credit Department(1)
2. Region No. 02(2) 4. A/R Department(1) 6. Vice-President (1)

COMMENTS:
1. Each Region receives list of its own accounts.
2. Also run on demand in customer name sequence.

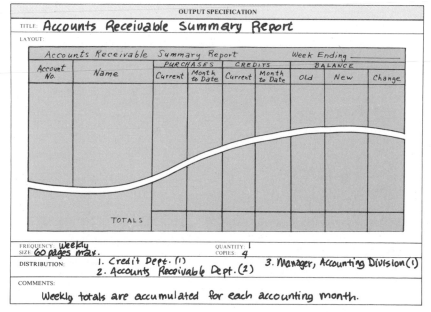

OUTPUT SPECIFICATION

TITLE: **Accounts Receivable Summary Report**

LAYOUT:

Accounts Receivable Summary Report		PURCHASES		CREDITS		BALANCE		Week Ending _____
Account No.	Name	Current	Month to Date	Current	Month to Date	Old	New	Change
TOTALS								

FREQUENCY: Weekly
SIZE: 60 pages max.
QUANTITY COPIES: 4

DISTRIBUTION:
1. Credit Dept. (1)   3. Manager, Accounting Division(1)
2. Accounts Receivable Dept. (2)

COMMENTS:
Weekly totals are accumulated for each accounting month.

Figure 12.7

Inputs to existing A/R system. The systems analyst reviews the inputs of the existing accounts receivable system. Many of these may be inputs for the new system.

CUSTOMER ACCOUNT APPLICATION

CUSTOMER ACCOUNT APPLICATION
ABCO Corporation
Walnut, California

FOR OFFICE USE
ACCOUNT NUMBER

FIRM NAME

EFFECTIVE DATE

INDICATE W for WHOLESALE or R for RETAIL   DATE   TELEPHONE

CREDIT CODE   DISCOUNT CODE

STREET ADDRESS

CITY   STATE   ZIP CODE

BANK REFERENCES

NAME	BRANCH	TELEPHONE	
ADDRESS	CITY	STATE	ZIP CODE
NAME	BRANCH	TELEPHONE	
ADDRESS	CITY	STATE	ZIP CODE

OTHER REFERENCES

NAME		TELEPHONE	
ADDRESS	CITY	STATE	ZIP CODE
NAME		TELEPHONE	
ADDRESS	CITY	STATE	ZIP CODE

FOR OFFICE USE
CREDIT APPROVED ☐   DISAPPROVED ☐
IF DISAPPROVED, REASON:

CUSTOMER ORDER

CUSTOMER ORDER
ABCO Corporation
Walnut, California

Name and Address

Invoice No. _____

Account No. _____

Date _____

QUANTITY	DESCRIPTION	UNIT PRICE	AMOUNT

Authorized Signature

Subtotal
Discount
Total

PAYMENT/CREDIT MEMORANDUM

PAYMENT/CREDIT MEMORANDUM

ACCOUNT NUMBER: _____   DATE: _____

PAYMENT AMOUNT: _____

CREDIT AMOUNT: _____

AUTHORIZED BY: _____

Figure 12.8
Accounts receivable cost summary (monthly cost per account at 10,000 accounts). The direct operating costs of the current system are a combination of computer time, personnel, keypunch, and supplies. These costs were determined during the initial investigation by examining current cost records.

accounts receivable cost summary (monthly cost per account at 10,000 accounts)	
computer time	$ .44
personnel	.93
keypunch	.20
supplies	.28
total	$1.85
total monthly cost is $18,500.00	

New system specific objectives We will assume that, after a review of the project directive and the results of the initial investigation, J. Herring prepared the following list of specific objectives for the on-line accounts receivable system (OARS). They are ranked in order of importance.

1. To establish billing cycles for each region
2. To mail customer statements no later than one day after the close of a billing cycle
3. To provide the customer with a billing statement two days after the close of a billing cycle
4. To speed up collections, reducing the float by 50 percent
5. To examine customer account balances through on-line inquiry at the time of order entry

Output Description

Output Identification and Description
As the final step in performance definition, the analyst must describe the outputs as they will appear to the user. For example, if the system outputs are to be in report format, the analyst prepares a layout and a data element description for each output. At this point in the life cycle of a system, it is not necessary to design the output report form. The output layout may be a neat sketch that the user can understand and comment upon. However, the sketch should conform to the general rules for good form design.

Figure 12.9
Project directive for OARS (on-line A/R system). The project directive reflects the conclusions reached by the principal user and the analyst as a result of the initial investigation. The system objectives, anticipated benefits, and descriptions of the outputs and inputs become the basis for the new system performance definition.

INFORMATION SERVICE REQUEST			Page 1 of 3	

INFORMATION SERVICE REQUEST Page 1 of 3

JOB TITLE: Study Phase for the On-line A/R System (OARS)	NEW ☑ REV. ☐	REQUESTED DATE: 11/7/XX	REQUIRED DATE: 12/19/XX

AUTHORIZATION

OBJECTIVE:		LABOR		OTHER	
		HOURS	AMOUNT	HOURS	AMOUNT
		600	$12,500		

ANTICIPATED BENEFITS: 1. faster billing   2. reduce A/R payment "float"

3. expand to 20,000 accounts   4. allow additional billing cycles

OUTPUT DESCRIPTION	INPUT DESCRIPTION
TITLE: Customer Monthly Statement	TITLE: Sales Order
FREQUENCY: monthly QUANTITY: 20,000	FREQUENCY: on-line
PAGES: 1 to 3 COPIES: 2	QUANTITY: 1,000 per day
COMMENTS: 2-part form; multiple billing cycles per month	COMMENTS: Sales transaction **generated**
TITLE: A/R Transaction Register	TITLE: Customer Account Application
FREQUENCY: daily QUANTITY: 1	FREQUENCY: daily
PAGES: 40 max. COPIES: 1	QUANTITY: 20 max.
COMMENTS: printed copy for A/R file; data available for on-line screen display	COMMENTS: application form submitted by customer batched and entered each day after approval

TO BE FILLED OUT BY REQUESTOR

REQUESTED BY: *G. Davis*	DEPARTMENT: 310	TITLE: head, A/R dept.	TELEPHONE: X3250
APPROVED BY: *Ben Franklin*	DEPARTMENT: 300	TITLE: manager, accounting div.	TELEPHONE: X3208

TO BE FILLED OUT BY INFORMATION SERVICES

FILE NO: ISR-310-24	ACCEPTED ☑ NOT ACCEPTED ☐		
SIGNATURE: *C. Hampton*	DEPARTMENT: 200	TITLE: manager, Info. services	TELEPHONE: X2670

REMARKS:
This is a project directive.
J. Herring is appointed project leader. Advise requestor when funds are 90% expended.

FORM NO: C-6-1	ADDITIONAL INFORMATION. USE REVERSE SIDE OR EXTRA PAGES

INFORMATION SERVICE REQUEST

JOB TITLE: Study Phase for the On-line A/R System (OARS)	NEW ☑ REV. ☐	REQUESTED DATE: 11/7/XX	REQUIRED DATE: 12/19/XX

OBJECTIVE:	AUTHORIZATION			
	LABOR		OTHER	
	HOURS	AMOUNT	HOURS	AMOUNT

ANTICIPATED BENEFITS:

OUTPUT DESCRIPTION	INPUT DESCRIPTION
TITLE: A/R Summary	TITLE: Payments
FREQUENCY: on demand QUANTITY: 20,000 max.	FREQUENCY: on-line
PAGES: —— COPIES: ——	QUANTITY: 500/day max.
COMMENTS: screen display; access by account number or customer name	COMMENTS: entered upon receipt of check at central site
TITLE: Aged A/R Report	TITLE: Credit Memo
FREQUENCY: monthly QUANTITY: 1	FREQUENCY: on-line
PAGES: 15/mo. max COPIES: 1	QUANTITY: 10/day max.
COMMENTS: exception report; past due accounts only	COMMENTS: transaction driven by need for billing adjustments

TO BE FILLED OUT BY REQUESTOR

REQUESTED BY: G. Davis	DEPARTMENT: 310	TITLE: head, A/R dept.	TELEPHONE: X3250
APPROVED BY: Ben Franklin	DEPARTMENT: 300	TITLE: manager, accounting div.	TELEPHONE: X3208

TO BE FILLED OUT BY INFORMATION SERVICES

FILE NO: ISR-310-24	ACCEPTED ☑ NOT ACCEPTED ☐		
SIGNATURE: C. Hampton	DEPARTMENT: 200	TITLE: manager, Info. services	TELEPHONE: X2670

REMARKS:

continued from page 1

FORM NO: C-6-1	ADDITIONAL INFORMATION: USE REVERSE SIDE OR EXTRA PAGES

Figure 12.9
continued

INFORMATION SERVICE REQUEST			Page 3 of 3

JOB TITLE: Study Phase for the On-line A/R System (OARS)	NEW ☑ REV ☐	REQUESTED DATE: 11/7/XX	REQUIRED DATE: 12/19/XX

OBJECTIVE:

AUTHORIZATION

	LABOR		OTHER	
	HOURS	AMOUNT	HOURS	AMOUNT

ANTICIPATED BENEFITS:

OUTPUT DESCRIPTION	INPUT DESCRIPTION
TITLE: Customer Account Data	TITLE:
FREQUENCY: on-demand QUANTITY: 20,000 accts.max.	FREQUENCY:
PAGES: COPIES:	QUANTITY:
COMMENTS: display only; access by account number or customer name	COMMENTS:
TITLE: Overcredit Notification	TITLE:
FREQUENCY: daily QUANTITY: 80 max.	FREQUENCY:
PAGES: 1 COPIES: 1	QUANTITY:
COMMENTS: printed and sent to customer (overcredit accounts only)	COMMENTS:

TO BE FILLED OUT BY REQUESTOR			
REQUESTED BY: *G. Davis*	DEPARTMENT: 310	TITLE: head, A/R dept.	TELEPHONE: X3250
APPROVED BY: *Ben Franklin*	DEPARTMENT: 300	TITLE: manager, accounting div.	TELEPHONE: X3208

TO BE FILLED OUT BY INFORMATION SERVICES			
FILE NO: ISR-310-24	ACCEPTED ☑ NOT ACCEPTED ☐		
SIGNATURE: *C. Hampton*	DEPARTMENT: 200	TITLE: manager, Info. services	TELEPHONE: X2670

REMARKS:

continued from page 2

FORM NO: C-6-1	ADDITIONAL INFORMATION: USE REVERSE SIDE OR EXTRA PAGES

To prepare an effective layout, the analyst has to make an initial judgment of the most likely output medium. The analyst should never undertake an assignment with a prejudgment that the solution requires a computer or computer-produced outputs. The outputs could be printed reports, CRT displays, microfilm, and so forth. The decision on output media will not be made until the feasibility analysis. However, if as a result of the initial investigation and the identification of specific objectives the analyst judges that the outputs are most likely to be computer-produced, sketches are prepared with the characteristics of the output device in mind. Thus, if the analyst were preparing a layout of a printed report, the characteristics of line printers would be taken into account.

If a computer is to be considered, the analyst's flexibility in output presentation may be increased because a variety of output media are available. Figure 12.10 displays examples of computer output media. In this figure, the flowchart symbol most appropriate to each output medium is indicated. Some of the output devices also can serve as input devices.

Reports are the most common computer outputs. These may be actual paper print-outs, called "hardcopy" reports, or they may be displays on cathode ray tubes (CRT's), called "softcopy" reports. This discussion of output specifications and data element lists is, in general, applicable to both types of outputs.

The initial specification of output formats and media is necessary not only to complete the system performance definition, but also to "trigger" the process of selecting the "best" of alternative systems. This selection process is described in chapter 13, Feasibility Analysis. The initially chosen outputs and media may or may not be those selected at the conclusion of the feasibility study. For the remainder of this section and throughout the discussion of OARS, which follows, we will assume that an initial decision was made to present the system outputs in the form of terminal displays whenever feasible, and in the form of printed reports where copies must be sent to customers or filed for future use.

Figure 12.11 is a output specification form that may be used to describe computer-generated output. It includes:

1. Output title
2. A layout, that is, a sketch or sample of the output form
3. An estimate of the frequency, size (number of pages), quantity (number of unique reports), and number of copies of the output
4. A distribution list, including the location of feasible distribution points
5. Special considerations, including identified constraints and controls

Figure 12.12 is an example of a data element list that should accompany the report specification. The *data element list* contains information describing each data element on the specified report. We will use this "typical entry" to illustrate the meaning of the column headings.

Figure 12.10
Typical computer output devices. If a computerized system is to be considered, the analyst must be familiar with available types of output media. Shown are (a) high speed printer; (b) electrostatic plotter; (c) document printer; (d) audio response unit; (e) CRT terminal; (f) hard copy terminal.

a

b

c

d

e

f

Figure 12.11
Output specification. The output specification form is used by the analyst to describe computer-generated outputs. The layout is a sketch of the proposed output.

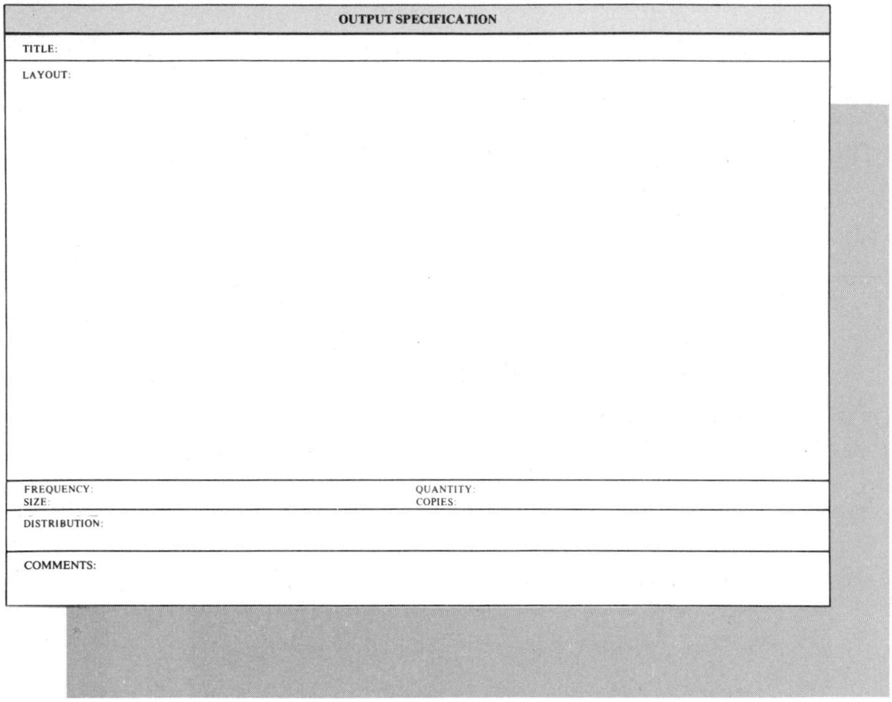

OUTPUT SPECIFICATION		
TITLE:		
LAYOUT:		
FREQUENCY: SIZE:	QUANTITY: COPIES:	
DISTRIBUTION:		
COMMENTS:		

1. Description is the name plus any other significant description of the report:
   Billing Cycle: Mo/Day/Yr
2. Format is an indication of the appearance of the output on a printed report. The format column should be used whenever any special punctuation or editing is required, for example, $12,456.00 or 02/12/83. *X*'s are used to represent numbers and characters. Editing symbols are used if they are to be part of the output:
   xx/xx/xx
3. Size indicates the number of characters to be printed:
   8 characters

By specifying exactly how an output item will appear, the data element list clarifies communication. It leads to an estimate of storage requirements in the study phase. In subsequent phases, it is a basis for file design and for a description of computer program report output.

Figure 12.12

Data element list—typical entry. Each data element shown in the output layout sketch is described in a data element list. The size, in characters, should be given for each data element. If the format of the data is other than a single string of characters, the format should also be shown on the data element list.

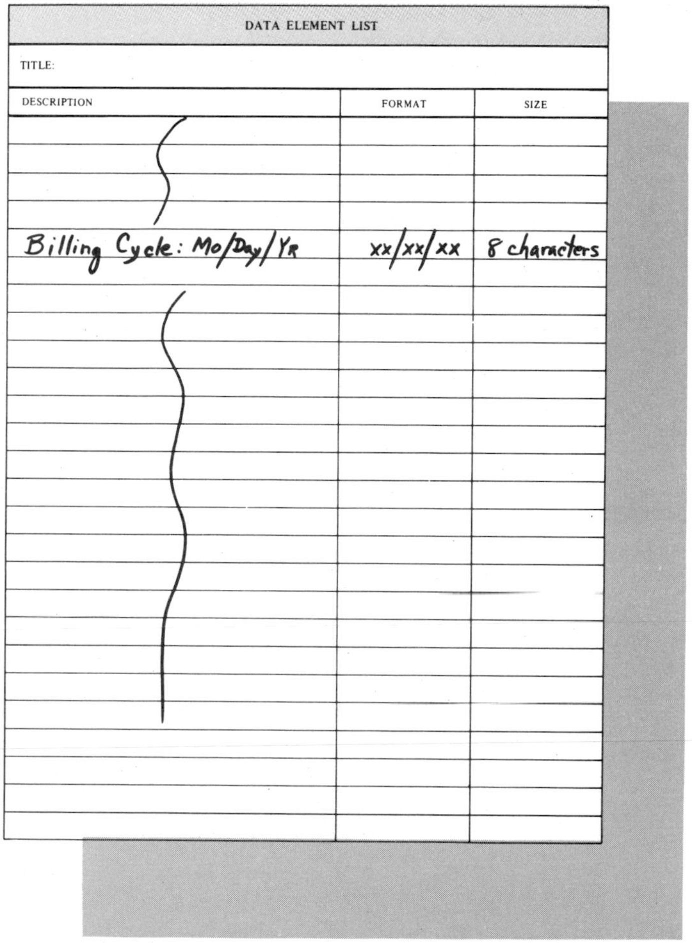

DATA ELEMENT LIST		
TITLE:		
DESCRIPTION	FORMAT	SIZE
Billing Cycle: Mo/Day/Yr	xx/xx/xx	8 characters

DETAILED DESIGN

Example System: Output Description

Sample outputs for the example A/R system are shown in figures 12.13 through 12.18. The first part of each figure is an output specification, the second, a corresponding data element list.

Figure 12.13, the customer monthly statement, is the same as the current system statement.

Figure 12.14, the A/R transaction register, contains the same data as the current system output, but is a CRT display as well as a printed report.

Figure 12.13

Sample customer monthly statement. The output specification and data element list of the customer monthly statement was developed by the user and the analyst. In this example, it is the same as the existing system's statement.

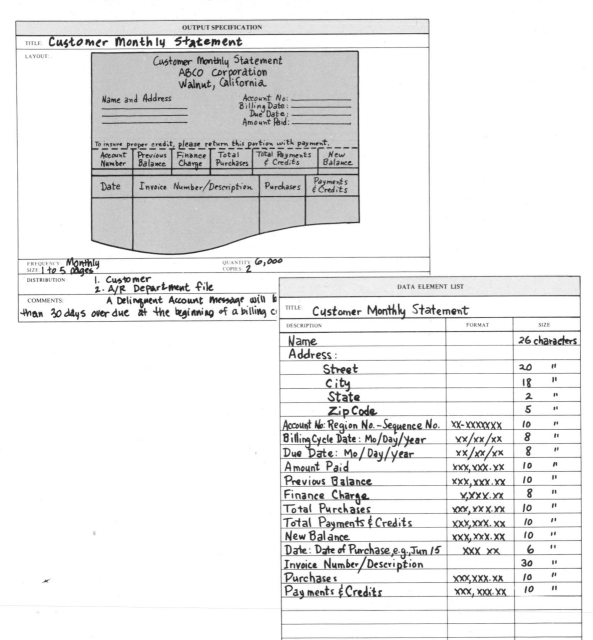

Figure 12.14

Sample A/R transaction register. The output specification and data element list of the accounts receivable transaction register for this system contain the same data as the current system output, but is to be a CRT display as well as a printed report.

OUTPUT SPECIFICATION

TITLE: Accounts Receivable Transaction Register

LAYOUT:

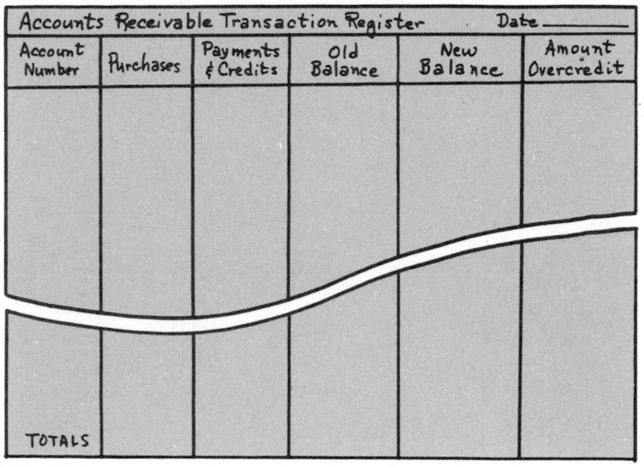

Accounts Receivable Transaction Register   Date _____

Account Number	Purchases	Payments & Credits	Old Balance	New Balance	Amount Overcredit

TOTALS

FREQUENCY: Daily   QUANTITY: 1
SIZE: 20 pages, max.   COPIES: 1

DISTRIBUTION:
1. A/R Department (As a display)
2. A/R File (1)

COMMENTS:
Account number sequence

DATA ELEMENT LIST

TITLE: Accounts Receivable Transaction Register

DESCRIPTION	FORMAT	SIZE
Date: Mo/Day/Year	xx/xx/xx	8 characters
Account No: Region No.-Sequence No.	xx-xxxxxxx	10 "
Purchases	xxx,xxx.xx	10 "
Payments & Credits	xxx,xxx.xx	10 "
Old Balance	xxx,xxx.xx	10 "
New Balance	xxx,xxx.xx	10 "
Amount overcredit	xx,xxx.xx	9 "
Totals	xx,xxx,xxx.xx	13 "

Figure 12.15
Sample A/R summary. The accounts receivable summary report has been modified from the current output to reflect current balances only. It will be a CRT display.

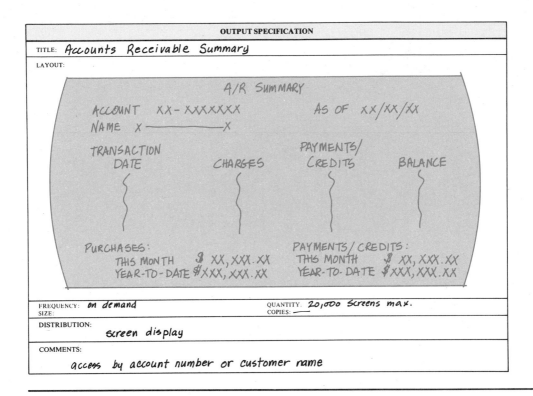

Figure 12.15, the A/R summary report, has been modified from the current output to reflect current balances only. It will be a CRT display showing the data for one customer on each screen.

Figure 12.16 is the aged A/R report. It is a monthly report produced at the end of each billing cycle when the customer statements are prepared. Current amounts, and 30, 60, 90, and over 90 days past due amounts are shown.

Figure 12.17 is the customer account list on demand in account number sequence. Each sales region will be able to view its customer account list on a terminal on demand. No printed copies will be available.

Figure 12.18 is the overcredit notification, which is prepared daily for each account that has exceeded its credit limit and for which the accounts receivable department decides not to grant additional credit. The printed notice is mailed to the customer.

DATA ELEMENT LIST		
TITLE: *Accounts Receivable Summary*		
DESCRIPTION	FORMAT	SIZE
Account Number	XX-XXXXXX	10 Char.
Current Date (as of)	MM/DD/YY	8 "
Customer Name		26 "
Transaction Date	MM/DD/YY	8 "
Charges	$XX,XXX.XX	10 "
Payments / Credits	$XX,XXX.XX	10 "
Balance	$XXX,XXX.XX	11 "
Purchases :		
This month	$ XX,XXX.XX	10 "
Year-to-date	$ XXX,XXX.XX	11 "
Payments / Credits:		
This month	$ XX,XXX.XX	10 "
Year-to-date	$XXX,XXX.XX	11 "

At this point, the analyst has identified the specific objectives of the system and has described the outputs required to meet those objectives. The analyst is now ready to evaluate the feasibility of alternative systems that might produce these outputs.

Summary

The system performance definition activities identify general constraints under which the system must operate. Subject to these constraints, the analyst identifies and ranks the specific objectives of the system. Whenever possible, the specific objectives should be stated in measureable terms. The final step in performance definition is the identification and description of the system outputs as they will appear to the user. Upon completion of the system performance activities, the analyst is ready to evaluate the feasibility of alternative systems that might produce the specified outputs.

Figure 12.16
Sample aged A/R report. The aged accounts receivable report is a printed monthly report produced at the end of each billing cycle. It is unchanged from the current system.

OUTPUT SPECIFICATION

TITLE: *Aged Accounts Receivable Report*

LAYOUT:

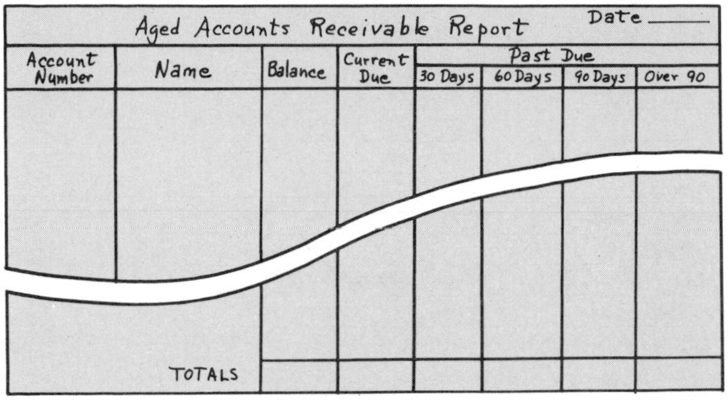

				Past Due			
Account Number	Name	Balance	Current Due	30 Days	60 Days	90 Days	Over 90

Aged Accounts Receivable Report   Date _____

TOTALS

FREQUENCY: *Monthly*   QUANTITY: *1*
SIZE: *5 pages max., each cycle.*   COPIES: *1*

DISTRIBUTION: *1. Accounts Receivable Department*

COMMENTS:
          Run by region on billing cycle date.

DATA ELEMENT LIST

TITLE: *Aged Accounts Receivable Report*

DESCRIPTION	FORMAT	SIZE
Date: Billing Cycle Date: Mo/Day/Yr.	xx/xx/xx	8 characters
Account No.: Region No. - Sequence No.	xx-xxxxxxx	10 "
Name		26 "
Balance	xxx,xxx.xx	10 "
Current Due	xxx,xxx.xx	10 "
Past Due		
30 days	XXX,XXX.XX	10 "
60 days	XXX,XXX.XX	10 "
90 days	XXX,XXX.XX	10 "
Over 90	XXX,XXX.XX	10 "
Totals	XX,XXX,XXX.XX	13 "

Figure 12.17
Sample customer account data. The new customer account data will be a CRT display.
Users will be able to view the customer account list at any time.

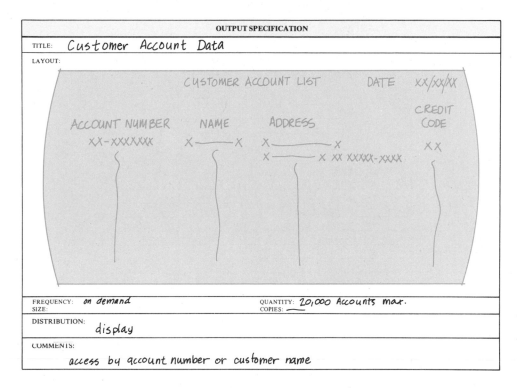

For Review

constraint   data element list   output specification
specific objective

For Discussion

1. What are the three steps that result in the definition of system performance?
2. What is a general constraint? A specific constraint?
3. What is a specific objective?
4. In what ways does the system performance definition process extend the information contained in the project directive?
5. What is the reason for ranking specific objectives in order of importance?
6. Why must the analyst make a tentative selection of output media?
7. What are the values of the data element list in the study phase? In subsequent phases?

Figure 12.17
continued

DATA ELEMENT LIST		
TITLE: **Customer Account List**		
DESCRIPTION	FORMAT	SIZE
Date: Mo/Day/Yr.	xx/xx/xx	8 characters
Acct. No.: Region and Seq.	xx-xxxxxxx	10 "
Name		26 "
Address:		
Street		20 "
City		18 "
State		2 "
Zip Code		5 "
Credit Code	xx	2 "
01 500 Limit		
02 2500 Limit		
03 10,000 Limit		
04 25,000 Limit		
05 Special		

Figure 12.18
Sample overcredit notification. An overcredit notification will be printed for each account that has exceeded its credit limit and for which the accounts receivable department decides not to grant additional credit to the customer.

OUTPUT SPECIFICATION

TITLE: Overcredit Notification

LAYOUT:

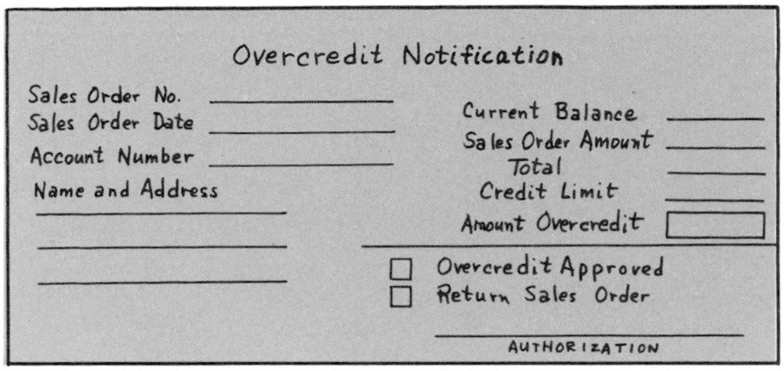

> **Overcredit Notification**
>
> Sales Order No. _____
> Sales Order Date _____
> Account Number _____
> Name and Address
> _____
> _____
> _____
>
> Current Balance _____
> Sales Order Amount _____
> Total _____
> Credit Limit _____
> Amount Overcredit [_____]
>
> ☐ Overcredit Approved
> ☐ Return Sales Order
>
> _____
> AUTHORIZATION

FREQUENCY: Daily   QUANTITY: 80 max.
SIZE: 1 page   COPIES: 2

DISTRIBUTION: 1. Accts. Receivable Dept.

COMMENTS:

DATA ELEMENT LIST

TITLE: Overcredit Notification

DESCRIPTION	FORMAT	SIZE	
Sales Order Number	XXXXXXXX	8	characters
Sales Order Date	xx/xx/xx	8	"
Account No: Region No.—Sequence No.	XX-XXXXXXX	10	"
Name		26	"
Address			
Street		20	"
City		18	"
State		2	"
Zip Code		5	"
Current Balance	XXX,XXX.XX	10	"
Sales Order Amount	XXX,XXX.XX	10	"
Total	XXX,XXX.XX	10	"
Credit Limit	XX,XXX.XX	9	"
Amount Overcredit	XX,XXX.XX	9	"

13
Feasibility Analysis

Preview

The previous chapter described the process of defining the performance required of a computer-based business information system. Next, a feasibility analysis is performed to choose the system that meets the performance requirements at the least cost. The most essential tasks performed by a feasibility analysis are: (1) the identification and description of candidate systems, (2) the evaluation of candidate systems, and (3) the selection of the best of the candidate systems.

Objectives

1. You will understand the purposes of a feasibility analysis.
2. You will learn the process for performing a feasibility analysis.
3. You will learn how to estimate storage requirements by performing an output data source analysis.
4. You will be able to prepare candidate system, candidate evaluation, and weighted candidate evaluation matrices.

Key Terms

"best" system   The system that meets the performance requirements at the least cost.

candidate system matrix   A table that lists functions to be performed and alternative systems for performing them.

candidate evaluation matrix   A table that lists evaluation criteria and rates alternative systems in terms of these criteria.

weighted candidate evaluation matrix   A table that weights the candidate evaluation matrix entries by their importance and applies a rating number; it is a means of calculating comparative total scores for each candidate.

Purposes of a Feasibility Analysis

Several activities have now been completed in the study phase:

1. A user has recognized a need.
2. An initial investigation has been completed to study the existing system and uncover problem areas.
3. The user and the analyst have cooperatively defined the problem and identified the general objectives, required outputs, and general constraints of the new system. This resulted in the project directive.
4. The system performance has been defined by ranking specific objectives in order of their importance, and by preparing detailed descriptions of the outputs.

At this point both the user and the analyst know what the problems are and the general characteristics of the system that will solve those problems.

The feasibility analysis activities will lead them, if they are successful, to a system that they have the capability to develop, that will work in their environment, and that the company can afford. The specific purposes of the feasibility analysis are (1) the selection and description of candidate systems; and (2) the selection of the "best" (that is, the most feasible) system.

The specific objectives of the new system have been identified through the system performance definition activities. The system that meets those objectives, at the lowest cost within the constraints, is the "best" system.

Steps in a Feasibility Analysis

The most difficult parts of a feasibility analysis are the identification of candidate systems and the evaluation of their performances and costs. This process is a highly creative one that requires imagination, experience, and a share of good luck. As with most complex tasks, however, a procedure can be followed. There are nine identifiable steps in performing a feasibility analysis:

1. Form the system team
2. Develop generalized system flowcharts
3. Develop the system candidates
4. Perform preliminary evaluation of candidates
5. Prepare detailed descriptions of candidates
6. Identify meaningful system characteristics
7. Determine performance and cost for each candidate
8. Weight the system performance and cost characteristics by importance
9. Select the "best" candidate

In the following discussion of the nine steps of the feasibility analysis, the examples presented are a continuation of the development of the on-line accounts receivable system (OARS) of the ABCO corporation.

Step 1: Form the System Team

The first step is to enlist qualified participants for the *system team*. The team should be a group of involved, interested people who can represent their respective areas to help define system problems and develop methods of solution. Typically, user organizations, management, and data processing should be represented on the team. The inputs from team members and their reactions to the inputs of others can tell the analyst much about what they really consider important in the system. Often, this information-gathering technique is the only way for the analyst to get the whole picture. In addition, team members usually supply excellent ideas for consideration.

Another benefit of the system team is that the involvement of users and management in planning a system makes the system "their" system. As was emphasized during the discussion of presentations in chapter nine, when the people who are going to use the system and/or pay for the system refer to it as "their" system rather than "your" system, the analyst most likely will have cooperation from the users at system reviews and during system conversion.

Step 2: Develop Generalized System Flowcharts

The second step of the feasibility analysis is to develop generalized, overview flowcharts for the system. These may include data flow diagrams, information-oriented system flowcharts and process-oriented system flowcharts. The value of the high-level system flowcharts, at this point, is that they draw attention to the system inputs, outputs, and data transformations rather than to detailed processing operations.

Developing the generalized system flowchart usually proceeds without difficulty. The team has the flowcharts that were prepared during the initial investigation to review. The team also has sketches and data element lists of all outputs identified during the system performance definition. Additionally, system inputs were identified on the information service request used as the project directive.

Returning to the example project, ABCO's OARS let us assume the following:

> The system team reviewed and discussed the outputs and inputs identified during the system performance definition activities. At this point the team prepared generalized flowcharts. Figure 13.1a is an information-oriented system flowchart that reflects the recommendations made by the system team and approved by Mr. Franklin, manager of the accounting division. Figure 13.1b is a data flow diagram drawn by the team and used to develop the flowchart of figure 13.1a.
>
> Figure 13.2 is a process-oriented flowchart that focuses on the data processing column of the flowchart in figure 13.1a.

Step 3: Develop the System Candidates

The third step of the feasibility analysis is to develop candidate systems that could produce the outputs identified in the generalized flowcharts. This process involves a transistion from logical system models, such as successively detailed data flow diagrams, to physical models, such as process-oriented flowcharts. The third step includes a consideration of hardware devices able to accomplish each of the four basic system functions of input, processing, storage, and output.

Figure 13.1a

Information-oriented flowchart and narrative—OARS. This information-oriented system flowchart reflects the recommendations made by the system team and approved by the principal user.

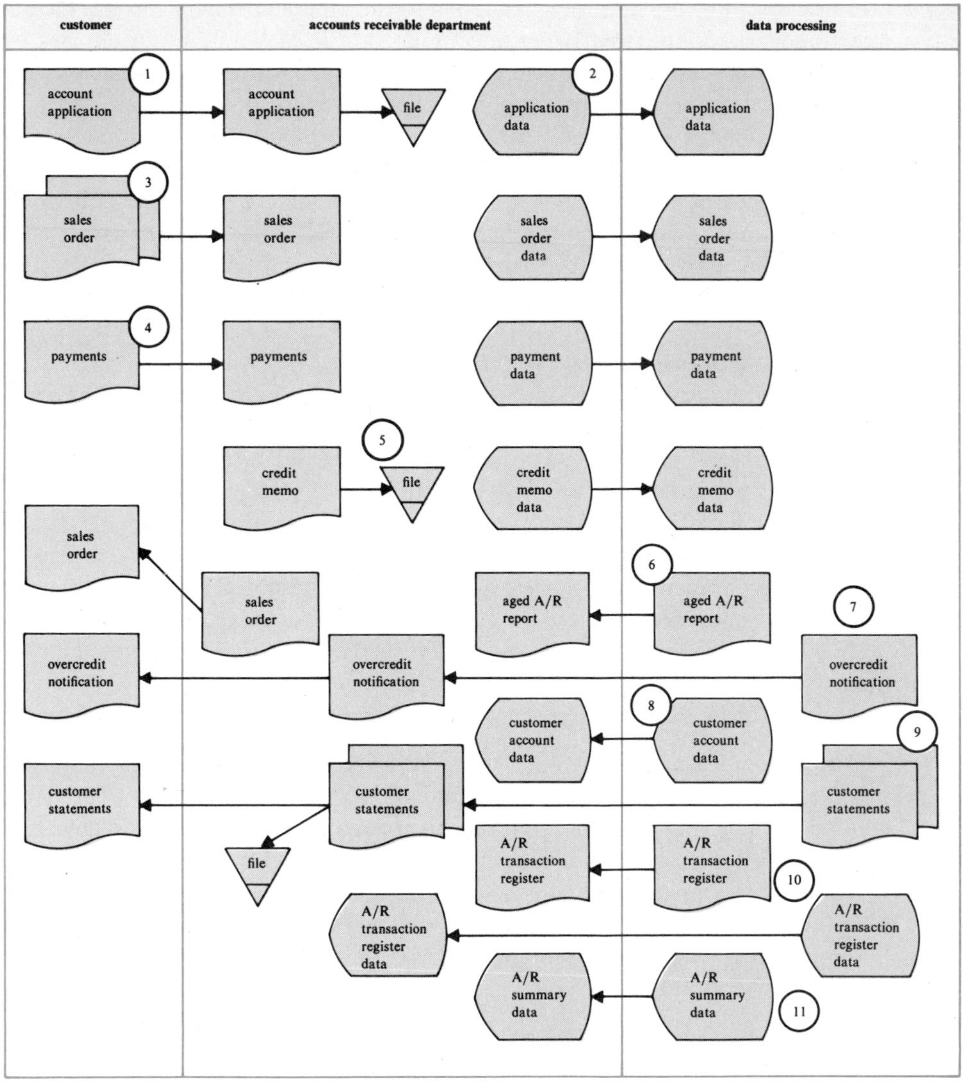

Figure 13.1a
continued

INFORMATION-ORIENTED FLOWCHART NARRATIVE

1. Customers submit account applications that are processed
    by the accounts receivable department.

2. If an application is accepted, it is keyed into the
    system. The application is filed.

3. Sales orders are entered through on-line order
    processing. The customer retains a carbon copy.

4. Payments are sent to the accounts receivable
    department and entered into the system.

5. Credit memos are generated and the data entered.
    The memo is filed.

6. An aged A/R report is sent to the accounts receivable
    department from data processing each month.

7. Data processing sends overcredit notices to the credit
    department whenever a new order would exceed the
    customer's credit limit and the additional credit is
    not approved. The order and notice are returned to the
    customer. If the additional credit is approved, the
    order is processed.

8. Customer account data is displayed on demand.

9. Customer statements are sent to accounts receivable
    in duplicate. The original copy is sent to the
    customer; the duplicate is filed. One-third of the
    statements are produced each ten days of the month,
    that is, on the 1st, 10th, and 20th.

10. A daily A/R transaction register is sent to the accounts
    receivable department as a file copy. The data is also
    available as a display.

11. The accounts receivable summary data is available as
    a display to the accounts receivable department.

Here again, the system team will be valuable to the systems analyst. The system team's task is to do some "brainstorming" by trying various hardware combinations for each of the four basic functions mentioned earlier. If some team members are not familiar with some of the potentially suitable devices, other members with knowledge of equipment, for example, the data processing representative, can make a presentation to the team.

Figure 13.3 shows a table that can be used in the development of system candidates. The table, called a candidate system matrix, lists functions to be performed and alternative systems for performing them. It is an effective means of presenting and comparing the basic functions of each candidate.

Figure 13.1b
OARS data flow diagram. This data flow diagram reflects the same system as the
previous information-oriented flowchart.

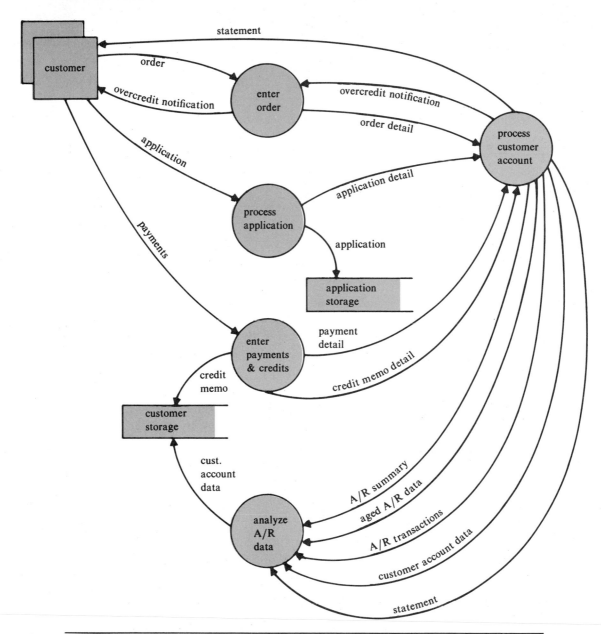

Figure 13.2
Generalized high-level process-oriented flowchart. The generalized high-level process-oriented flowchart focuses on the data processing column of the information-oriented flowchart. It is generalized in that it does not identify input or output media.

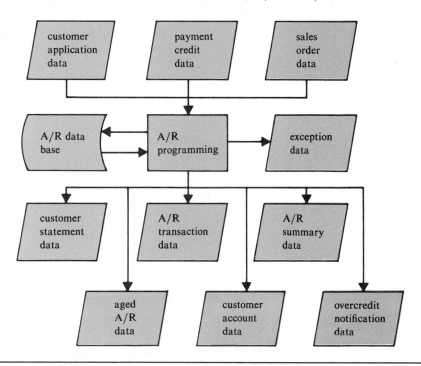

We can illustrate the use of a **candidate system matrix** by continuing with our example system.

In the ABCO corporation's OARS, the output media initially chosen was computer-printed reports or CRT displays, depending upon the nature of the output. These choices were made by the users and the analyst, Ms. J. Herring, during the system performance definition activities because they judged them the most likely output media. Those initial assumptions were made so that sketches of the outputs could be made with the characteristics of the output device in mind. Those initial assumptions, however, do not now prevent the system team from considering other media, several of which were shown in figure 12.10 in the preceding chapter.

Possible input media are depicted in figure 13.4.

Figure 13.5 displays examples of storage media to be considered for the system's data base and for intermediate storage. Each of these potential media has performance and cost advantages and disadvantages.

Figure 13.3
Candidate system matrix format. This table can be an effective means of presenting and comparing the basic functions of each potential system candidate.

CANDIDATE SYSTEM / FUNCTIONS	I	II	III	IV	V	VI
OUTPUT						
INPUT						
STORAGE						
PROCESSING						

The last function to be considered is processing. The general processing choices are manual processing and computer processing. Methods of manual processing can range from pencil and paper to desk calculators to manually operated equipment such as bookkeeping machines. If a potential processing medium is a computer, there are a large number of computers with differing processor capabilities, processor speeds, main storage size ranges, auxiliary storage handling capabilities, and levels of software support. Computer vendors will be more than happy to provide the system team with information on their equipment and to make recommendations. Representatives of the team, however, should always make it a point to verify data from vendors by making visits or contacts with other users or user groups. Each potential processor has its own advantages and disadvantages. Compromise may be necessary; however, know what you are giving up for what you are getting.

Figure 13.4

Typical computer input devices. To develop potential system candidates, the analyst must be familiar with available input devices. Shown are (a) magnetic ink character reader; (b) point of sale terminal; (c) information display system with light pen; (d) supermarket checkout system; (e) portable data terminal.

a

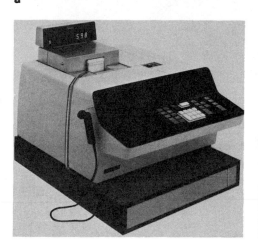

b

c

Step 4: Perform Preliminary Evaluation of Candidates

Usually the team has far too many candidates to evaluate each one in detail. Hence, the fourth step in the feasibility analysis is to make a preliminary evaluation of the system candidates and thus narrow down the number of candidates to a manageable number.

In developing system candidates, the idea is to "brainstorm" as many candidates as possible, without attempting any evaluation. In the preliminary evaluation, any system that would not be practical because of its obvious high cost or its "overkill" for the task at hand is eliminated. Candidate systems that require technical knowledge beyond that available to the company or that do not fit the corporate philosophy should also be dropped from consideration.

Figure 13.4
continued

d

e

The process of elimination should continue until the number of candidates is reduced to a manageable size. The actual number of systems to be considered in detail is, of course, a function of the amount of time and the resources available to the team. The systems to be evaluated in detail are entered into a candidate system matrix.

For example:

The system team for the OARS project, headed by Ms. J. Herring, generated a candidate system matrix for OARS. This matrix is shown in part in figure 13.6. All of the candidates are computer-based systems. Some of the candidates use a central-site large computer; others distribute the processing to the sales regions with on-site minicomputers that transmit summary data to the central-site computer.

Figure 13.5
Typical computer storage devices. To develop potential system candidates, the analyst must be familiar with storage media to be considered for the system's data base and for intermediate storage. Typical devices are (a) magnetic tape unit; (b) disk pack mounted into disk drive; (c) removable disk pack and data cartridges; (d) honeycomb.

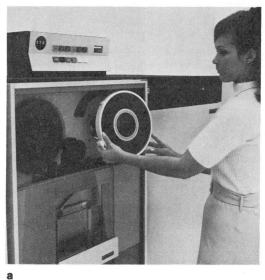

a

b

c

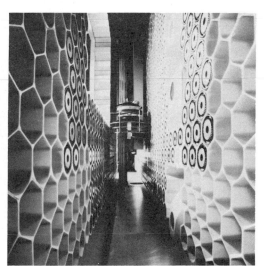

d

Figure 13.6
Candidate system matrix. This candidate system matrix depicts, in part, the system candidates generated by the system team. Each candidate is described by its output, input, storage, and processing functions.

candidate system / functions	I	II	III	
output	all outputs printed at central site	all outputs generated at central site; all outputs printed	all outputs generated locally; printed and/or displayed as appropriate	
input	dumb terminal each store	intelligent terminal each store	intelligent terminal attached to minicomputer located at each store	
storage	data base on magnetic disk at central site	master file on magnetic disk at central site; diskette local storage for batch transmission to central site	master file on magnetic disk at local store	
processing	all processing at central site	local data editing; all other at central site	all A/R system processing done locally	

Step 5: Prepare Detailed Descriptions of Candidates

The fifth step in the feasibility analysis is to prepare detailed descriptions of the systems appearing in the candidate system matrix. The detailed descriptions should include flowcharts and narratives, specific constraints, identified inputs, processing requirements, and storage requirements.

The OARS example illustrates this step with the following assumed situation:

> In the OARS project, Ms. Herring and the system team have reduced the number of candidates for detailed study to three:
>
> 1. Candidate I is a computer system utilizing a larger (than the existing) central-site computer with dumb terminals located in each region. All outputs would continue to be produced at the central-site location.
> 2. Candidate II requires a larger central-site computer with intelligent terminals located in each region. The intelligent terminals allow data editing to be performed at the data entry point. All outputs continue to be produced at the central site.

3. Candidate III uses the current central-site computer with minicomputers located in each region. This candidate distributes the accounts receivable processing to each region, with summary data transmitted back to the central site. All outputs, therefore, are generated at the local region.

Both candidates I and II require a central-site computer that is larger than the existing one. Candidate III requires of a minicomputer for each region, and this candidate is used as the example system for the discussions that follow. Vendor proposals will be solicited and evaluated early in the design phase, assuming the project gets approval to proceed from the users and management. The high-level information-oriented flowchart for this candidate is shown in figure 13.1a. The flowchart of 13.1a shows the source of input documents, the number of required copies for each document, the path of each document through the organization, and the generation and distribution of all outputs.

Parts 1 through 4 of figure 13.7a, and also the figure 13.7b drawn for candidate III, illustrate detail of the description that the team prepared for each candidate. Part 1 of figure 13.7a is an intermediate-level process-oriented system flowchart with accompanying narrative. It is the high-level flowchart in figure 13.2 modified to reflect specific media to be used in the input, storage, and output functions. Figure 13.7b is a data flow diagram that was used to derive figure 13.7a. It emphasizes the inputs, outputs, and data transformations of the new accounts receivable system. These flowcharts present a comprehensive overview of the candidate system.

Specific constraints are unique to each candidate. They must either be consistent with the general constraints of the project directive or must identify areas where the candidate does not meet the general constraints. If the specific constraints are too restrictive, the candidate will be eliminated or the general constraints of the project directive will have to be modified. Returning to our example system:

Part 2 of figure 13.7a lists the specific constraints for OARS candidate III. Part 3 identifies the four major OARS inputs and specifies, in general, the equipment to be required by the system. In this case the equipment required is the existing central-site computer and a minicomputer system for each region.

Part 4 presents the estimated number of characters of storage for each master record in the data base, the number of characters for the master file at current customer levels, and the size of the master file at the end of a five-year projection of customer growth. A general constraint affecting the storage estimates was the capacity to process 20,000 customers within five years.

The gross storage estimates can be developed from a worksheet like the form used in the OARS example, shown in figure 13.8. This worksheet is called an *output data element source analysis sheet,* and its purpose is to help determine the source of the data elements in the outputs of a system. The data element lists developed to accompany each output sketched during the system performance definition activities are referred to. The completed worksheet identifies the origin of the output data elements as master file, transaction file, or calculated result. In addition, it provides an overview of the output documents and their data element relationships.

Figure 13.7a

OARS candidate III description. Candidate III is described in terms of its process-oriented system flowchart, specific constraints, identified inputs, processing requirements, and storage requirements. Although not shown here, each of the other candidates were described in a similar manner.

PART 1

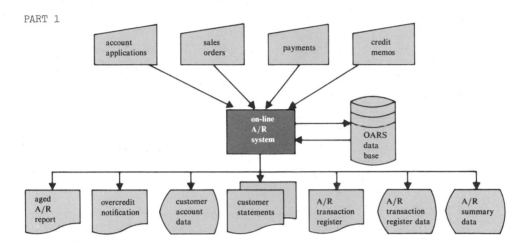

SYSTEM FLOWCHART NARRATIVE

1. Customer account applications, payments, credit memos, and sales orders are the four major system inputs.

2. The A/R program inputs the transaction data and reads/ updates the master file on magnetic disk.

3. The seven major outputs of the system are the aged A/R report, overcredit notification, customer account data display, customer statements, A/R transaction register, A/R transaction register data display, A/R summary data display.

PART 2

SPECIFIC CONSTRAINTS

1. This candidate does not require replacement of the central site computer.

2. This candidate requires the installation of an in-store minicomputer at each store. These minicomputers will off-load the central site computer. This will tend to extend the useful life of the currently installed equipment.

3. The implementation schedule for OARS allows very little time for the acquisition and installation of the required minicomputers. A letter of intent must be sent to vendors as soon as possible to get an appropriate delivery date for the computers.

4. The minicomputers will require program modification and/or new programs for existing A/R system outputs.

5. Dial-up communication capability between each store and the central site is required.

PART 3

IDENTIFIED INPUTS

1. Customer Account Applications - maximum of 20 per day

2. Sales Orders - maximum of 1,000 per day

3. Payments - maximum of 500 per day

4. Credit memos - maximum of 10 per day

PROCESSING REQUIREMENTS

1. The current central site medium-size CPU or its equivalent must be retained.

2. Each store must have a minicomputer system including:
   a. line printer of approximately 600 lpm
   b. magnetic disk storage
   c. backup capability for files - magnetic tape or diskette
   d. display terminals
   e. communications hardware/software to communicate with the central site computer system.

PART 4

STORAGE REQUIREMENTS

1. Each master record consists of 15 data fields for a total of 167 characters.

2. Assuming approximately 8,000 characters of data per disk track:
   a. 213 tracks will be required for the current 10,000 customers.
   b. 426 tracks will be required for the expected 20,000 customers in five years.

3. The 426 tracks of storage to be required in five years will be distributed among five regions. Thus, each region will have no more than approximately 85 tracks of customer master file data.

Figure 13.7b

OARS data flow diagram. The OARS data flow diagram emphasizes the inputs and outputs of the new accounts receivable system. This diagram conveys the same information as the intermediate-level system flowchart in figure 13.7a, except that it does not indicate the media to be used for the inputs or outputs.

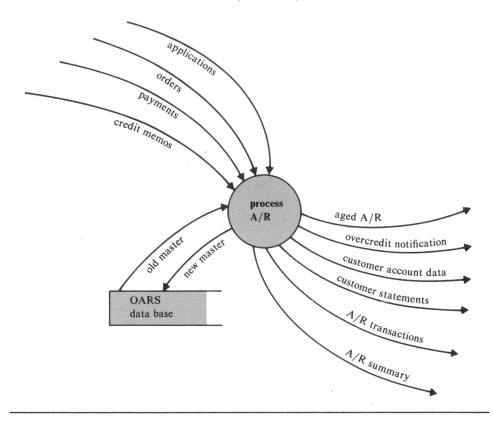

The following are the steps used in performing an output data element source analysis:

1. Complete the heading information.
2. List all output data elements found on the data element lists. Do not list a data element more than one time. Watch for single data elements listed by two or more different data element names. The names used on the data element lists are names that were appropriate and meaningful to that particular output. A slightly different name for the same data element might have been more meaningful on a different output. For example, in figure 13.8 the data element "previous balance" is shown as being used in five of the seven outputs. The data element lists refer to the same data element as "previous balance" on one output, as "old balance" on two outputs, as "balance" on one output, and as "current balance" on one output. It is obviously the same field in each case, regardless of its name on any one data element list.

Figure 13.8
Output data element source analysis. The data element source analysis form may be useful in identifying the sources of data required for system outputs. In this example it is used to identify data elements that must be held in the OARS master file. This form may also be used to identify required transaction files.

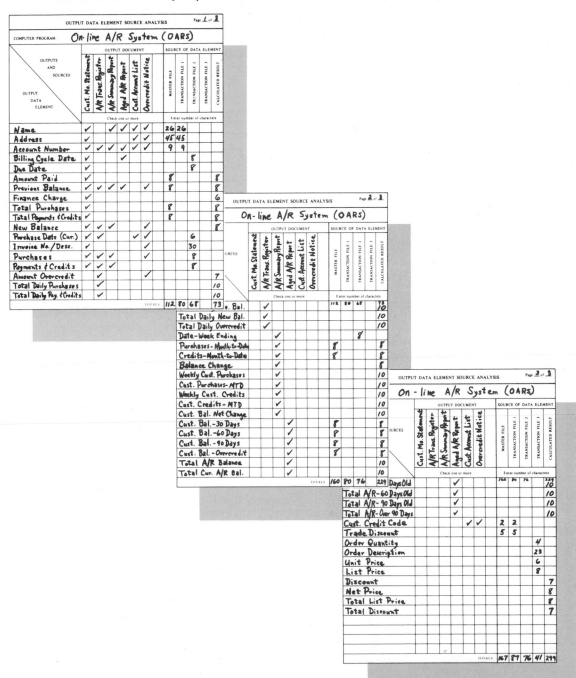

3. In the column below each output document, place a check mark opposite each data element appearing on the document. The frequency of check marks opposite the data elements will be valuable information when designing the master record layout.

4. With the frequency-of-use information, decide which data elements should be maintained in a master file or a transaction file, or which ones are to be calculated by the computer program. In the appropriate source of data element column, enter the number of characters for the data element. Data in the master file must originate either in transaction files or as calculated results. Such data is entered in two columns. The number of characters for each data element can be found in the appropriate data element list. Watch for inconsistent data element sizes on the data element list and make adjustments to the lists where required. Note that the numbers of characters entered in the source of data element columns do not include editing symbols, that is, dollar signs, commas, and decimal points, as these would not be stored in the master file or transaction file records.

5. Total each source of data element column. The master file total is a usable estimate of the master record size. It does not take data formats, for example zoned or packed decimal, into consideration, but it is adequate for study phase planning.

Step 6: Identify Meaningful System Characteristics

Step six of the feasibility analysis is to select the criteria for evaluating the candidate systems. The candidates are evaluated by two major categories of criteria: performance characteristics and costs.

The performance evaluation criteria relate to the satisfaction of specific objectives identified and ranked in the process of system performance definition. Typical criteria are accuracy, control capability, flexibility, growth potential, response time, storage requirements, and usability. Often these characteristics do not lend themselves to quantitative measurements and must be described qualitatively. In any event, qualitative measurements (that is, best, good, bad, worst) can be used to gauge the relative performance of candidate systems. These measures are based upon the collective judgement of the systems team.

System costs include the costs of developing the system and operating it after its implementation. Cost factors that may be particularly important in evaluating a system are those for equipment, facilities, and training. Equipment costs are important when additional equipment, such as computers and word processing machines, must be acquired or existing equipment must be modified. Existing equipment costs usually need not be considered in the evaluation of a system if these costs would remain the same whether or not the system was adopted. Facility costs reflect the costs of additional buildings or rooms, or the modification of existing facilities. Computer installations requiring additional air conditioning, subflooring to allow for cables, alterations to fire sprinkler systems, or installation of security devices are examples of such costs. Training costs are usually not collected unless they can be easily identified. They can be identified if employees must be sent to equipment vendor schools or if classes are to be held in-house with overtime being paid to either the employee attending classes

Figure 13.9
Candidate evaluation matrix. The candidate evaluation matrix is used to record the
performance characteristics and cost of each candidate. The performance and cost
evaluation criteria shown in this example are typical of the criteria used to evaluate
computer-related systems.

candidate system / evaluation criteria	candidate I	candidate II	
performance			
accuracy			
control			
flexibility			
growth potential			
response time			
storage requirements			
usability			
costs			
system development			
system operations			
payback			

or to his substitute on the job. However, normal training on the job is usually too difficult to separate from regular job duties to be counted as system costs.

Figure 13.9 depicts the system evaluation criteria to be used to evaluate the three remaining OARS candidates. This table is called a **candidate evaluation matrix.** When completed, it lists evaluation criteria and rates alternative systems in terms of this criteria.

Step 7: Determine Performance and Cost for Each Candidate
The next step in the feasibility analysis is to develop the entries for the candidate evaluation matrix. Although the performance ratings are often subjective, the analyst must be fair in making appraisals of system performance and should use consistent units of measure for each candidate.

The "accuracy" of a candidate refers not to the accuracy of the equipment, but of the system. One computer is not likely to be more accurate than another, but a system using one computer can be more accurate than another system using the same computer. Accuracy, therefore, relates to the steps involved in getting source data into the system and the steps that are taken to keep the data as error free as possible. No system will be completely free of bad data. "Control capability" relates to the security of the system; it also provides for auditing of the system. Control capability provides protection from mistakes made by humans and from fraud or illegal data manipulation. "Flexibility" refers to the ease of making adjustments to the system, such as modified or new outputs. "Growth potential" is a measure of how much the system can continue to grow without extensive modification to the system or a major component, such as a computer. Such changes are often costly. A system should be expandable for at least a few years. The analyst's confidence that the system can produce the desired outputs cycle after cycle is the "reliability" rating. Weak points in the system, or in equipment utilized within the system that could bring the operation to a halt should be reflected in the reliability rating. A system's "response time" is the total elapsed time between submission of data by a user and its return as a computer output. In a batch processing system, response time includes the system activities of data collection and output distribution, as well as the time required to process the data. In an on-line, transaction processing system using a CRT terminal, the response time is the time from entering the data (or request for data) and seeing the output on the screen. "Storage requirements" refers to computer-based systems and refers to both main storage size and required auxiliary storage for all files. One of the most important criteria is "usability." Usability, as we have emphasized, is the worth of a system as evaluated by the persons who must use it. It is a measure of the "user friendliness" (or ease of use) of the system and it is the final measure of its acceptance.

Costs can be developed most easily when the system benefits are relatively tangible. The most tangible cost comparisons are related to their actual cost savings. The next most tangible comparisons are those related to cost avoidance, usually the case when large growth factors are involved. To illustrate:

> For the example OARS, it has been stated that the growth in the number of A/R accounts will double within the next five years. An estimate was made that sales and accounts would increase to 12,000 accounts from the current 10,000 accounts before any new system could be developed and implemented. In addition, it is expected that accounts will increase by approximately 2,000 accounts per year for the next five years. Figure 13.10 depicts this estimated growth. Using this projection, the number of accounts can be estimated for any year over the next five years. Knowing the number of accounts to be processed, the analyst can break the total costs down into "per account" cost for comparison purposes.

The OARS example cost calculations that follow are for OARS candidate III only. Similar calculations were made for the other candidates, and the results are tabulated.

> Figure 13.11 illustrates the calculations required to determine the OARS candidate III cost of operation. The figure includes the number of accounts, as well as the monthly cost. This information, plus the fact that the number of regions will increase to four in year two and to five in year three of the new system will allow us to calculate the operating costs on a per account basis. Candidate III

Figure 13.10
Account projection. The system team estimated that the number of A/R accounts will
increase from 12,000 accounts at the beginning of the year to 20,000 accounts by the
end of five years.

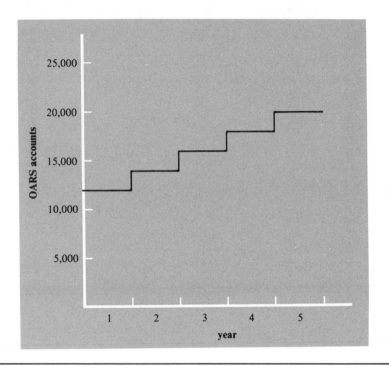

Figure 13.11
Category III—monthly cost calculations. The system team used collected account
projection data and estimates of the communication, computer, system personnel, and
central-site costs to determine the total monthly costs of each candidate system.

year	1	2	3	4	5
number of accounts	12,000	14,000	16,000	18,000	20,000
per region costs:					
communications	$ 150	$ 150	$ 150	$ 150	$ 150
*computer	2,000	2,000	2,000	2,000	2,000
system personnel	2,500	2,750	3,000	3,250	3,500
totals	$4,650	$4,900	$5,150	$5,400	$5,650
central-site costs:	$6,000	$6,000	$6,000	$6,000	$6,000

*Note: The computer costs include a minicomputer, printer, disk storage, and
    terminals (2).

Figure 13.12
Candidate III—operating costs per account. The data shown in figure 13.11 was used to calculate the per-account costs. As an example, the total cost during year 1 was $13,950 ($4,650 per region times three regions) plus $6,000 at the central site for a total of $19,950. That total, divided by the 12,000 accounts to be serviced during year 1, gives a per-account cost of $1.67 (rounded up). Note that the number of regions will increase to four in year 2 and to five in year 3.

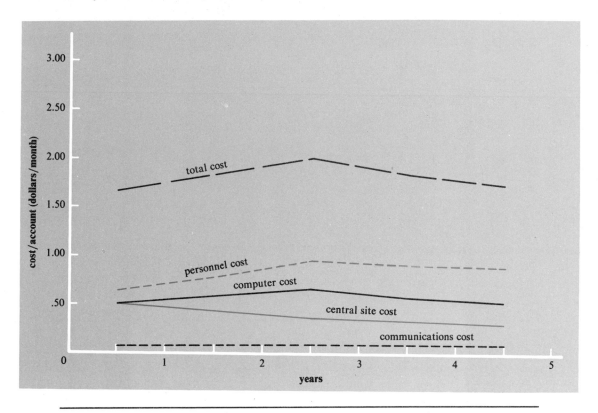

distributes the processing of the accounts receivable system to each of the regions. Each region, as well as the central site, will therefore incur operating costs. The per region cost includes the cost of communications, the computer, and the system personnel. The central-site cost includes the communications hardware/software and central-site system personnel.

Figure 13.12 summarizes the per account costs. As new regions are created and minicomputers are installed, the total cost per account fluctuates from $1.67 up to $2.00 and down to $1.72. Clearly, the personnel cost is the major component of the total cost.

Figure 13.13
Candidate III—project cost estimate. The system team estimated the total personnel and computer time costs for the study, design, and development phases. This figure summarizes these costs for candidate III. Note the increase in cost as the project proceeds from study to design to development.

	per week	total	
study phase: (10 weeks)			
senior analyst	$700	$7,000	
analyst	550	5,500	
			$12,500
design phase: (20 weeks)			
senior analyst	$700	$14,000	
analyst	550	11,000	
senior programmer	500	10,000	
			$35,000
development phase: (30 weeks)			
senior analyst	$700	$21,000	
analyst	550	16,500	
senior programmer	500	15,000	
programmers (5)	450	67,500	
computer time	250	7,500	
			$127,500
			$175,000

The estimated cost to design and develop the OARS candidate III is shown in figure 13.13. This cost is the total of the study, design, and development phase costs. Some analysts choose not to include the study phase costs, since they are considered "sunk" costs (or costs already spent). In this example the study phase cost was included in the total cost of all candidates. Figure 13.14 depicts the rate of spending and the cumulative cost of the study phase, design phase, and development phase over a sixty-week period.

For the purpose of comparison, the cost of operating the current A/R system was determined. Figure 13.15 is the same as figure 12.8, except for the added cost due to "float." Float is the cost incurred due to delays in receiving payment from credit customers. The estimated delay in collecting an average two million dollars per month at current interest rates results in an estimated monthly cost of $31,992. This float adds a cost of $.27 per account each month, raising the cost from $1.85 to $2.12.

Figure 13.14
Candidate III—project cost estimate chart. The system team used estimated costs, such as those shown in figure 13.13 for candidate III, to chart each system candidate's costs. In this example for candidate III, the estimated cumulative study phase, design phase, and development phase cost is $175,000. Actual project costs incurred will be compared with this estimate as the project proceeds.

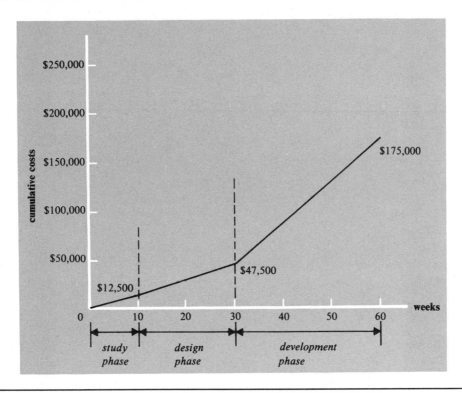

Figure 13.15
Operating costs of current A/R system. Monthly operating costs of the current system were collected during the initial investigation. These operating costs, plus the hidden cost of float, total $50,492 per month, or $2.12 per account for 10,000 accounts.

monthly operating costs:		
computer time	$ 4,400	$ .44/account
personnel cost	9,300	.93/account
keypunch	2,000	.20/account
supplies	$ 2,800	$ .28/account
operating totals	$18,500	$1.85/account
monthly float costs	$31,992	$ .27/account
total cost of system	$50,492	$2.12/account

Figure 13.16

Candidate cost calculations. This table of estimated development and operating costs was prepared by the system team to compare the cost of each candidate. As an example, the first operational year (year + 1) costs for candidate III were calculated as follows: (1) The per region cost of $19,950 per month (figures 13.11 and 13.12) was multiplied by 12 to give an annual cost of $239,400. (2) This cost was added to the development cost of $175,000 (figure 13.13) to give a cumulative cost of $414,400. Costs for successive years were calculated by a continuation of this process. The existing system, while not a candidate, is included as a reference for comparing the candidates. Costs for the current system were calculated as follows: (1) The monthly cost per account for the expanded current system would increase from $2.12 to $2.37. (2) $2.37/account × 12,000 accounts × 12 months/year = $341,280 per year. Similar costs for candidates I and II, details not provided, are also shown.

year	*ref.-existing*		*I*		*II*		*III*	
	ann.	*cum.*	*ann.*	*cum.*	*ann.*	*cum.*	*ann.*	*cum.*
dev. cost	- 0 -	- 0 -	118,500	118,500	145,000	145,000	175,000	175,000
+1	341,280	341,280	274,000	352,500	239,000	384,000	239,400	414,400
+2	398,160	739,440	270,000	622,500	277,000	661,000	307,200	721,600
+3	455,040	1,194,480	307,000	929,500	316,000	977,000	381,000	1,102,600
+4	511,920	1,706,400	312,000	1,241,500	321,000	1,298,000	396,000	1,498,600
+5	568,000	2,275,200	317,000	1,558,500	326,000	1,624,000	411,000	1,909,600

Figure 13.16 depicts the costs for each candidate. Even though the current system is not a viable candidate, its costs has been included in figure 13.16 to show the relationships between the costs of replacement systems and the cost of maintaining the existing system. The data for figure 13.16 was determined from the following:

1. The number of accounts for each year was taken from the account projection in figure 13.10.
2. The number of regions will increase from three to four in the second year of system operation and to five in the third year.
3. The cost of float will increase as the number of accounts increases. This cost was determined to be $31,992, $37,332, $42,660, $48,000, and $53,328 for each of the projected five years.
4. The float cost from number 3 was added to the annual operating costs of candidates I and II.
5. It was estimated that the expanded existing system would increase the per account cost each month from $2.12 to $2.37. It was assumed that this per account rate would continue over the next five years.
6. The annual operating costs for candidate III were calculated using the data in figure 13.11.

Figure 13.17
OARS payback analysis. The cumulative costs for the existing system and each of the three candidates, shown in figure 13.16, are plotted to show the payback period of each candidate. The payback point occurs where the candidate's cost plot goes below the reference line of the existing system.

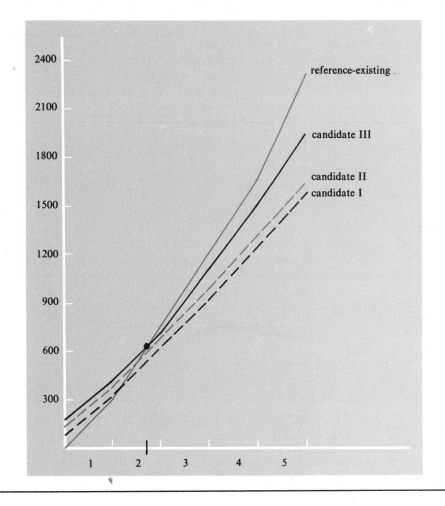

Figure 13.17 is called a *payback analysis*. Its purpose is to show the point in time at which the investment in the new system is recovered—as a result of cost savings or cost avoidance. The payback occurs where the candidate's accumulated cost crosses the accumulated cost of the reference existing system.

The performances and costs of each candidate are summarized in figure 13.18, which is a completed candidate evaluation matrix for candidate III. As the analyst proceeds through the feasibility analysis, additional system criteria may be identified, and/or criteria identified previously may be discarded. In this figure, all of the criteria identified in figure 13.19 have been retained.

Figure 13.18
Candidate evaluation matrix. This candidate evaluation matrix summarizes the estimated
performance and costs of each of the three candidate systems. It is a combination of
quantitative and qualitative evaluations.

candidate system / evaluation criteria	candidate I	candidate II	candidate III
performance			
accuracy	good	very good	very good
control	good	good	very good
flexibility	very good	good	very good
growth potential	poor	poor	very good
response time	good	good	very good
storage requirements	very good	very good	fair
usability	fair	good	very good
costs			
system development	$118,500	$145,000	$175,000
system operation	$1.26 to 1.09	$1.30 to 1.13	$1.45 to 1.16
payback	13 months	18 months	21 months

Step 8: Weight the System Performance and Cost Characteristics

In some cases the performance and cost data collected for each candidate will show
one candidate as the obvious choice. When this occurs, the task of the feasibility analysis is completed. Many times, however, the "best" candidate is still not clearly identified. The eighth step of the feasibility analysis is to prepare a **weighted candidate
evaluation matrix.** This is a matrix that weights the candidate evaluation entries by
their importance and then applies a rating number; it is a means of calculating total
numeric scores for each candidate.

A weighted candidate evaluation matrix is prepared in five steps:

1. Divide the evaluation criteria into categories of importance, for example, very
   important, moderately important, important.
2. Assign a weighting factor to each category. The weighting factors should be in
   proportion to each criterion's effect on the success of the selected candidate
   system.

3. Rate each candidate for each criterion relative to the other candidates. This relative rating is often with a scale of 1 to 5, with 5 being the best and 1 the lowest.
4. Calculate the candidate's score for each criterion by multiplying the relative rating by the weight assigned to the category.
5. Add the score column for each candidate to determine its total score.

Figure 13.19 is a weighted candidate evaluation matrix that was developed for OARS by following the previous steps. The relative weights for each criterion were determined cooperatively by the system team, and the ratings of the candidates were developed from data summarized in figure 13.18.

Step 9: Select the "Best" System

The last step is to select the best candidate system. If the analyst has been objective and has judged each of the evaluation criteria for each candidate without prejudice, the weighted candidate evaluation matrix summarizes the facts collected, as well as the subjective "impressions" about each candidate. The candidate with the highest total score *probably* is the best system. The analyst must select a candidate to present to the users and to management for their acceptance or rejection. If the total scores of two or more candidates are close, the analyst should reexamine the weights and ratings that differ the most or in which the team has the least amount of confidence. Finally, the analyst should select the candidate with the highest total score that is consistent with the team's confidence in the data. The analyst usually should not present several candidate choices to management for their selection, for they do not have the background to evaluate system performance as well as the analyst. The analyst has just completed a period of concentrated study of the candidates and is therefore in the best position to recommend a selection.

In the example ABCO corporation OARS, the criterion with the biggest effect on the total scores was the system growth potential. Candiates I and II were not expected to be able to grow as easily as candidate III because of the requirement for a large central-site computer to support them. In addition, the response time and the usability of candidate III are far superior to the other candidates. Candidate III will also allow ABCO to pursue its profit-center concept of organization. Therefore, candidate III was selected for presentation to management and the users.

The number of candidates that could be considered in the preceding example was necessarily limited. Actual situations usually are more complex, and choices are more difficult. However, there is one important practical guideline that the systems analyst always should keep in mind: The **"best" system** usually is the system that meets the performance requirements at the least cost.

This "best" system is not necessarily the system that provides the best performance, or even the best performance-to-cost ratio, if cost is a serious consideration. Overdesign frequently disguised as "the most bang for the buck" is liable to be an expensive luxury in a competitive, cost-conscious business environment. For example,

Figure 13.19
Weighted candidate evaluation matrix. The evaluations in this matrix are quantitative. The system team converted each of the performance and cost evaluations to a relative rating on a scale of 1 to 5. In addition, the evaluation criteria were assigned a weight to reflect their relative importance. The recorded scores are equal to the product of the assigned weights and ratings. The candidate with the highest total score that is consistent with the team's confidence in the data is the "best" system candidate.

evaluation criteria \ candidate system	candidate I		candidate II		candidate III	
	rating	score	rating	score	rating	score
performance						
accuracy (wt 2)	3	6	5	10	5	10
control (wt 4)	3	12	3	12	4	16
flexibility (wt 2)	5	10	3	6	4	8
growth potential (wt 4)	1	4	1	4	5	20
response time (wt 5)	3	15	3	15	5	25
storage requirements (wt 2)	5	10	5	10	2	4
usability (wt 5)	2	10	3	15	5	25
costs						
system development (wt 5)	5	25	4	20	2	10
system operation (wt 2)	5	10	4	8	3	6
payback (wt 2)	5	10	4	8	3	6
total score		112		108		130

a medium-to-large-scale computer may have a better performance-to-cost ratio than a small computer. However, if a small computer will meet the performance requirement at a lower cost, the additional dollars might be more effectively utilized elsewhere in the corporation.

Summary

The purpose of the feasibility analysis is to identify candidate systems; reduce these systems to a reasonable number; and evaluate the systems on the basis of relative cost and performance. The nine steps in performing a feasibility analysis are:

1. Form the system team.
2. Develop generalized flowcharts.
3. Develop the system candidates.

4. Perform preliminary evaluation of candidates.
5. Prepare detailed descriptions of candidates.
6. Identify meaningful system characteristics.
7. Determine performance and cost for each candidate.
8. Weight the system performance and cost characteristics by importance.
9. Select the "best" candidate.

Three tables, called matrices, are useful tools in proceeding from step 4 to step 9. These are:

1. A candidate system matrix: A table that lists functions to be performed and an alternative system for performing them.
2. A candidate evaluation matrix: A table that lists evaluation criteria and rates alternative system in terms of these criteria.
3. A weighted candidate evaluation matrix: A table that weights the candidate evaluation matrix entries by their importance and then applies a rating number.

The weighted candidate evaluation matrix is a means of calculating total scores for each candidate. The candidate with the highest total is identified as the "best" system. The best system usually is the system that meets the performance requirements at the least cost. After completing a period of concentrated study, the analyst should be prepared to discuss all candidates with the principle user and make a recommendation.

For Review

"best" system
system team
output data
system candidates
matrix
source analysis

system performance
system costs
payback analysis
**candidate system
  matrix**

**candidate evaluation
  matrix**
**weighted candidate
  evaluation matrix**

For Discussion

1. What are the two specific purposes of the feasibility analysis?
2. What is the purpose of the systems team? Who should be on the team?
3. What are the four basic functions to be performed by any system candidate? How does considering these functions aid in developing candidates?
4. Why is it necessary to perform a preliminary evaluation of system candidates prior to preparing detailed descriptions?
5. What should be included in a detailed description of a system candidate?
6. What is the purpose of the output data element source analysis worksheet?
7. Is the feasibility analysis in the study phase the only study of alternatives in the life cycle? Explain.
8. What is the purpose of:
   a. A candidate system matrix?
   b. A candidate evaluation matrix?
   c. A weighted candidate evaluation matrix?
9. What is the relationship among the three matrices cited in question 8?
10. Discuss the statement: "The system that meets all the performance criteria at the least cost is the 'best' system."

14
Study Phase Report
and Review

Preview

At the conclusion of the study phase activities the analyst prepares a report and reviews it with users of the computer-based information system. The central element of the study phase report is the performance specification, which is a user-oriented document. The study phase review is attended by the principal user and managers who will be affected by the system, and the outcome of the review is a decision whether or not to proceed with the design of the proposed system.

Objectives

1. You will become familiar with the content of a performance specification.
2. You will learn how to prepare a study phase report.
3. You will understand the purposes of a study phase review.

Key Terms

study phase report   A comprehensive report prepared for the user-sponsor of the system and presented at the conclusion of the study phase.

performance specification   A baseline specification that describes what the computer-based system is to do.

study phase review   A review for presenting the results of the study phase activities and determining future action.

Performance Specification

The *study phase* is concluded by a study phase review. Prior to this review, the performance specification is completed and a study phase report is prepared. The performance specification is contained in the study phase report. This specification is the first major baseline document. It is a complete, stand-alone specification that can be extracted from the study phase report as visible evidence of life-cycle progress.

The **performance specification** is the system performance communication link between the user and the analyst. Figure 14.1 illustrates the content of this specification. It is divided into two parts. The first part describes the interaction of the system with its operating environment, that is, inputs, outputs, interfaces with other systems, and resource needs. The resource needs include personnel, facilities, and equipment. These interactions are external to the computer program around which the computer-based business information system is built. For this reason, this part of the specification is called the *external performance description.* It includes an *information-oriented flowchart* and/or a *data flow diagram.*

The second part of the performance specification, the *internal performance description,* describes the internal environment of the system. It includes a *process-oriented flowchart* and/or *data flow diagram* and describes the related data processing operations and storage requirements.

The performance specification is the central element of the study phase report. It is a user-oriented document, written in the language of the user. In subsequent phases of the system life cycle, both the external and internal parts of the performance specification are expanded. The greatest expansion in documentation occurs in the internal part of the specification. This part becomes more technical and more comprehensive as the system moves through the design and development phases.

Figure 14.1
Performance specification outline. The performance specification is the technical core of the study phase report. It has two major sections: the external (to the computer program component) performance specification, and the internal (to the computer program component) performance specification.

performance specification	
a. *external performance description:*	1. flowchart: information-oriented and/or data flow diagram
	2. system output description
	3. system input description
	4. system interface identification
	5. system resource identification
b. *internal performance description:*	1. flowchart: process-oriented and/or data flow diagram
	2. data storage description

Study Phase Report

Structure and Content

The **study phase report** is a carefully prepared document. It is a management-oriented report that must be free of computer jargon so that it can be understood by senior managers who may not have a computer-oriented background. The structure and content of the study phase report are shown in figure 14.2.

The discussion of system scope is based on the project directive and on the performance definition activities by which specific objectives were identified. The problem statement and purpose section is a brief discussion of the problem and a statement of

Figure 14.2

Study phase report outline. The study phase report, completed at the end of the study phase, has five major sections: system scope; conclusions and recommendations; performance specification; plans and cost schedules; and appendices.

study phase report	
i. *system scope*	a. system title
	b. problem statement and purpose
	c. constraints
	d. specific objectives
	e. method of evaluation
ii. *conclusions and recommendations*	a. conclusions
	b. recommendations
iii. *performance specification*	a. external performance description
	1. flowchart—information-oriented and/or data flow diagram
	2. system output descriptions
	3. system input descriptions
	4. system interface identification
	5. system resource identification
	b. internal performance description
	1. flowchart—process-oriented and/or data flow diagram
	2. data storage description
iv. *plans and cost schedules*	a. detailed milestones—study phase
	b. major milestones—all phases
	c. detailed milestones—design phase
v. *appendices*	as appropriate

the system's general objective and anticipated benefits. The results of the initial investigation, which preceded the formulation of the project directive, may be referred to in this area.

The constraints are of the type referred to in chapter 12 as general constraints. They are "ground rules" that apply to all the alternative means by which the general objective of the system may be accomplished.

The specific objectives are derived from the general objective and anticipated benefits. They should be complete and quantitative wherever possible. The method of evaluation should describe how the accomplishment of the specific objectives is to be measured during the operation phase of the life cycle.

The conclusions and recommendations are presented next in the report in order to emphasize them and to accommodate the executive who may not need to read the entire report. However, the conclusions and recommendations must be substantiated in the other sections of the study phase report. The conclusions reflect the significant results of the system performance definition and feasibility analysis activities. The recommendations relate to the user's decision either to proceed with a design phase or to terminate the project.

The performance specification follows the conclusions and recommendations. Major elements of this baseline document were outlined in figure 14.1.

The plans and cost schedules prepared at the onset of the study phase are updated to report actual progress and cost versus schedule for the entire study phase. The report also includes two additional sets of project plans and cost schedules. A chart of *major milestones* is prepared for the entire project. The purpose of such a chart is to make visible to the reviewers the key activities, called milestones, to be completed and the costs to be incurred in order for the proposed system to become operational. Figure 14.3 is a list of key design phase and development phase tasks that are common to the life cycle of most computer-based business systems. A cost schedule is prepared to accompany the major milestone project plan.

A *detailed milestone* project plan also is prepared for the design phase. This plan is required since, after a successful study phase review, the authorization to proceed applies to the entire design phase. There is another review at the end of the design phase, when similar detail is given for the development phase. Figure 14.4 lists the milestones that will be described in chapter 15. A cost schedule is prepared to accompany the plan.

Appropriate appendices are included in the study phase report. Typically, they contain the project directive, the significant results of the initial investigation, the feasibility analysis, and pertinent memoranda.

Example Study Phase Report

Exhibit 1 on the following pages shows the study phase report that might be prepared for the example system OARS. It includes all the features outlined in figure 14.2. However, many illustrations that have already been presented in the text are not repeated here but simply referred to. In addition, the content of the appendix is limited to a listing of items typically included.

Figure 14.3
Major milestones in the design and development phases. The major design and development phase milestones are included in the study phase report to provide the user-sponsor of the information system with an overview of the activities and associated costs required to complete the project.

major milestones	
design phase	allocation of functions
	computer program functions
	test requirements
	design specifications
	design phase report
	design phase review
development phase	implementation plan
	equipment acquisition
	computer program development
	system tests
	personnel training
	changeover plan
	system specifications
	development phase report
	development phase review

Study Phase Review

The **study phase review** is held to present the user with the results of the study phase activities and with recommendations for future action. If the recommendation of the analyst is to proceed with the design of the system, the analyst should remember that the user will be making a commitment that transforms the system effort from a study to a formal project.

The analyst should be factual, yet not hesitate to sell a recommendation; after all, it is the analyst who has been working on the study phase for weeks, even months. The analyst *is* knowledgeable about the system and would be less than human in not forming strong convictions. If the analyst is enthusiastic, it should show.

The study phase review should be attended not only by the *principal user,* but also by senior representatives of other affected organizations, including the information services organization. All attendees should be provided with a copy of the study phase report in advance of the review meeting.

Figure 14.4
Detailed milestones for the design phase. The detailed milestones for the design phase are included in the study phase report to provide the user-sponsor with specific information about the activities and associated costs that are to be authorized for the subsequent phase.

design phase—detailed milestones	
allocation of functions	
manual functions	task definition
	reference manual identification
equipment functions	function definition
	equipment specification
computer program functions	data base design
	computer programs
	design (for each program)
test requirements	system test requirements
	computer program test requirements
design specifications	
design phase report	
design phase review	

If the outcome of the study phase review is a decision to proceed with the design phase, the user-sponsor issues a written approval to proceed. This approval is an example of the renewed, or incremental, commitment that occurs at the end of each life-cycle phase. It extends the project directive by authorizing the expenditure of resources for the design phase.

Summary

At the conclusion of the study phase activities, a study phase report is completed. This report is the basis for a study phase review with the principal user and managers affected by the computer-related business information system. The study phase report has five principal parts:

1. System scope
2. Conclusions and recommendations
3. Performance specification

4. Plans and cost schedules
5. Appendices

The performance specification is the central element of the study phase report. It is written in language that the user can understand, and it describes both the external (to the computer program), and the internal performance of the business information system. The study phase report includes completed project plan and cost schedules for the study phase, major milestones for all phases to provide visibility for the entire project, and detailed milestones for the design phase, since that is the next major activity. At the study phase review, the systems analyst presents the results of the activities completed and makes recommendations for future effort. If the recommendation of the systems analyst is to proceed, and if that recommendation is approved, the project moves into the design phase.

For Review

study phase
study phase report
performance
 specification
external performance
 definition

internal performance
 definition
study phase review
principal user
major milestone
data flow diagram

detailed milestone
information-oriented
 system flowchart
process-oriented
 system flowchart

For Discussion

1. What is the purpose of the study phase report?
2. What is the content of the study phase report?
3. Why are the conclusions and recommendations presented early in the study phase report?
4. Discuss the importance of the performance specification.
5. What are the differences between the external performance specification and the internal performance specification?
6. How many sets of project plans and schedules are presented at the study phase review? How do they differ?
7. What is the purpose of the study phase review? Who should attend?

Exhibit 1: The Study Phase Report

A case study that follows the life-cycle road map is introduced in chapter 11, Initial Investigation. In the previous edition, the case study resulted in a Modified Accounts Receivable System (MARS), which was an improvement over a predecessor manual system. MARS operated in a batch processing mode and depended upon punched card data preparation. In this edition MARS is the "old" system; the case study updates MARS to an On-line Accounts Receivable System (OARS). A principal goal of OARS is to establish geographically separate profit centers. The evaluated candidate systems are distributed data processing systems, which range from "dumb" terminals to a substantial local data processing capability, and which provide for direct data entry and several communications options.

Exhibits of example study phase, design phase, and development phase reports are included in chapter 14 (Study Phase Report and Review); chapter 20 (Design Phase Report and Review); and chapter 24 (Development Phase Report and Review). The attachment to this chapter is Exhibit 1: The Study Phase Report.

OARS Study Phase Report

I. System Scope
   A. System Title
      On-line Accounts Receivable System (OARS)
   B. Problem Statement and Purpose
      The ABCO corporation's present accounts receivable system is at its maximum capacity of 10,000 accounts. The number of accounts is expected to double to 20,000 accounts over a five-year period. The present system cannot meet this projected growth and satisfy the corporate goal of distributing

information processing resources to regional profit centers. Serious problems already have been encountered in processing the current volume of accounts. Specific problems that have been identified are:

1. Saturation of the capacity of the present computer system, causing difficulties in adding new accounts and obtaining information about the status of existing accounts.
2. Processing delays in preparing customer billing statements because of the batch-oriented design of the current accounts receivable system.
3. Excessive elapsed time between mailing of customer statements and receipt of payments, which creates a high-cost, four-day float.
4. Inadequate control of credit limits.
5. Inability to provide regional centers with timely customer-related information.

Therefore, the purpose of the OARS project is to replace the existing accounts receivable system with one that can eliminate the stated problems and meet ABCO's growth and regional accountability goals.

C. Constraints

The OARS constraints are:

1. Development of the on-line accounts receivable system is to be completed within fourteen months.
2. OARS is to have a growth potential to handle 20,000 customer accounts.
3. OARS is to interface with the existing perpetual inventory system.
4. OARS is to be designed as an on-line system operating in a distributed data processing environment.
5. The design must be compatible with corporate plans to install regional profit centers.

D. Specific Objectives

The specific objectives of OARS are:

1. To establish billing cycles for each region.
2. To mail customer statements no later than one day after the close of a billing cycle.
3. To provide the customer with a billing statement two days after the close of a billing cycle.
4. To speed up collections, reducing the float by 50 percent.
5. To examine customer account balances through on-line inquiry at the time of order entry.

E. Method of Evaluation

After OARS has been operational from sixty to ninety days:

1. A statistical analysis will be made of customer account processing to verify the elapsed time between the close of the billing cycle and the mailing of customer statements.
2. The float time will be measured, and the cost of the float will be calculated at three-month intervals.
3. Periodically, random samples of customer accounts in each region will be audited for accuracy and to validate the effectiveness of on-line inquiry.

4. The validity of OARS transactions that affect the inventory system will be measured by random sampling and physical count.
5. Personal evaluations of the effectiveness of the system will be obtained from its principal users.

II. Conclusions and Recommendations

A. Conclusions

The feasibility analysis of the on-line accounts receivable system (OARS) involved the evaluation of three candidate systems and led to the conclusion that the best system would be one that required minimal modification of the existing central-site computer system and that utilized this system as a host for a network of minicomputers. The minicomputers, one for each region, would be programmable terminals. They would provide a local capability for transaction processing and would be used to transmit summary information to the central site. All outputs would be available as CRT displays wherever feasible, with printed reports provided or available when necessary.

The projected monthly operational cost for the selected system varies from $1.66 per account at a volume of 12,000 accounts in three regions, to $1.71 per account for a volume of 20,000 accounts in five regions. The current system costs $2.37 per account at a volume of 12,000 accounts. It can expand to handle 20,000 accounts at a cost of $2.29 per account; however, it would continue to operate in a batch-oriented mode and would not meet the on-line, distributed data processing goals of the corporation.

The projected development cost for OARS is $175,000. The savings in the operating cost of the selected system, when compared with the cost of expanding the existing system, will result in recovery of the development costs in twenty-one months of operation.

B. Recommendations

It is therefore recommended that the OARS project be approved for the design phase.

III. Performance Specification

A. External Performance Description

1. *Flowcharts*   Figure E1.1a is an information-oriented system flowchart for OARS. The accompanying narrative appears as figure E1.1b. Figure E1.2 is the logically equivalent data flow diagram.
2. *System output descriptions*   The six OARS outputs are:
   a. Customer monthly statement
   b. Accounts receivable transaction register/display
   c. Accounts receivable summary display
   d. Aged accounts receivable report
   e. Customer account list display
   f. Overcredit notification

   An output specification and a data element list for an output are presented as figure E1.3.

Figure E 1.1a
OARS information-oriented flowchart.

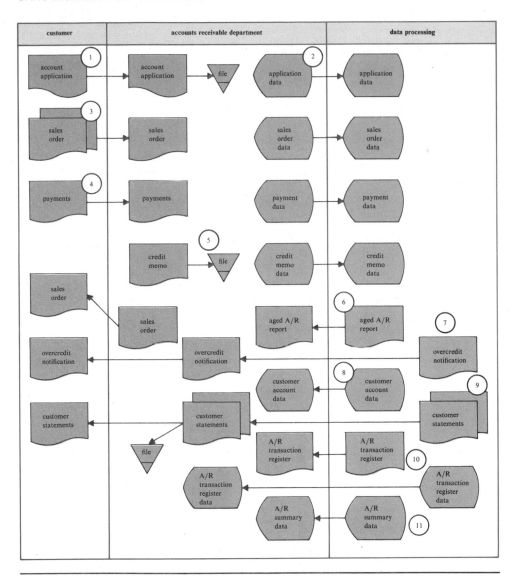

3. *System input descriptions*   The three OARS inputs are:
   a. Customer account application
   b. Customer order
   c. Payment/credit
   An example of the system inputs is included as figure E1.4.
4. *System interface identification*   The OARS must interface with (that is, transfer data to and from) the existing inventory system. Inventory

Figure E1.1b
Narrative for OARS information-oriented flowchart.

```
INFORMATION-ORIENTED FLOWCHART NARRATIVE


  1.  Customers submit account applications that are processed
      by the accounts receivable department.

  2.  If an application is accepted, it is keyed into the
      system.  The application is filed.

  3.  Sales orders are entered through on-line order
      processing.  The customer retains a carbon copy.

  4.  Payments are sent to the accounts receivable
      department and entered into the system.

  5.  Credit memos are generated and the data entered.
      The memo is filed.

  6.  An aged A/R report is sent to the accounts receivable
      department from data processing each month.

  7.  Data processing sends overcredit notices to the credit
      department whenever a new order would exceed the
      customer's credit limit and the additional credit is
      not approved.  The order and notice are returned to the
      customer.  If the additional credit is approved, the
      order is processed.

  8.  Customer account data is displayed on demand.

  9.  Customer statements are sent to accounts receivable
      in duplicate.  The original copy is sent to the
      customer; the duplicate is filed.  One-third of the
      statements are produced each ten days of the month,
      that is, on the 1st, 10th, and 20th.

 10.  A daily A/R transaction register is sent to the accounts
      receivable department as a file copy.  The data is also
      available as a display.

 11.  The accounts receivable summary data is available as
      a display to the accounts receivable department.
```

must be on hand for shipment prior to billing the customer, and inventory quantities should be reduced as merchandise is committed for shipment.

5. *System resource identification*   The current central-site computer system will be augmented by regional minicomputers, each with the following characteristics:
   a. CPU with 128K positions of main memory.
   b. One 600 line:-per-minute printer.
   c. One 130 megabyte magnetic disk drive.
   d. One console keyboard/printer.
   e. Two CRT keyboard/display terminals, expandable to eight stations.
   One additional programmer will be required prior to start of the development phase. No other additional resource requirements are anticipated.

Figure E1.2
OARS data flow diagram.

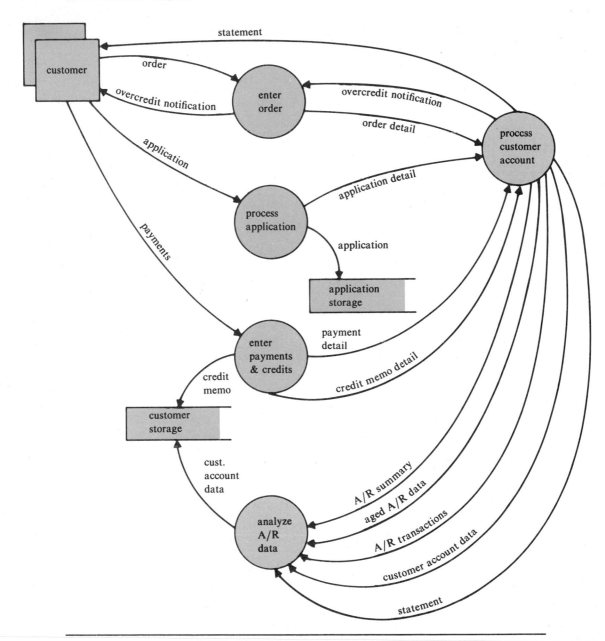

Figure E1.3
OARS output description: overcredit notification.

OUTPUT SPECIFICATION

TITLE: Overcredit Notification

LAYOUT:

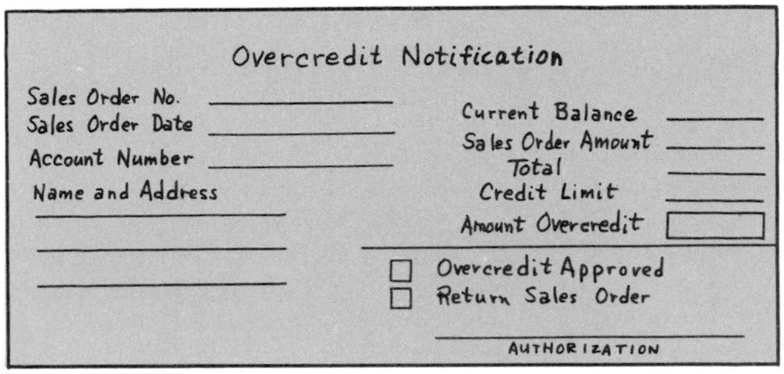

> ### Overcredit Notification
>
> Sales Order No. _____   Current Balance _____
> Sales Order Date _____   Sales Order Amount _____
> Account Number _____   Total _____
> Name and Address   Credit Limit _____
> _____   Amount Overcredit [_____]
> _____
>
> ☐ Overcredit Approved
> ☐ Return Sales Order
>
> AUTHORIZATION

FREQUENCY: Daily   QUANTITY: 80 max.
SIZE: 1 page   COPIES: 2

DISTRIBUTION: 1. Accts. Receivable Dept.

COMMENTS:

DATA ELEMENT LIST		

TITLE: Overcredit Notification

DESCRIPTION	FORMAT	SIZE
Sales Order Number	XXXXXXXX	8 characters
Sales Order Date	XX/XX/XX	8 "
Account No: Region No. – Sequence No.	XX-XXXXXXX	10 "
Name		26 "
Address		
Street		20 "
City		18 "
State		2 "
Zip Code		5 "
Current Balance	XXX,XXX.XX	10 "
Sales Order Amount	XXX,XXX.XX	10 "
Total	XXX,XXX.XX	10 "
Credit Limit	XX,XXX.XX	9 "
Amount Overcredit	XX,XXX.XX	9 "

Figure E1.4
OARS input description: sketch of customer account application.

CUSTOMER ACCOUNT APPLICATION ABCO Corporation Walnut, California			
FIRM NAME	WHOLESALE ☐ RETAIL ☐	DATE	
STREET ADDRESS		TELEPHONE	
CITY	STATE	ZIP CODE	
BANK REFERENCES:			
NAME	BRANCH	TELEPHONE	
ADDRESS	CITY	STATE	ZIP CODE
NAME	BRANCH	TELEPHONE	
ADDRESS	CITY	STATE	ZIP CODE
OTHER REFERENCES:			
NAME		TELEPHONE	
ADDRESS	CITY	STATE	ZIP CODE
NAME		TELEPHONE	
ADDRESS	CITY	STATE	ZIP CODE
FOR OFFICE USE:	APPROVED ☐		
ACCOUNT NUMBER	DISAPPROVED ☐		
EFFECTIVE DATE	AUTHORIZATION SIGNATURE		
CREDIT CODE			
DISCOUNT CODE			
IF DISAPPROVED, REASON:			

Figure E1.5
OARS process-oriented flowchart and narrative.

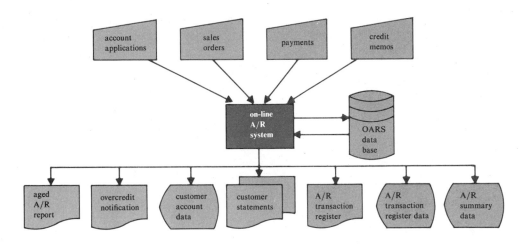

SYSTEM FLOWCHART NARRATIVE

1. Customer account applications, payments, credit memos,
    and sales orders are the four major system inputs.

2. The A/R program inputs the transaction data and reads/
    updates the master file on magnetic disk.

3. The seven major outputs of the system are the aged A/R
    report, overcredit notification, customer account data
    display, customer statements, A/R transaction register,
    A/R transaction register data display, A/R summary data
    display.

B. Internal Performance Specification
   1. *Flowcharts*   The process-oriented system flowchart is shown in the upper
       part of figure E1.5. The narrative description of the flowchart appears in
       the lower part of the same figure. Figure E1.6 is the logically equivalent
       data flow diagram from which figure E1.5 was derived.
   2. *Data storage description*   Each master record contains 15 data ele-
       ments, for a total of 167 characters. The master file will require a max-
       imum of 388 tracks (20,000 customers). This is less than 100 tracks per
       region, allowing ample storage for applications growth.

Figure E1.6
OARS data flow diagram—processing detail.

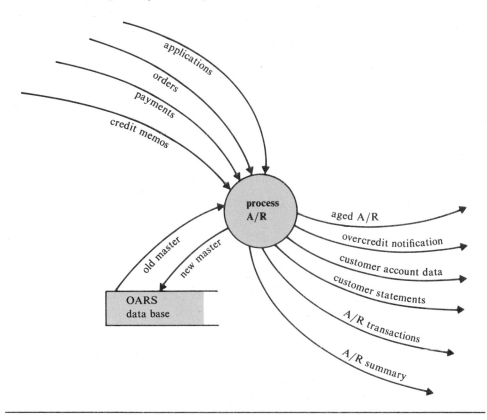

IV. Project Plans and Schedules
   A. Study Phase

   The study phase was scheduled for a ten-week period, beginning 9/1/xx (this year) and ending 11/10/xx. The funding authorized for the study phase was $12,500. As shown in figure E1.7, the project is on schedule. Only the study phase review remains to be completed.

   As shown in figure E1.8, expenditures are slightly under budgeted cost.

Figure E1.7
Project plan and status report—study phase.

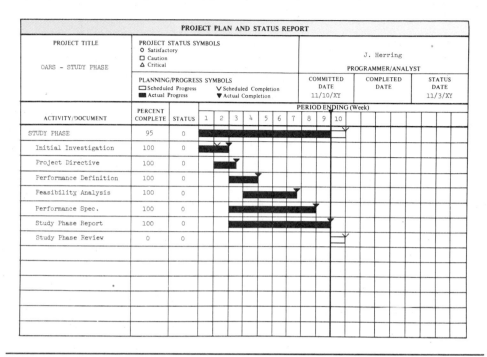

B. Major Milestones—All Phases

Figure E1.9 is a schedule for the entire project. The design phase is scheduled for twenty weeks, and the development phase is scheduled for thirty weeks. If we proceed as planned, the design phase will be completed 3/30/xy (next year), and the development phase will be completed on 11/19/xy.

The estimated cumulative cost for the entire project (study, design, and development phases) is graphed in figure E1.10. The total cost is estimated to be $175,000.

C. Detailed Milestones—Design Phase

Since the next phase to be undertaken is the design phase, detailed projections are presented for that phase. Figure E1.11 displays the specific milestones to be achieved in the course of a twenty-week design phase effort.

Figure E1.12 presents the accompanying cumulative cost estimate. The total design phase cost is estimated to be $35,000.

Figure E1.8
Project cost report—study phase.

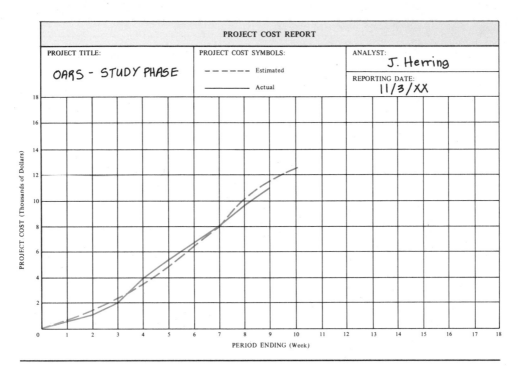

V. Appendices

Note: The appendices include all supporting data for the conclusions of the study. For purposes of this exhibit, such detail is not included. However, the following is a list of typical appendices, with in-text references.

*Calculations (not developed in the text) should also be included for other candidates as well as for candidate 3.

Figure E1.9
Project plan and status report—OARS major milestones.

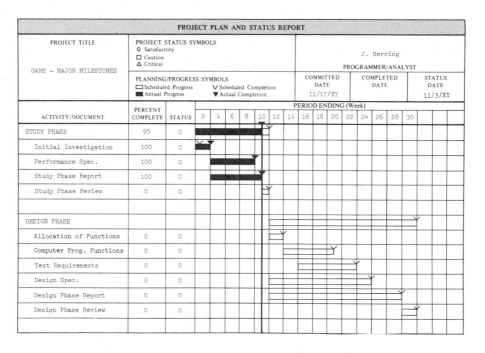

PROJECT PLAN AND STATUS REPORT																		

PROJECT TITLE: OARS – MAJOR MILESTONES
PROGRAMMER/ANALYST: J. Herring
COMMITTED DATE: 11/17/XY **STATUS DATE:** 11/3/XY

PROJECT STATUS SYMBOLS: O Satisfactory, □ Caution, △ Critical
PLANNING/PROGRESS SYMBOLS: □ Scheduled Progress, ■ Actual Progress, ∨ Scheduled Completion, ▼ Actual Completion

ACTIVITY/DOCUMENT	PERCENT COMPLETE	STATUS	32	34	36	38	40	42	44	46	48	50	52	54	56	58	60
DEVELOPMENT PHASE	0	O															∨
Implementation Plan	0	O			∨												
Equipment Acquisition	0	O								∨							
Computer Program Dev.	0	O												∨			
Personnel Training	0	O													∨		
System Tests	0	O													∨		
Changeover Plan	0	O													∨		
System Spec.	0	O												∨			
Dev. Phase Report	0	O														∨	
Dev. Phase Review	0	O															∨

Figure E1.10
Project cost report—total project.

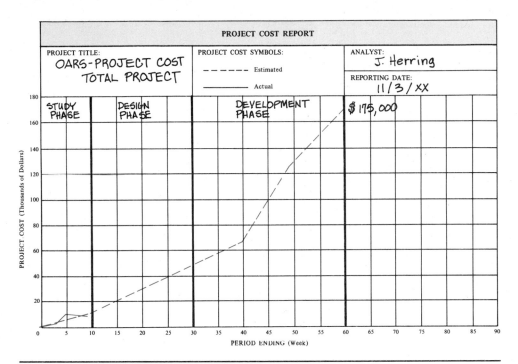

Figure E1.11
Project plan and status report—design phase.

PROJECT PLAN AND STATUS REPORT																		

PROJECT TITLE: OARS – DESIGN PHASE

PROJECT STATUS SYMBOLS
O Satisfactory
□ Caution
△ Critical

PLANNING/PROGRESS SYMBOLS
□ Scheduled Progress   V Scheduled Completion
■ Actual Progress   ▼ Actual Completion

J. Herring — PROGRAMMER/ANALYST

COMMITTED DATE 3/30/XY
COMPLETED DATE
STATUS DATE 11/3/XY

ACTIVITY/DOCUMENT	PERCENT COMPLETE	STATUS
DESIGN PHASE	0	0
Allocation of functions	0	0
Manual Functions	0	0
Task Definition	0	0
Ref. Manual Iden.	0	0
Equipment Functions	0	0
Function Def.	0	0
Equipment Spec.	0	0
Computer Prog. Functions	0	0
Data Base Design	0	0
Data Edit Program	0	0
A/R Program	0	0
Customer Program	0	0
Overcredit Program	0	0

ACTIVITY/DOCUMENT	PERCENT COMPLETE	STATUS
DESIGN PHASE (CONT'D)	0	0
Test Requirements	0	0
System Test Req.	0	0
Computer Prog. Test Req.	0	0
Design Spec.	0	0
Design Phase Report	0	0
Design Phase Review	0	0

Figure E1.12
Project cost report—design phase.

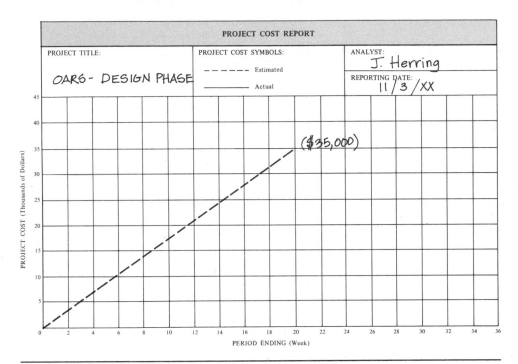

The Design Phase

Unit 4

15
Design Phase
Overview

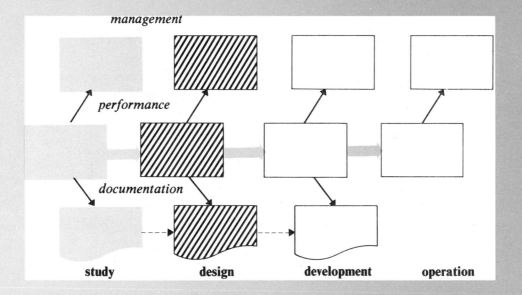

Preview

In the design phase the detailed design of the system selected in the study phase is accomplished, and a user-oriented performance specification is converted into a technical design specification. Principal activities performed during the design phase include the allocation of functions between computer programs, equipment, and manual tasks; data base design; and test requirements definition. The design phase concludes with a design phase report and a user review.

Objectives

1. You will become familiar with the major design phase activities.
2. You will acquire a perspective of the design phase that will serve as a reference and guide as you study specific topics in the chapters in this unit.

Key Terms

design phase   The life-cycle phase in which the detailed design of the system selected in the study phase is accomplished.

design specification   A baseline specification that serves as a blueprint for the construction of a computer-based business information system.

design phase report   A report prepared at the end of the design phase; it is an extension of the study phase report that summarizes the results of the design phase activities.

design phase review   A review held with the user organization at the conclusion of the design phase to determine whether or not to proceed with the development phase.

Design Phase Organization

The **design phase** is the life-cycle phase in which the detailed design of the system selected in the study phase is accomplished. In the course of the design phase, the performance specification is expanded into the design specification. The user-oriented baseline document prepared in the study phase becomes a baseline document oriented to the needs of the programmers and other professional personnel who will actually develop the system. A smooth transition from the study phase to the design phase is necessary because the design phase continues the activities begun in the earlier phase. However, the project becomes enlarged in scope, and personnel are added. As examples:

1. The user organization assigns additional personnel to participate in the project as intermediaries between the project and the user organization. These persons are particularly concerned with defining user requirements and developing the resources (for example, training manuals, procedures, and personnel) required by the user organization to insure successful operation of the system.
2. The information service organization assigns additional technical personnel, such as analyst/programmers and equipment specialists, to the project. These persons develop the technical requirements for computer programming and operations support. They are particularly concerned with the effective translation of system performance requirements into computer program design requirements. They aid in the selection of the best techniques for utilizing existing computer hardware and software. They also aid in the development of specifications on which to base the selection of new computer systems or components.

Design Phase Activities

The flowchart of figure 15.1 is a pictorial overview of the design phase. Each of the activities shown in this flowchart is described briefly as follows:

1. *Allocation of functions*   The *information-oriented* system flowcharts, high-level *process-oriented system flowcharts,* and *data flow diagrams* prepared during the study phase are reviewed and expanded in order to allocate functions between *manual tasks, equipment functions,* and *computer program functions.* The alternatives are analyzed until all functions have been allocated. With respect to the computer program in particular, the expanded flowcharts identify the inputs, outputs, and files accessed by the set of programs that are the building blocks of the overall computer program component of the computer-related business information system. The expanded flowcharts also identify the controls required to ensure valid system performance.

   Interfaces between the computer-based component of the system under development and other computer-based systems and equipment are defined by specifying the data that must be made available to or accessed from other systems.

Figure 15.1
Design phase activity flowchart. After the allocation of functions, the principal activity
sequences of the design phase relate to the definition of manual tasks, equipment
functions, and the definition of test requirements.

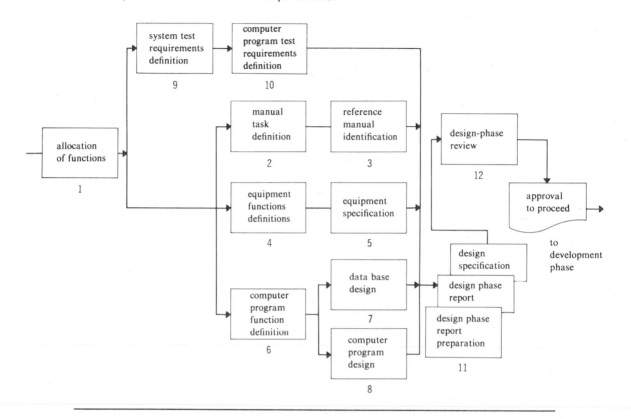

2. *Manual task definition*   Requirements resulting from human interfaces with
   the computer-based component of the system are described. Human interfaces
   include preparation of source documents, operation of equipment (for example,
   machine operation, display console operation, data entry), and other input- and
   output-related activities.
3. *Reference manual identification*   Reference manuals required by user
   personnel, programmers, and equipment operators are identified.
4. *Equipment functions definition*   The functions to be performed by hardware
   (rather than by computer programs or manual operations) are defined. Special
   functions unique to the application (for example, console displays) are described
   in detail.
5. *Equipment specification*   The hardware configuration used to convert input
   data to meaningful output information is described. If existing hardware is not
   adequate, alternatives, which may range from adding special equipment to

procuring an entire computer system, must be considered. Since computer hardware may be a long lead-time item, it may be necessary to initiate procurement of critical equipment during the design phase.[1]

6. *Computer program function definition* The specific functions of the computer program component of the overall system are defined, and design requirements for external system inputs are established. For example, when data entry is required from an input source document, a CRT screen layout is made. In addition, input frequency, quantity, and source are specified. Similarly, the design requirements for system outputs are established. Thus, if a printed report is an output, a printer output layout is prepared. Report frequency, size, quantity, and number of copies are stated. Control requirements are specified. These requirements include input preparation, input acceptance, and processing controls.

The above activities lead to establishing requirements for the data base and for each of the computer programs that comprise the overall computer program component of the system.

7. *Data base design* Relationships between data elements, functions to be performed, and techniques for file organization are studied in detail so the most appropriate storage device(s) can be selected and an efficient data base design can be achieved.

The storage requirements for all the data elements on which the computer programs operate are calculated, taking into account the size and volume of the records to be stored and the methods of file organization and access.

The interfaces between the system data base and other data bases are identified by specifying the data that must flow between them.

8. *Computer program design* The computer programs, which make up the overall computer program component, share the system data base. However, they have their own output, input, and processing requirements, which must be specified for each one. Special hardware and supporting software requirements also are identified. As necessary, additional expanded system flowcharts are prepared at the computer program and subprogram levels. Narratives, equations, algorithms and decision tables or trees may be developed as aides in defining the functions of the computer programs. Control requirements are also extended to the computer programs.

9. *System test requirements definition* Requirements are established for the tests necessary to verify the performance of the entire computer-based system. This is accomplished in parallel with the activities associated with system design.

10. *Computer program test requirements definition* Requirements also are determined for the tests necessary to verify the performance of the major computer programs. This is done after the definition of the system test requirements. (However, these tests are performed prior to the system tests.)

---

1. Manufacturers of computers will frequently cooperate by accepting a letter of intent, subject to subsequent final verification, to establish a production and delivery schedule.

11. *Design phase report preparation* A **design phase report** is prepared at the conclusion of the design phase. This report is the outgrowth of the data acquired and added to the project file during the design phase. An extension of the study phase report, it contains a summary of the results of all significant activities undertaken during the design phase. An important element of the design phase report is a recommendation relative to proceeding with the development phase. If the recommendation is to proceed, a detailed project plan is provided for the remainder of the project.

Included in the design phase report is the *design specification*—the second major baseline document. It represents an expansion of the *performance specification* into a blueprint for the development of the computer-based business system. It is a "build to" specification.

12. *Design phase review* The system design is reviewed at the conclusion of the design phase by the management of the user organization and by representatives of the information systems organization. The principal documents upon which the review is based are the study phase report, including the performance specification; and the design phase report, including the design specification. Any changes to the performance specification as a result of the design phase activities are identified and discussed. The detailed progress plan and the cost schedule for the development phase are reviewed, as are the estimates of operational costs.

After the conclusion of a successful *design phase review,* manpower and other resources are committed. Written approval to proceed with the development phase is provided by the user organization.

Summary

The design phase is the second of the four life-cycle phases. It is the phase in which the detailed design of the system selected in the study phase is accomplished. The computer-related business information system project is expanded to include the addition of personnel from the user and information services organizations.

The principal activities performed during the design phase include: the allocation of functions between computer programs, equipment, and manual operations; the design of the data base used by the computer programs; specification of the requirements for input, processing, and output; and the definition of system and computer program test requirements.

A design phase report is prepared after the completion of the design phase activities, and a review is held with the user organization in order to determine whether or not the computer-based business information system project is ready to proceed to the development phase.

For Review

design phase
information-oriented
  system flowchart
process-oriented
  system flowchart
manual task

equipment function
computer program
  function
performance
  specification

design specification
design phase review
design phase report

For Discussion

1. Define design phase.
2. What is meant by "allocation of functions"?
3. Why is it important to establish test requirements in parallel with the on-going design activities?
4. What is the purpose of the design phase report? The design phase review?
5. What is the design specification?

16
System Design

Preview

At this point in the life cycle the design phase begins. The problem has been identified, solutions have been studied, and a management review has ended with an authorization to proceed with a recommended solution. The first design phase step, system design, involves selection of those system functions to be performed by either the people in the system, the equipment, or computer programs. These functions are usually identified by drawing expanded system flowcharts.

In addition to the allocation of functions, system and computer program test requirements must be identified. Structured walk-throughs are useful techniques in detecting and eliminating errors as the computer-based business information system is being designed and developed.

Objectives

1. You will learn how system functions are allocated among humans, equipment, and computer programs.
2. In addition, you will learn the importance of identifying test requirements.
3. You will understand the purpose and use of structured walk-throughs.

Key Terms

expanded system flowchart   A system flowchart that has been expanded in its detail until each of the processing functions can be identified.

structured walk-through   A technical review to assist the people working on a project; used to discover errors in logic of a computer program or in other system components.

Allocation of Functions

Expanded System Flowcharts

The first major activity of the design phase is the allocation of functions to manual operations, equipment, or *computer programs*. This is accomplished by expanding the results of the study phase activities to greater levels of detail. The flowcharts prepared in the study phase are reviewed and decomposed until all functions the system must perform are evident. Because of the importance of identifying system functions, the analyst may draw HIPO charts instead of or in addition to the process-oriented flowcharts or data flow diagrams. The OARS example illustrates the use of expanded flowcharts, which identify each of the processing functions.

Figure 16.1a is an expanded process-oriented system flowchart, with an accompanying narrative, prepared for the example OARS system. It displays all the major data processing functions that must be performed by individual computer programs. In addition, the major processing control functions are shown.

Figure 16.1b is an expanded data flow diagram that was used to derive the major data processing functions. Figures 16.1a and 16.1b are expansions of figures 13.7a and 13.7b, which were drawn during the study phase.

Figure 16.1a
Expanded A/R process-oriented flowchart and narrative. This intermediate-level system flowchart was prepared by the system team to emphasize the major data processing functions that must be performed by individual computer programs. This flowchart is an expansion of the flowchart shown in figure 13.7a, drawn during the study phase.

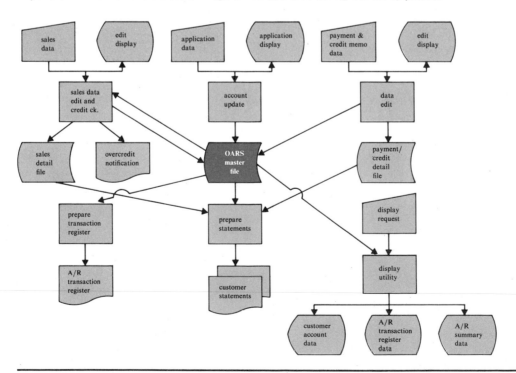

Figure 16.2 is an example of a hierarchy chart and an IPO chart used to show system functions. A *HIPO chart* set should include an IPO chart for each block of the hierarchy chart, but in this example, only the IPO for the data edit function is shown. Note that the HIPO charts show the same computer programs as the *expanded process-oriented system flowchart*. The major differences in the HIPO charts are the chart structure and the emphasis on system functions.

Identification of Controls

The purpose of controls is to minimize incorrect data, and so they are an indispensable part of any information system. To ensure that the appropriate controls are built into the system, the analyst should plan for them early in the design phase. Later in the development phase there will be pressures from programming schedules, test schedules, and storage limitations that will tend to distract from the importance of controls.

Although controls help to prevent or detect theft and fraud, the vast majority of problems detected through controls arise from honest mistakes. The principal controls used relate to input, processing, and output. Figure 16.3 identifies some commonly

```
OARS EXPANDED PROCESS-ORIENTED FLOWCHART NARRATIVE

    Sales data edit and credit check:

       The sales data is entered via terminal keyboard.  A
    numeric edit is performed.  Bad data discovered during the
    edit is displayed to the terminal operator.  If the data is
    valid, a check is performed to determine whether or not the
    customer will exceed their credit limit.  An overcredit
    notification is produced if the limit is exceeded and the
    data is recorded in the sales detail file.

    Account update:

       Application data is entered via terminal keyboard.  The
    OARS master file is updated to reflect the new account.

    Payment/credit memo data edit:

       The payment and credit memo data is entered via terminal
    keyboard.  A check is made to verify account numbers and
    valid numeric data.  Good data updates the master file and
    is recorded in the payment/credit detail file.

    Prepare transaction register:

       At the end of each month, a transaction register is printed
    using data from the OARS master file.

    Prepare customer statements:

    Display utility:

       Requests to display customer account data, A/R transaction
    register data, or A/R summary data are made through any
    terminal with proper password clearance.  All data comes from
    the OARS master file.
```

Figure 16.1b

Expanded data flow diagram. This expanded data flow diagram is a further decomposition of the diagram shown in figure 13.7b, drawn during the study phase. It was used by the systems team to display system logic, and to lead to the physical implementation implied by the system flowcharting symbols used in figure 16.1a.

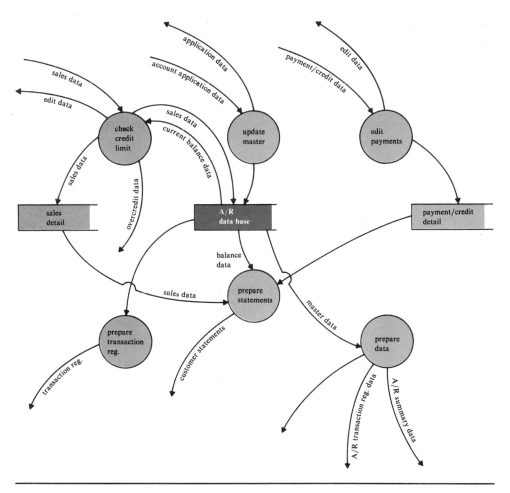

encountered controls with which the systems analyst should be familiar. Some of the controls are simply standards to be established and used; some involve manual operations; others are accomplished through computer program components. In the OARS example:

The expanded A/R system flowchart in figure 16.1 and the HIPO charts in figure 16.2 illustrate two control techniques. First, the OARS utility program is used to protect the master file with magnetic tape backups for recovery of the master file data. It also provides the only general access to the master file. All updates involving new customers added to the file, address changes, or credit

Figure 16.2
OARS HIPO charts. HIPO charts consist of a hierarchy chart (H) and a series of input-process-output (IPO) charts. The hierarchy chart identifies the major system functions. Each IPO chart shows the inputs, processing, and outputs of one system function, such as the data edit module shown in this figure.

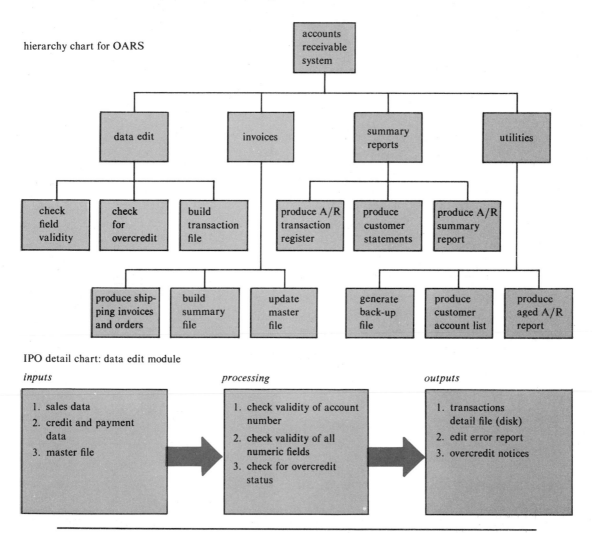

hierarchy chart for OARS

IPO detail chart: data edit module

Figure 16.3
Example control techniques. Controls are used to minimize incorrect data and should be a part of any information system. Control techniques may be applied to inputting, processing, and outputting functions.

control techniques	
input controls	1. Define responsibility for input preparation and delivery.
	2. Verify data for significant numeric fields.
	3. Utilize self-checking numbers and codes where practical.
	4. Sequence number input data.
	5. Number and take hash totals for each batch. Keep a transmittal tape for each batch.
processing controls	1. Establish a comprehensive program of computer program and system testing.
	2. Centralize the authority for all computer program changes or modifications.
	3. Protect files with backups and limited accessibility.
	4. Provide standards for utilizing hardware checks.
	5. Provide standards for the development and use of software checks.
	6. Log all machine time usage.
output controls	1. Assign responsibility for output handling and distribution.
	2. Insist upon an auditable trail from outputs back to inputs.
	3. Inspect all outputs for reasonableness.

limit changes must be a part of a processing run separate from the daily transaction runs. Accessibility of the file can be limited through controlling the use of the utility program. A second control, the data edit program, utilizes software checks for data validity, for example, valid inventory codes, customer account numbers, and numeric fields. It also prevents customers from exceeding their credit limit without specific authorization.

Allocation Process

Key project personnel, as a team, determine the most effective means of performing each function. This determination is made by studying and evaluating alternatives. The process follows steps similar to those of the feasibility analysis, but alternatives are studied at a level of greater detail. For example, during the study phase it might have been found that a remote terminal was the most cost-effective means of entering data into the system. Now, in the design phase, a particular terminal must be selected from the many models or types available to perform the input function. As another example, it might be decided that certain data could be evaluated more effectively by manual techniques than by a computer. This could be the case if human judgment or particularly rapid handling of an exceptional situation were required.

Manual Tasks

Manual Task Definition

Typically, humans interface with computer-based systems in at least six principal ways, each involving *manual tasks:*

1. Humans prepare source documents.
2. Humans enter data from source documents for conversion to a machine-readable format.
3. Humans prepare and use control documents.
4. Humans write computer programs.
5. Humans operate equipment.
6. Humans use computer-produced outputs.

All humans who are involved with a computer-based system must know certain aspects of the system. This information is provided by reference manuals.

Reference Manual Identification

All the reference manuals to be used by programmers, users, and operators are identified in the design phase. Their principal content may be outlined. There may be a large number of these manuals because of the varying needs of many different types of users. The manuals are prepared during the development phase, and they are an essential part of personnel training. Training plans and programs, including reference manuals, are discussed in chapter 22, Preparing for Implementation.

Equipment Functions

Equipment Functions Definition

The process of allocating functions between manual tasks, equipment, and the computer program took into account the computer equipment resources currently available, and additional resources that might be required. The range of equipment options includes:

1. Acquire a computer system
2. Current computer equipment is adequate
3. Add some components to the current equipment
4. Greatly modify the current equipment
5. Replace the current equipment by a new computer system

Within each of these options there may be many equipment choices. The process for evaluating specific equipment model choices is similar to that used during the feasibility analysis activities of the study phase.

As an example, consider a situation in which it has been found necessary to re-place the current equipment with a computer system in order to handle an anticipated increase in workload. The equipment selection process would involve the following steps:

1. Define the *equipment functions* to be performed by the computer system
2. Identify the required capabilities of the computer system
3. Evaluate the candidate computer systems
4. Select the computer system to be acquired on the basis of its performance and cost

The definition of the functions to be performed by the system begins with an analysis of the workload that the system is expected to process. This includes the con-tinuing current workload plus the estimated additional future workload. The purpose of analyzing workload is to develop representative samples that can be used to test the performance of various computer configurations. These samples are commonly drawn from either actual programs currently in use, a composite made up to represent the planned workload, or they are a simulation of the workload and computer.

Whichever technique is used, the sample workload should be representative of the tasks the computer must perform. This is called "benchmark" testing because iden-tical series of tasks can be used to compare two or more computers. Typical tasks are compilations, updates, sorts, edits, report generation, and input and output handling. If it is cumbersome to assemble an adequate and representative sample from problems currently being processed, then a composite can be made of tasks currently being per-formed and expected to be performed. This composite is not an actual production pro-gram. It is a series of tasks of the types described above. It can, however, be used as a benchmark for measuring the relative performance of proposed hardware and soft-ware.

Another alternative is the use of a simulation program, which makes one com-puter appear to perform like another computer. The inputs to the simulation program are the input, processing, and output tasks desired, and the frequency of their occur-rence. The result of the simulation is data on timing, price, and performance.

Equipment Specification

After the functions to be performed have been converted to a representative workload sample, the additional requirements for the computer equipment can be stated, and the candidate systems can be evaluated. Figure 16.4 depicts performance and cost fac-tors that typically are considered in evaluating products of vendors.

Workload analysis and measurement were discussed in the preceding section on the definition of equipment functions.

Growth is the ability of the equipment to meet not only the current workload but also future workloads. Three common ways of providing for growth are (1) equipment with reserve power, (2) equipment with add-on capability, and (3) equipment that is compatible for upward conversion.

Reserve power usually is the least economical way of providing for growth, since it involves paying for capability not being used. Add-on capability is attractive because it offers an increase in capability when needed. Compatibility for upward conversion

Figure 16.4

Vendor evaluation matrix. A vendor evaluation matrix is useful to the systems team in comparing the performance and cost factors of equipment to be acquired. Vendor evaluation is another example of a study of alternatives, and it follows the steps discussed in chapter 13 for performing a feasibility analysis.

FACTOR / VENDOR	Vendor A	Vendor B	Vendor C
PERFORMANCE:			
Workload			
Growth			
Hardware			
Software			
Support Services			
COSTS:			
Rental			
Third-Party Lease			
Purchase			

gives the potential for replacing the equipment by a more powerful system on which all of the existing programs will run without modification. Since most vendors provide upward-compatible equipment, this, too, often is a feasible route to follow. However, possible pitfalls, such as subtle differences between operating systems, should be evaluated carefully.

Hardware performance is measured by factors such as memory size, internal speed of the central processor, and idle time waiting for input or output. Software performance depends on the efficiency of the operating system, the quality and quantity of utility programs, the programming languages provided, and the applications packages that the vendor may supply. Support services include field engineering, for example, maintenance, back up, and response to problems; systems (that is, software) support; and education and training of customer employees.

The three most common ways of acquiring computer equipment are rental, lease, and purchase. If the equipment is to be used for an extended period of time, both purchase and lease plans are less expensive than rental. Many lease plans provide an option to purchase. The lease of equipment from someone other than the vendor is called a third-party lease. There are many leasing companies, and a variety of lease plans are available. Third-party leasing can result in cost savings because the leasing company, anticipating a continuing market, often can spread the cost of the equipment over a longer period of time. If the equipment is to be retained for a long period of time, usually in excess of four years, then outright purchase may be the most economical means of acquisition. Among the most important factors that affect cost decisions are changes in technology (which still appear to be producing more performance for less cost), confidence in workload projections, interest rates, flexibility, and lead times for equipment delivery.

The equipment acquisition process has not been described in detail. Rather, by identifying the factors that must be considered, we have emphasized the need for evaluation procedures. Systems analysts must realize they cannot rely entirely upon vendors. They must know their own needs, be able to determine their own performance requirements, and be able to evaluate vendor responses to those requirements.

Computer Program Functions

Computer Program Functions Definition

The computer program component of an information system can be broken down into the functions of its individual computer programs. Each program performs one or more functions. These are typical functions provided by programs:

1. Master file load or creation
2. Master file update
3. Master file back-up
4. Master file maintenance
5. Input data editing
6. Report/screen generation

While systems are unique in their overall functions, these six functions of programs are common to most systems.

Interface Management

An important design phase task is the review of interfaces between the new system and existing or proposed future systems. *Interface management* is particularly important, since the success of the integrated data base system concept depends on the planning of each system so that it is capable of accepting data from, and providing data for, related data-base-oriented systems. OARS provides an example of interface management.

An interface with another system is a concern in the OARS system. The invoice program of OARS produces a shipping invoice and shipping orders containing a detailed listing of all items ordered and shipped. In order to avoid printing shipping invoices and then discovering that some items are out of stock during the distribution process, the OARS is to interface with an existing inventory system. The invoice program is to determine the availability of merchandise prior to printing the invoice. If the merchandise is short or out of stock, that condition should be noted on the shipping invoice, and the customer charges should represent goods actually shipped. If the merchandise is not out of stock, the invoice program is to reduce the inventory by the amount to be shipped.

The inventory system already exists. The OARS project must interface with it in a manner that will not require extensive redesign or modification of the inventory system.

Test Requirements

Identification of Test Requirements

The requirements for tests are established after the allocation of functions is completed. They are established in the following sequence:

1. System test requirements
2. Subsystem test requirements
3. Component test requirements

The requirements for overall system performance are established first because they will determine the requirements for each subsystem. The requirements for testing the components of each subsystem are defined last.

A computer-based business system has manual, equipment, and computer program subsystems. Typically, the most important subsystem is the computer program subsystem. Its components are computer programs.

In the design phase, the tests that must be performed are identified in the sequence indicated above, progressing from the system level to the component level. In the development phase, test plans are prepared to correspond to the test requirements. In this phase actual testing also takes place. The tests are performed in a planned sequence, progressing from the component level to the system level. Development phase testing activities are discussed in chapter 22, Preparing for Implementation.

Structured Walk-throughs

Structured reviews are a technique used in developing efficient and reliable systems. A **structured walk-through** is a technical review to assist the technical people working on a project. It is a "structured" review because it is one of a series of reviews that are a planned part of the system design and development activities. It is referred to as a "walk-through" because the project is reviewed in a step-by-step sequence. Structured walk-throughs can be a valuable tool in the design and development of any system component. They are especially valuable in the design and development of computer programs.

The purpose of a structured walk-through is to discover errors in logic of a computer program or problems with other system components. The underlying philosophy is that others often can see errors that are not obvious to the programmer or analyst. The structured walk-through is a very powerful "test-as-you-go" technique.

The review team consists of selected peers of the project developer. For example, if the project is a computer program, the review group consists of other programmers. The project developer "walks" the review group through the logic of the project. If any errors, omissions, or discrepancies are uncovered, they are recorded by one of the group members. The structured walk-through allows problems to be discovered early when it is easier and less costly to correct them. If the structured walk-through technique is used consistently, fewer errors will be found during system testing, and less debugging time required. It should be noted that successful structured walk-throughs must be positive and nonthreatening experiences for the project developer; therefore, management does not attend, and the review must not be a basis for employee evaluation.

The structured walk-through technique is consistent with the entire life-cycle approach to the design and development of systems. The study, design, and development phase reviews are structured reviews. Structured walk-throughs are meaningful supplements to those reviews for the purpose of examining the technical logic of system components. Structured walk-throughs are particularly valuable in the development phase, which usually is much longer and costlier than the study and design phases. Therefore, it is essential that periodic structured management and technical reviews be held as work progresses.

Summary

The first step in the design phase is system design. System design consists of three major steps:

1. Drawing expanded system flowcharts to identify all of the processing functions required.
2. The allocation of the identified functions between the people in the system, the equipment to be used, and the computer programs.
3. The identification of the test requirements for the system and each of the system components.

Decomposition of data flow diagrams and high-level system flowcharts leads to expanded process-oriented system flowcharts. Manual tasks include the identification of the reference manuals needed by programmers, operators, and users of the system. Equipment function definition includes the development of representative workload samples and the evaluation of vendor products. The functions to be performed by the individual programs that make up the computer program component of the information system are identified, and the interfaces with other systems are reviewed. Test requirements are established at the system, subsystem, and component levels. Structured walk-throughs are a powerful technique for eliminating errors as the computer program component is designed and developed.

For Review

computer program
manual tasks
equipment functions

**expanded system
flowchart**

interface management
**structured walk-
through**
HIPO chart

For Discussion

1. How do the expanded system flowchart and HIPO charts aid in the allocation of functions?
2. Why is it important to plan controls as early as possible in system design?
3. Who determines the most effective means of performing each system function?
4. Name and discuss some ways in which humans interface with computer-based systems.
5. What studies of alternatives may be required in the equipment functions definition activities?
6. What is a "workload sample"? How is it obtained?
7. What are some typical computer program functions?
8. What is interface management?
9. What test requirements must be identified?
10. What is a structured walk-through? What is its purpose?

17
Output Design

Preview

Output design has been an ongoing activity almost from the beginning of the project. In the study phase outputs were identified and described generally in the project directive. A tentative output medium was then selected and sketches made of each output. In the feasibility analysis, a "best" new system was selected; its description identified the input and output media. In the design phase the system design process has included an evaluation and selection of specific equipment for the system. The design team must now refine the sketches into detailed descriptions of the outputs. To do this, we must plan the output with a specific medium. The most common output media are computer printers and CRT screens. Computer print charts and CRT display layout sheets are used for the detailed description of outputs.

Objectives

1. You will learn how to read and draw computer print charts.
2. You will understand the special requirements for preparing CRT display layouts.

Key Terms

display layout sheet   A form used to design CRT screen layouts. The form is divided into 24 lines of 80 characters each to simulate the possible display positions on a screen.

print chart   A form used to design computer printer outputs. Each line on the form is divided into printer print positions to allow for the detailed design of titles, column headings, detail lines, and so forth.

Computer Output

With few exceptions, output designs describe "lines of characters." The most common output medium is the line printer. However, data often is displayed on other devices, such as cathode ray tubes (CRT) and typewriter-like terminals. In all cases, the format descriptions are similar. Chapter 18, Input Design, will also discuss the design considerations of the cathode ray tube and typewriter terminals, which are both input and output devices.

Computer Printer Output

Computer Print Charts

The detailed description of outputs includes the identification of the print positions to be used for the title, column headings, detail data, and totals. Figure 17.1 depicts a typical form used to make this detailed description. The example **print chart** allows for 144 possible printing positions. Computer printers typically have a maximum of either 120 or 132 print positions, and may use forms of much narrower width. A vertical line should be drawn to indicate the desired width of the form. Form length also varies with the needs of users. Common form lengths are 3 1/3, 3 1/2, 3 2/3, 5 1/2, 6, 7, 7 1/2, 8 1/2, and 11 inches.

Older computer printers had a physical device that could read carriage control data from a closed-loop, called a *carriage control tape*. The carriage control data was recorded by punching holes within areas called channels. This tape is described on the left end of print charts. Modern printers no longer have a physical carriage control device. Instead, they store the carriage control information within a small memory buffer. The data in the carriage control buffer functions just as the carriage control tape did on older equipment.

Carriage control data are most often used to identify the first and last line of printing for a form on a computer. Two standard uses of channels are *channel 1*, which indicates the position of the first line to be printed on a page, and *channel 12*, which indicates the position of the last line of print on a page. Channels 2 through 11 are used at the discretion of the programmer.

When printer forms are moved one line at a time, the carriage control tape unit is used only to detect the channel 12 punch—the bottom of the form. The process of moving the form one, two, or three lines at a time is called *spacing*. *Skipping* is the process of releasing the form, or moving it rapidly through the printer carriage until a designated channel punch is detected by the carriage control tape reader. The skipping technique is useful because most printers skip faster than they space when the form is to be moved four lines or more.

As examples, let us consider some of the four OARS printer outputs.

The two parts of figure 17.2 depict the sketch and the data element list of the overcredit notification developed in the OARS system performance definition. This information is the basis for preparing the output design of figure 17.3. The field sizes and formats come from the data element list, and the layout comes from the sketch. The width and length of the form are outlined on the print chart.

Figure 17.1

Print chart. Print charts are used to prepare detailed descriptions of computer-printed outputs. Each of the small squares on the chart represents a possible printing position. The title, column headings, detail, and total lines can be shown exactly as they will appear in the final output.

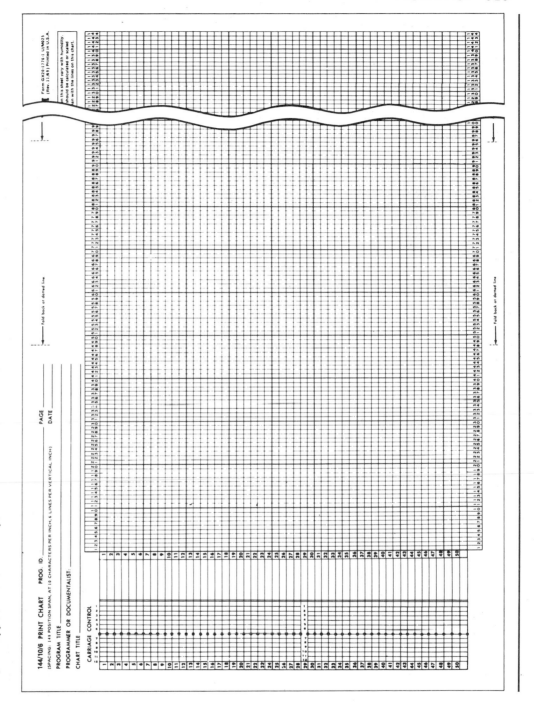

Figure 17.2
Sample overcredit notification. Output design is based upon the output specification and data element list developed during the system performance definition activities of the study phase (figure 12.18).

OUTPUT SPECIFICATION		
TITLE: Overcredit Notification		
LAYOUT:		

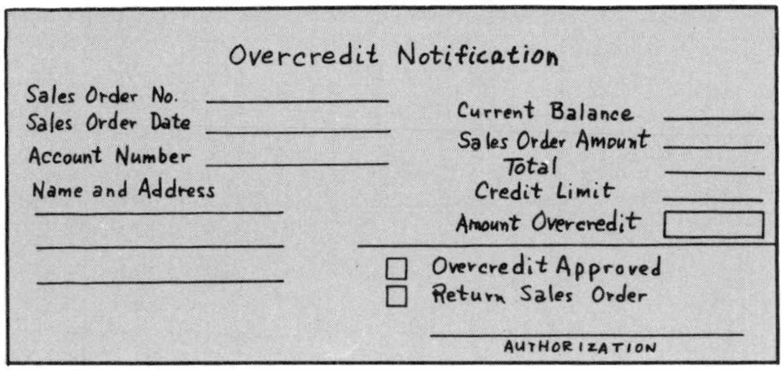

Overcredit Notification

Sales Order No. _____   Current Balance _____
Sales Order Date _____   Sales Order Amount _____
                                                        Total _____
Account Number _____   Credit Limit _____
Name and Address   Amount Overcredit [_____]
_____
_____   ☐ Overcredit Approved
_____   ☐ Return Sales Order

AUTHORIZATION

FREQUENCY: Daily
SIZE: 1 page   **QUANTITY:** 80 max.   **COPIES:** 2
DISTRIBUTION: 1. Accts. Receivable Dept.
COMMENTS:

DATA ELEMENT LIST		
TITLE: Overcredit Notification		
DESCRIPTION	FORMAT	SIZE
Sales Order Number	XXXXXXXX	8 characters
Sales Order Date	xx/xx/xx	8 "
Account No: Region No. – Sequence No.	XX–XXXXXXX	10 "
Name		26 "
Address		
Street		20 "
City		18 "
State		2 "
Zip Code		5 "
Current Balance	XXX,XXX.XX	10 "
Sales Order Amount	XXX,XXX.XX	10 "
Total	XXX,XXX.XX	10 "
Credit Limit	XX,XXX.XX	9 "
Amount Overcredit	XX,XXX.XX	9 "

Figure 17.3

OARS overcredit notification. The OARS overcredit notification print chart was developed from the overcredit notification output specification and data element list shown in figure 17.2.

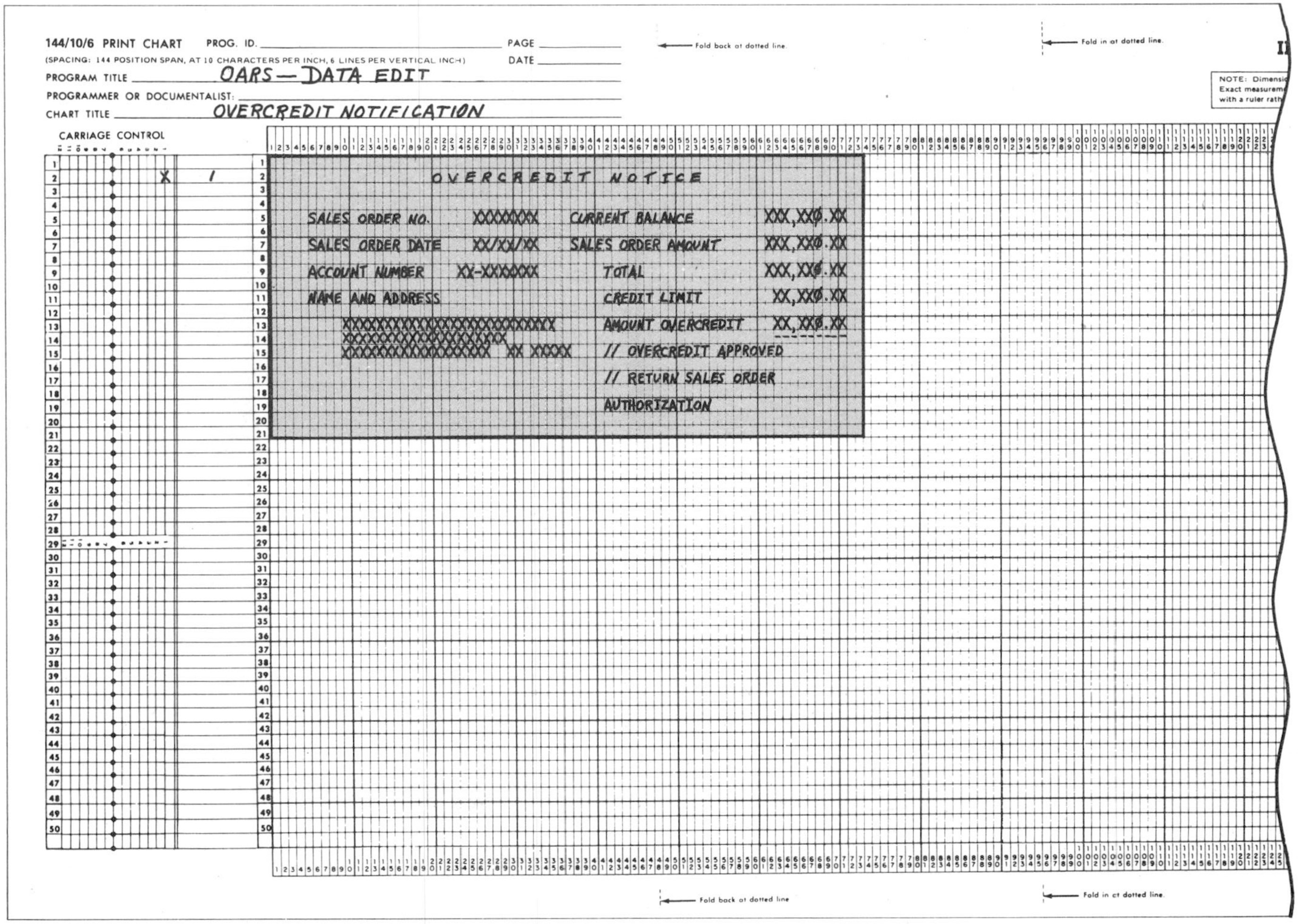

The location of the channel 1 punch in the carriage control tape is on the same line as the first line of printing, the title line. A channel 12 punch to indicate the last line of printing is not required because each line on this form is different from all others. Thus, the computer program logically can determine when the bottom of the form is reached.

Figure 17.4 is an example of a form with several detail lines. When several detail lines of the same format are to be printed, it is not necessary to show all the lines on the print chart. The first two detail lines should be shown to illustrate the spacing of the body. If the report has totals, the last detail line also is required, to show the spacing between detail lines and total lines. The wavy lines between the second and last detail lines indicate that the format is repeated without change.

Edit characters, for example, dollar signs, commas, and decimal points, should be included in field descriptions. For example, if a numeric field could have a value of zero, the number of zeros to be printed should be depicted on the print chart. The placement of a zero in a data field indicates the first significant digit to print. The format XXX,XXO.XX indicates to the programmer that 0.00 should be printed if the field value is zero, 0.99 if the field value is 99 cents.

Specialty Forms

Forms manufacturers can provide a wide variety of specialty forms on a custom-designed basis. Figure 17.5 depicts the OARS customer monthly statement. These are the advantages of this type of form:

1. Column headings can be much more flexible with different sizes and styles of type.
2. Numeric fields may be edited by the form itself, for example, dashed vertical lines to separate dollars and cents.
3. Fewer lines need to be printed by the computer, saving computer time.

The obvious disadvantages are higher cost and the need to maintain a larger forms inventory.

Specialty printer forms include continuous forms with built-in carbon copies (called multi-part forms); envelopes; mailing labels; and forms presealed into envelopes, which are printed through the envelope by carbon copy techniques.

CRT Screen Output

The principles used in designing CRT screen output are similar to those of form design and computer printer output. CRT displays typically include titles, column headings, detail data, and, possibly, totals. These are the same requirements identified with computer print charts. The major differences are: (1) the size of the screen, and (2) the amount of data to be outputted as a record. The typical screen size is 80 characters across and 24 lines down. The CRT display may consist of many detail lines arranged

Figure 17.4

OARS transaction register. The OARS transaction register print chart was developed from the output specification and data element list developed in the study phase (figure 12.14).

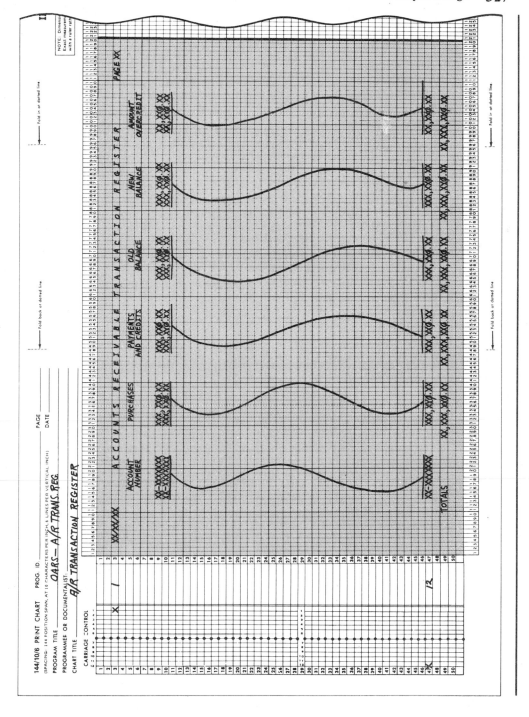

Figure 17.5

Preprinted form. Some computer-printed outputs may use forms that contain preprinted information. These specialty forms are obtainable from forms manufacturers on a custom-designed basis. This figure is a layout for a specialty form on which the titles and column headings will be preprinted.

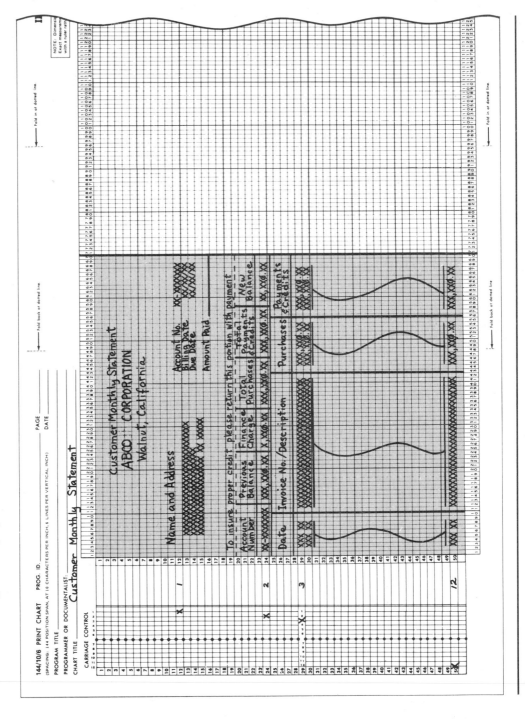

Figure 17.6
Display layout sheet.

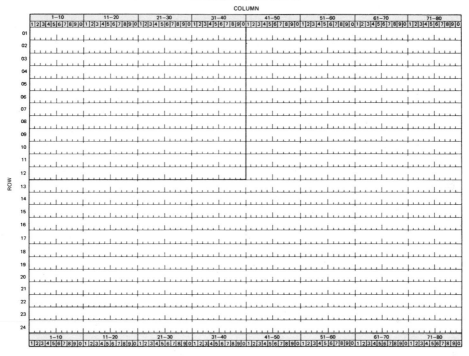

in columns much like a printed output, yet many times only a single record will be displayed. This is especially true with systems that randomly access data. Figure 17.6 is an example of the form for designing CRT displays. This form is called a **display layout sheet,** and it is very similar to the computer print chart in that each display position and line are numbered.

A detailed example of the use of display layout sheets is included in chapter 18, Input Design.

Summary

The output design, an ongoing activity almost from the beginning of the project, follows the principles of form design. Both printed outputs and CRT displays can include a title, column headings, detail data, and totals; they must be described in detail for programmers. Forms called print charts and display layout sheets are used to communicate this design detail. Specialty forms can be designed for printed reports. CRT display layout design differs from that of computer print charts because the area is limited by the size of the screen, as is the amount of data that can be displayed in a record.

For Review

print chart   channel 12   edit characters
carriage control tape   spacing   **display layout sheet**
channel 1   skipping

For Discussion

1. What is the most common output medium?
2. What is the purpose of a print chart?
3. What is the difference between spacing and skipping?
4. What are the "standardized" uses of channels on a carriage control tape?
5. Why should at least two detail lines of a report be shown on a print chart?
6. Under what conditions should an analyst consider the use of a preprinted specialty form?

18
Input Design

Preview

The most common source of data processing errors is inaccurate input data. Effective input design minimizes errors made by data entry operators. This chapter extends the previous discussions of form design and output design to important input devices, particularly the cathode ray tube (CRT) station.

Objectives

1. You will learn input design principles.
2. You will learn the design techniques for CRT displays.
3. You will become aware of the input design considerations for key-to-tape devices, key-to-disk devices, optical readers, and punched card machines.

Key Terms

CRT   Cathode Ray Tube: a television-like input or output station that displays information.

display layout sheet   A form used to design CRT screen layouts. The form is divided into 24 lines of 80 characters each to simulate the possible display positions on a screen.

The Input Design Process

Input design is the process of converting a user-oriented description of the inputs to a computer-based business system into a programmer-oriented specification.

Inaccurate input data is the most common cause of data processing errors. If poor input design—particularly where operators must enter data from source documents—permits bad data to enter a computer system, the outputs produced are of little value. The input design process was initiated in the study phase where, as a part of the feasibility study:

1. Input data were found to be available for establishing and maintaining master and transaction files and for creating output records.
2. The most suitable types of input media, for either off-line or on-line devices, were selected after a study of alternative data capture techniques.

In the design phase, as discussed earlier in this unit, the input design process was continued. Specifically:

1. The expanded system flowchart identified master files (the data base), transaction files, and the computer programs.
2. The input media selected in the study phase were reviewed. Additional studies of alternatives were performed as required, and tasks were allocated among equipment, manual operations, and computer programs.

In this chapter we will describe the rest of the input design process. We will examine cathode ray tube (CRT) display stations and then extend the discussion to other output media, including key-to-tape and key-to-disk devices, optical readers, and punched cards.

Input Design Considerations

In addition to the general form design considerations such as collecting only required data, grouping similar or related data, input design requires consideration of the needs of the data entry operator. Three data entry considerations are:

1. The field length must be documented. The field length must be known to the data entry operator, so that the data entered will not exceed the allocated space and/or numeric data may be right justified where appropriate.
2. The sequence of fields must match the sequence of the fields on the source document. The data entry operator must be able to scan the source document in a logical sequence.
3. The data format must be identified to the data entry operator. That is, a data field that is to be entered in an edited format must be documented. Data choices might be mm/dd/yy or mmddyy.

Figure 18.1

Display station and buffer. Each display station has its own storage called a buffer. The buffer holds the data being displayed on the display station screen.

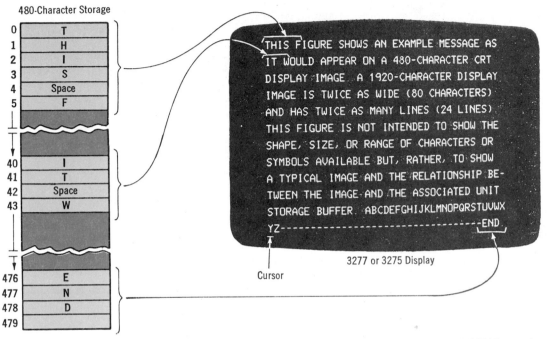

480-Character Storage

0	T
1	H
2	I
3	S
4	Space
5	F
40	I
41	T
42	Space
43	W
476	E
477	N
478	D
479	

THIS FIGURE SHOWS AN EXAMPLE MESSAGE AS IT WOULD APPEAR ON A 480-CHARACTER CRT DISPLAY IMAGE. A 1920-CHARACTER DISPLAY IMAGE IS TWICE AS WIDE (80 CHARACTERS) AND HAS TWICE AS MANY LINES (24 LINES). THIS FIGURE IS NOT INTENDED TO SHOW THE SHAPE, SIZE, OR RANGE OF CHARACTERS OR SYMBOLS AVAILABLE BUT, RATHER, TO SHOW A TYPICAL IMAGE AND THE RELATIONSHIP BE-TWEEN THE IMAGE AND THE ASSOCIATED UNIT STORAGE BUFFER. ABCDEFGHIJKLMNOPQRSTUVWX YZ----------------------------------END

Cursor

3277 or 3275 Display

Courtesy of IBM Corporation

CRT Screen Design

CRT terminals are becoming increasingly important as input devices. Effective screen design not only can reduce data entry errors, it also can increase productivity and user satisfaction. Many of the on-line data entry stations are CRTs that provide both a visual verification of input data and a means of prompting the user. As data is entered it is echoed, or displayed, on the screen. The user can modify or delete any data display before sending it to the computer system for storage and processing. Each *display station* has its own memory, called a buffer, for storing data. In figure 18.1 the size of the example buffer is 480 characters (12 rows of 40 characters each). As the message in the figure indicates, a typical larger-size display is 1,920 characters (24 rows of 80 characters each). The most common size display screen is 24 rows or lines of 80 characters each.

The input design goal is to input data as accurately as possible. To avoid data entry errors, it is common to prompt the operator with concise but clear labels on the screen. In addition, we must communicate to the operator the maximum length of each data element to be entered. Figure 18.2 is an example of a design form used to communicate the required layout to programmers.

Figure 18.2
Display layout sheet. The display layout sheet is used to plan the layout of a CRT screen.
It allows up to 24 lines of 80 characters each. Each small square on the form is the
position of one character. This form is similiar to the print chart introduced in chapter 17.

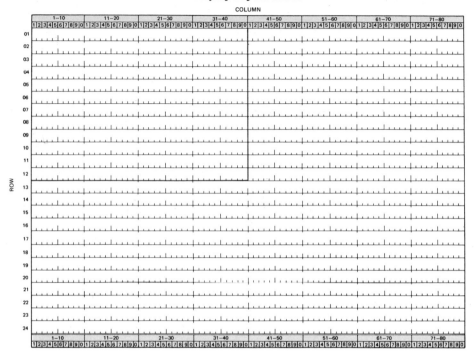

To illustrate the use of a **display layout sheet,** we will use the customer account application data as sample input data. Figure 18.3 depicts the source document for this data. Note that it is not necessary to input all data collected on the form. Only the data blocked off in small squares needs to be inputted. The remaining data are for the use of the personnel department and are not to be entered into the system. Figure 18.4 illustrates the display layout for the customer account application. The screen display is titled and each data element has a clearly stated prompt.

Many modern computer systems have software utilities to assist in the design and development of screen layouts. As an example, IBM has a utility called Screen Design Aid (SDA). With this utility, it is possible to eliminate filling out the display layout sheet. The screen layout can be designed at the terminal. The utility allows the designer to add, delete, or move the display components. When the design is completed, SDA outputs a printed record of the design that looks much like the display layout sheet. In addition, the utility generates the program code to produce the design display. Figure 18.5 is an example of SDA documentation.

Figure 18.3
Customer account application. The customer account application form is a source document for OARS input data. A display layout must be designed to enter the data shown in small squares.

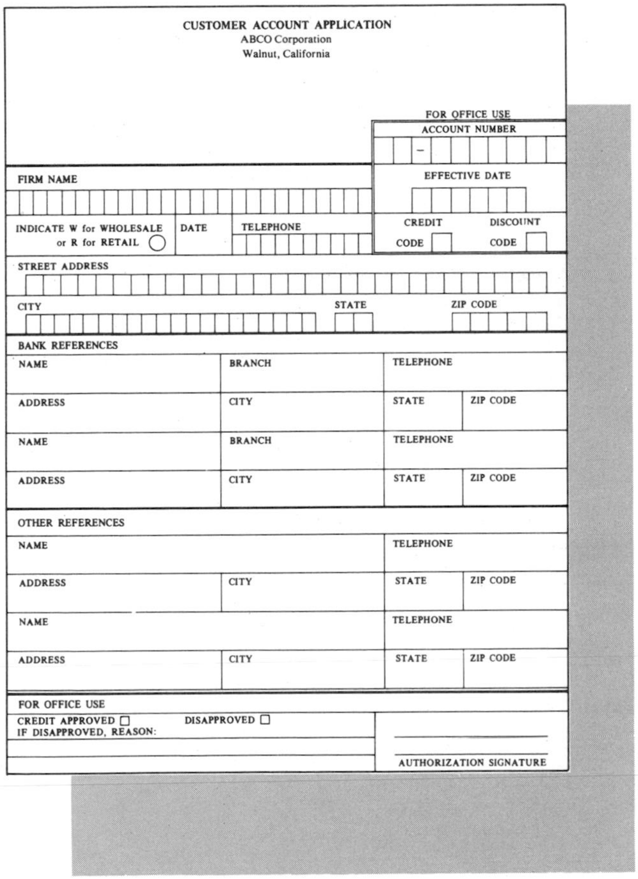
CUSTOMER ACCOUNT APPLICATION
ABCO Corporation
Walnut, California

FOR OFFICE USE
ACCOUNT NUMBER
FIRM NAME
EFFECTIVE DATE
INDICATE W for WHOLESALE or R for RETAIL
DATE
TELEPHONE
CREDIT CODE
DISCOUNT CODE
STREET ADDRESS
CITY
STATE
ZIP CODE
BANK REFERENCES
NAME | BRANCH | TELEPHONE
ADDRESS | CITY | STATE | ZIP CODE
NAME | BRANCH | TELEPHONE
ADDRESS | CITY | STATE | ZIP CODE
OTHER REFERENCES
NAME | TELEPHONE
ADDRESS | CITY | STATE | ZIP CODE
NAME | TELEPHONE
ADDRESS | CITY | STATE | ZIP CODE
FOR OFFICE USE
CREDIT APPROVED ☐ DISAPPROVED ☐
IF DISAPPROVED, REASON:
AUTHORIZATION SIGNATURE

Some screen design rules that are important for user satisfaction are:

1. Use the same formats with related screens. Users expect to find the same data in the same place.
2. Do not overcrowd the screen. Often two neat, eye-pleasing screens look better and are easier to use than one screen filled with too much data.
3. Provide instructions. Tutorial information that is easily accessible is helpful to users, particularly novices or infrequent users.
4. Use consistent terminology. Changing names of terms (for example, "erase" instead of "delete") can be confusing.

Figure 18.4

Customer application display layout sheet. This display layout sheet illustrates the completed design of the input screen for the data collected from the customer account application (figure 18.3).

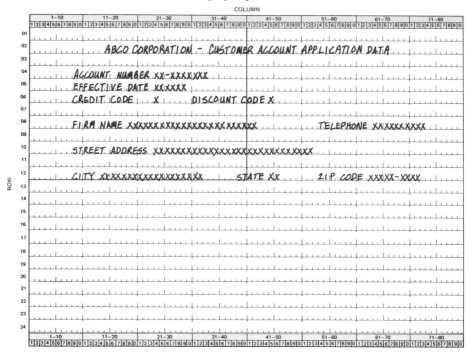

5. Keep instructions brief and grammatically similar. Nouns and action words are useful. Some examples are: inventory record update, customer record add, and sales amount.

6. Coordinate forms and screen design. This is particularly important for data entry, where data elements are common to the source document and to the screen.

Other Input Media

Optical Readers

Optical readers are examples of input devices that can capture data directly. Three important types of optical devices are mark readers, bar-code readers, and character readers. *Optical mark readers* are able to accept data in the form of pencil marks on paper.

Figure 18.5
Screen design aid. Many modern computer systems include utility programs to aid in the design of screens. This example is an IBM utility called screen design aid (SDA). It allows the screen to be designed at a terminal and produces the equivalent of a display layout sheet on a computer printer.

```
      1...+....10....+....20....+....30....+....40....+....50....+....60....+....70....+....80
 1 **                                                                                  **  1
 2 **             ABCO CORPORATION--CUSTOMER ACCOUNT APPLICATION DATA                  **  2
 3 **                                                                                  **  3
 4 **        ACCOUNT NUMBER XX-XXXXXXX                                                 **  4
 5 **        EFFECTIVE DATE XXXXXX                                                     **  5
 6 **        CREDIT CODE   X    DISCOUNT CODE X                                        **  6
 7 **                                                                                  **  7
 8 **        FIRM NAME XXXXXXXXXXXXXXXXXXXXXXXXX      TELEPHONE XXXXXXXXXX              **  8
 9 **                                                                                  **  9
10 **        STREET ADDRESS XXXXXXXXXXXXXXXXXXXXXXXXXXXXXX                             ** 10
11 **                                                                                  ** 11
12 **        CITY XXXXXXXXXXXXXXXXXXX     STATE XX     ZIP CODE XXXXX-XXXX             ** 12
13 **                                                                                  ** 13
14 **                                                                                  ** 14
15 **                                                                                  ** 15
16 **                                                                                  ** 16
17 **                                                                                  ** 17
18 **                                                                                  ** 18
19 **                                                                                  ** 19
20 **                                                                                  ** 20
21 **                                                                                  ** 21
22 **                                                                                  ** 22
23 **                                                                                  ** 23
24 **                                                                                  ** 24
      1...+....10....+....20....+....30....+....40....+....50....+....60....+....70....+....80

          FORMAT . . . . APPLICAT
```

Optical bar-code readers detect combinations of marks by which data is coded. These systems usually are complex to design; the most widely known is the Universal Product Code (UPC), which appears on most retail packages. Figure 18.6 shows several of the variety of sizes and shapes in which bar codes can be printed. The human-readable characters are printed alongside.

Optical character reader (OCR) devices have been designed for applications that can make use of special, optically readable symbols. A typical design application is embossed credit cards, which produce an imprint that can be read by optical scanners. Documents that use special type fonts are in common use; a typical application is in customer billing.

Optical readers are good examples of the expanding trend toward using technology to minimize the role of error-prone humans in creating large volumes of input transaction data. To the extent that human operations can be replaced by machine operations, the integrity of input data, and therefore of system output, can be improved.

Figure 18.6
Bar codes. Optical bar-code readers detect combinations of marks by which data is coded. Bar code systems are commonly found in many types of retail stores.

Punched Cards

For many years punched cards were the common method of preparing data for computer processing. With the increase in cost-effective, on-line, transaction-oriented computer systems, the use of punched cards is declining rapidly. However, punched cards remain an alternate off-line method for data preparation and are useful in special situations. A discussion of punched cards is therefore included in this text.

A punched card contains data or instructions coded in machine-readable format. This data is not read directly by the computer. A device, such as a card reader, converts the data coded on the card into electrical impulses that are transmitted to the computer. The data is recoded and recorded in main memory. Many different types of cards and card readers are available. The most common type of card is called a transcript card.

A transcript card is punched from information previously recorded on another document. It is the most frequently encountered type of punched card. Figure 18.7 displays the punched card equivalent of a cut form. Specialty transcript cards are available from card manufacturers. As is shown in figure 18.8, specialty cards contain preprinted information, such as card name, column headings, and column dividing lines.

Figure 18.7
Stock IBM 5081 card. The most common punched card face is one with printed column numbers and with the row numbers printed for rows 0 through 9.

Figure 18.8
Typical specialty transcript card. Transcript cards that have a custom printed face are called specialty transcript cards. It is common practice to print the company name and the names of the data fields on the card.

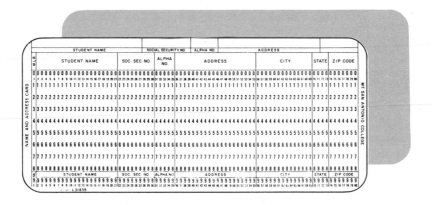

There are no invariable rules for card design. However, some useful guidelines that have emerged from trial-and-error experience can provide a "ballpark" preliminary design. Thereafter, the card, the source document, or both can usually be modified easily if necessary. The result is a card that can be punched rapidly with minimum errors caused by the form.

One OARS source document was shown in figure 18.3. Although the OARS system is designed to use CRT stations for on-line data entry, the source document in

Figure 18.9
Card design aid for customer account card I. Card design aid forms are useful to the analyst when designing the layout of a punched card.

CARD DESIGN AID

CARD NAME *Customer Account Card I*   SOURCE DOCUMENT *Customer Account Application*

DESIGNER *J. Herring*   DATE *2/17/xx*   CARD TYPE *Transcript*

DATA ELEMENTS REQUIRED	Columns in other cards	Sequence on source document	FIELD FORMAT	CARD FIELD SIZE			FINAL DESIGN		REMARKS
				Trial	Trial	Final	Card Columns	Sequence	
Firm Name		1	Alpha	26	26	26	11-36	2	Card 1
Street Address		2	Al-Num	30	—	—			Card 2
Telephone		3	Num	10	10	10	43-52	4	Card 1
City		4	Alpha	20	—	—			Card 2
State		5	Alpha	2	—	—			Card 2
Zip Code		6	Num	5	—	—			Card 2
Effective Date		7	Num	6	6	6	37-42	3	Card 1
Account Number		8	Num	10	10	10	1-10	1	All Cards
Credit Code		9	Alpha	1	1	1	53	5	Card 1
Discount Code		10	Num	1	1	1	54	6	Card 1
Card Code			Num		1	1	80	7	All Cards; not on source document
		TOTALS →		111	55	55			

figure 18.3 can be used to discuss punched card design. Not all the data on the customer account application need to be keypunched for computer processing. For example, the application date and credit references are not stored as part of the data base after the application has been approved. However, the fields required as inputs for accounts receivable processing or data storage must be punched into a customer account card. The fields to be punched are shown in their source document sequence in figure 18.9, which is called a *card design aid.*

    Inspection of the trial field size column of figure 18.9 reveals that all the data cannot be made to fit on one card. The account number is a key field that must appear on all cards. Hence, it is assigned to columns 1 through 10 on the punched card. Since all the address information (street, city, state, and code) will not fit along with firm name, the decision was made to enter this information on the second card. These decisions are reflected in the remarks column of figure 18.9. In addition, a card code field

Figure 18.10

Card design aid for customer account card II. It is not unusual for the amount of data to be recorded to exceed the 80 character limit of the punched card. In such cases, multiple cards must be designed.

DATA ELEMENTS REQUIRED	Columns in other cards	Sequence on source document	FIELD FORMAT	CARD FIELD SIZE			FINAL DESIGN		REMARKS
				Trial	Trial	Final	Card Columns	Sequence	
Account Number	1-10	8	Num	10		10	1-10	1	All Cards
Street Address		2	Al-Num	30		30	11-40	2	
City		4	Alpha	20		20	41-60	3	
State		5	Alpha	2		2	61-62	4	
Zip Code		6	Num	5		5	63-67	5	
Card Code	80		Num	1		1	80	6	All Cards; not on source document
			TOTALS →	68		68			

CARD DESIGN AID

CARD NAME Customer Account Card II   SOURCE DOCUMENT Customer Account Application

DESIGNER J. Herring   DATE 2/17/xx   CARD TYPE Transcript

is assigned to distinguish between the two cards. Although it is not to be shown on the application form, the necessary directions are to be supplied to the keypunch operator. In this case the second trial field size column becomes the final field size column.

Figure 18.10 illustrates the design of the card into which the address portion of the input data is to be punched. The fields previously assigned columns on the first card design aid are entered under the heading "Columns on other cards." In this case the first trial field size column becomes the final field size column.

Figure 18.11 is prepared in part from the data on the card design aids for the customer account card. The customer account card layouts are shown in the upper two nine-rows. Figures 18.12 and 18.13 illustrate the specific instructions given to the keypunch operator.

Figure 18.11

A/R multiple-card layout form. This form is used to document the layout of punched cards. The layouts for up to six cards may be shown on a single form. The form is prepared from the data on the card design aid forms.

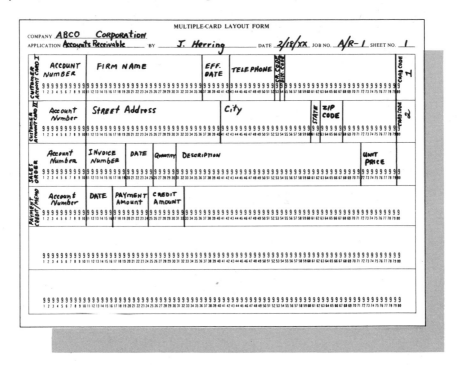

Key-to-Tape and Key-to-Disk Devices

The data rates (that is, the rate at which data can be transferred to the computer memory) of on-line, data entry terminals and machines that read punched cards are much slower than those of magnetic tape and magnetic disk devices. For these reasons, keyboard equipment that can transfer operator keystrokes directly to magnetic tape or magnetic disk often is used.

Key-to-tape devices record data magnetically on reels of tape, cassettes, or cartridges. The data is then entered into the computer at relatively rapid (compared with card readers) data rates.

Key-to-disk devices are often used as input stations on multi-terminal, shared processor systems. Typically, eight or more stations are linked to a minicomputer that provides magnetic disk storage for each station. This computer, in turn, can act as an "intelligent" terminal for a larger, central-site machine. It is able to perform local editing, validation, and correction of data, greatly speeding up the entry of error-free data. Some input data stations are equipped with a cathode ray tube (CRT) screen, on which the data that has been "keyed in" can be inspected visually and verified before it is recorded magnetically.

Figure 18.12
Keypunch instructions for customer account card I. Keypunch instruction forms provide specific instructions to the keypunch operator. These instructions include information on field position, field type, justification requirements, and zero fill requirements.

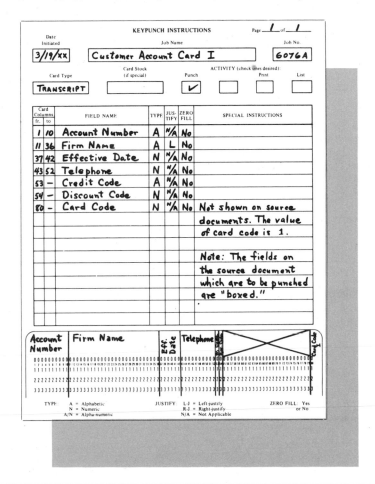

Although key-to-tape and key-to-disk devices provide a means of entering data into a computer without using data terminals or punched cards, operators of these devices work with source documents in much the same way as CRT terminal or key punch operators do. There is an interface with a human, so source documents still must be designed to take into account readability and rapid, accurate keying of data.

Figure 18.13

Keypunch instructions for customer account card II. If multiple cards must be punched for each record, each card layout must have a keypunch instruction sheet. This example is for the second card required for the customer account data.

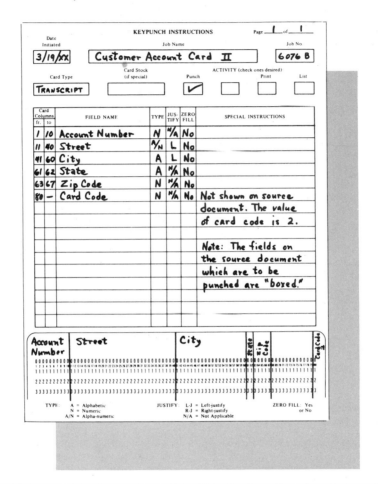

Summary

The most common source of data processing error is inaccurate input data. The objective of input design is to create an input layout that is easy to follow and does not induce operator errors. One of the most common input devices is the CRT terminal. The screen design must include appropriate labels, or prompts, for the data entry operator. The layout of the screen is designed and documented with the use of a display layout sheet or its equivalent. Many computer systems have software packages that assist in the design and documentation process.

Other input media, such as key-to-tape and key-to-disk devices, optical readers, and punched cards are also used and have particular design considerations.

For Review

key-to-tape	cathode ray tube **(CRT)**	optical character
key-to-disk	optical reader	reader (OCR)
display station	optical mark reader	**display layout sheet**
	optical bar code	punched card
		card design aid

For Discussion

1. What is the most common cause of data processing errors?
2. Describe the input design process.
3. What is a display layout sheet? How is it used?
4. What is the purpose of a card design aid?
5. Under what conditions would key-to-tape or key-to-disc devices be selected as input media?

19
File Design

Preview

So far in the design phase the analyst has (1) developed expanded process-oriented flowcharts identifying each task to be accomplished by the system; (2) allocated each task as a manual, equipment, or computer program function; (3) prepared layouts for all system outputs; and (4) designed the inputs for the system. We are now ready to complete the system design by designing the files to be used in the system. The files will be designed by describing the file structure and the records that make up the file.

Objectives

1. You will learn file design concepts, including data formats, record formats, file access methods, and file organizations.
2. You will be able to design sequential or indexed sequential files.
3. You will become familiar with the principles of data base management systems (DBMS).

Key Terms

file   The collection of logically related records, for example, records of a single type, such as employee records, payroll records, and so forth.

logical record   The unit of data to be processed at one time, for example, the payroll data for one employee.

physical record   The unit of data to be inputted to or outputted from the computer's memory at one time. This may be one or more logical records. The physical record is also referred to as a block.

Data Base Management System (DBMS)   Software that allows data descriptions to be independent from computer programs. This system provides the capability for describing logical relationships between files to facilitate efficient maintenance and access to the data base.

Objectives of File Design

In the design of a system's **files,** two types of files must be designed—master files and transaction files. Master files contain relatively permanent data, such as customer names, addresses, balances, and year- or month-to-date information. Transaction files contain data with a limited useful life, typically one processing cycle. Transaction information on sales or payments, for example, is of value only during the current billing cycle.

The objectives of file design are to provide effective auxiliary storage and to contribute to the overall efficiency of the computer program component of the business information system. The auxiliary storage medium must provide efficient access to the data. The data, in turn, must be in a form that minimizes the need for computer program instructions to change data formats.

File Design Concepts

The following concepts are presented as a review of principles that are important in file design, but not as a detailed presentation on data management concepts. The discussion assumes the use of a computer utilizing an eight-bit byte and the Extended Binary Coded Decimal Interchange Code (*EBCDIC*) as the internal storage code.

Data Formats
There are four basic data formats: (1) EBCDIC characters; (2) packed decimal; (3) fixed point; and (4) floating point. The fixed point and floating point formats are used mainly in scientific applications. EBCDIC character and packed decimal formats are, then, the primary concern of the analyst developing a business-oriented system.

EBCDIC characters include all letters, numbers, and special symbols that can be punched into a card or printed. Each character has a four-bit zone and a four-bit digit indication and occupies one byte of storage. Figure 19.1 depicts the zone-digit combinations used to record each character. Note that many of the zone-digit combinations are used for control purposes or are unused. The unused combinations are available for further expansion of the number of characters.

EBCDIC characters 0 through 9 are also referred to as *zoned decimal*. The number 123 would be represented as F1F2F3. Each decimal digit is associated with an F zone; hence, the term of zoned decimal. (Note: Internally the number 123 would be stored in three bytes as 11110001 11110010 11110011. However, the hexadecimal digits 0 through F are used as a shorthand notation for binary 0000 through 1111. Hence, 123 can be referred to as F1F2F3.) One problem with zoned decimal is that most computers do not use zoned decimal numbers in arithmetic operations. They require another format, such as *packed decimal*. Figure 19.2 shows the relationship of packed decimal to zoned decimal and the conversion process used. All zone indications are dropped with the exception of the zone in the low-order byte, that is, the byte occupying the position of lowest place value. The zone of the low-order byte is retained as the sign of the number. A hexadecimal C is a standard plus sign, hexadecimal D is a standard minus sign, and hexadecimal F indicates an unsigned number.

Figure 19.1
EBCDIC graphic characters. Each EBCDIC character is recorded as a zone-digit combination. This chart shows the zone-digit combinations assigned to each graphic character. Many zone-digit combinations do not have an assigned graphic character, but are used as control characters or are available for further expansion.

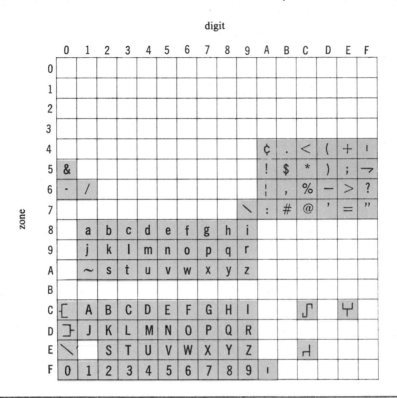

The packed decimal format is important in file design for its contribution to efficiency. First, note that the packed decimal format in figure 19.2 requires fewer bytes of storage than does zoned decimal. This allows more data to be stored in a given amount of space. In addition, input and output operations are more efficient because data records can be read or written in a shorter time. The rule for calculating the number of bytes of storage required for a number in packed decimal is: Add 1 (for the sign) to the number of digits in the number and divide by 2. Always round upward if the division has a remainder.

Second, since numbers in packed decimal are ready to be handled arithmetically, instructions in the computer program to convert from zoned decimal to packed decimal can be eliminated. This contributes to smaller, faster-running computer programs.

Numeric data that is not to be involved in arithmetic operations, dates, for example, are usually left in zoned decimal format unless the space saved by packing is significant.

Figure 19.2
Zoned and packed decimal. Zoned decimal is converted to packed decimal by dropping all zones except the zone holding the sign of the number. A plus sign is indicated with a zone of C; a minus sign is a D; and a sign of F indicates that the value is unsigned.

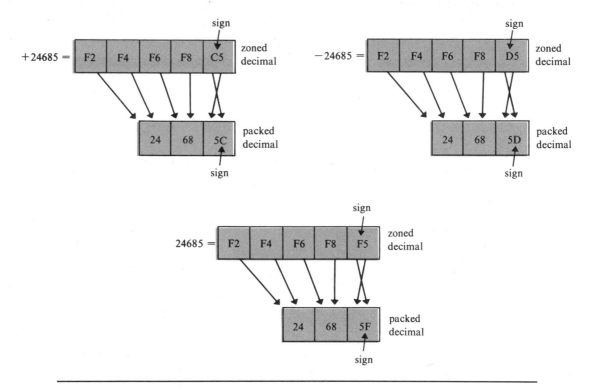

Record Formats

There are two "records" involved with auxiliary storage—logical records and physical records. A **logical record** is defined as the unit of data that is operated upon by a computer program. It is a collection of logically related fields, for example, one customer's accounts receivable data. A **physical record** is the unit of data transferred by one input or output operation. It is made up of one or more logical records. When a physical record contains one logical record, the record is said to be unblocked. If the physical record contains more than one logical record, the records are said to be blocked. The number of logical records in each physical record is called the *blocking factor*. Figure 19.3 illustrates both format choices and identifies the logical and physical records.

The spaces between physical records depicted in figure 19.3 are an important consideration in file design. If the storage medium is magnetic tape, the space is needed as room to start and stop the tape between each physical record that is read or written. This space, called an interrecord gap, is approximately 3/4 inch long. If the blocking factor were increased from one to five, four interrecord gaps totaling three inches of

Figure 19.3

Blocked and unblocked records. Blocked records are physical records that contain more than one logical record. The number of logical records within the physical record is called the blocking factor.

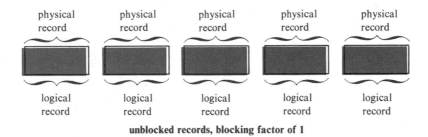

unblocked records, blocking factor of 1

blocked records, blocking factor of 5

tape (4 × 3/4″) would be eliminated. The file length in inches of tape would be reduced three inches for each five logical records in the file, greatly increasing the capacity of the reel of tape. In general, the larger the blocking factor, the more efficiently the magnetic tape is used. Reducing the length of tape to be passed reduces the time required for input or output operations. Unfortunately, the larger the blocking factor, the larger the physical record. Since the entire physical record must be read or written at one time, adequate main storage must be available to hold it. Magnetic tape blocking factors are selected as a compromise between tape efficiency and the amount of main storage available for input or output areas.

Direct access storage devices, such as magnetic disk, have additional limitations. A physical record must be wholly contained on one *track,* the magnetic surface covered by one read-write head with the access arms in one position. Also, address, key length, and data length information are recorded, in addition to gaps, for each physical record. As with magnetic tape, the blocking factor used with magnetic disk is a compromise between disk efficiency and main storage usage. One difference between tape and disk storage is that the disk's efficiency varies as the blocking factor is increased. This variation is due to wasted space at the end of a track, which occurs when the space is too small to hold an additional physical record. Figure 19.4 illustrates the typical change in disk efficiency with changes in blocking factor. The most efficient blocking factors correspond to the charted peaks. The blocking factor to use is the "peak" for which the physical record size does not exceed the available input/output areas.

Figure 19.4

Efficiency of blocking factors. The capacity of a magnetic disk varies with changes in the blocking factor of the physical records. The selected blocking factor should be at one of the charted peaks for the most efficient use of the magnetic disk. In this example, the charted disk capacity is for a 133-byte record. A different logical record size or disk storage system will produce a different plot.

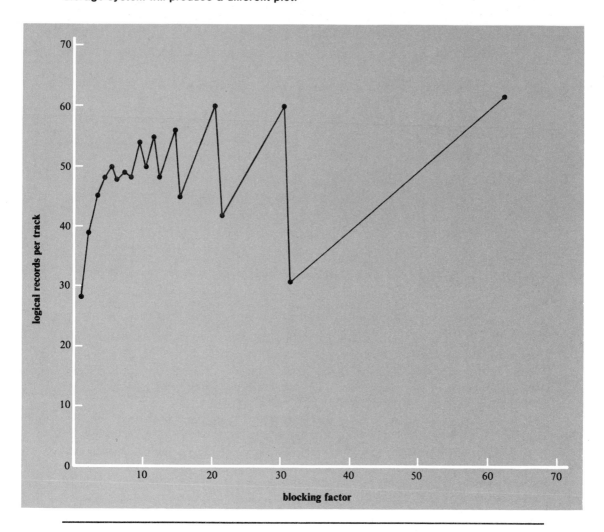

A common installation standard is the specification of a blocking factor for which the physical record does not exceed approximately 3,000 bytes. The examples presented in this chapter assume the use of an IBM 3340 disk drive. The largest physical record that does not exceed 3,000 bytes is a record of 2,678 bytes. Three records of this size will fit on each track. This blocking factor would therefore result in third-track blocking. The OARS example has a logical record size of 133 bytes. A blocking factor of 20 (2,678 ÷ 133 = 20.13) would result in physical records of 2,660 bytes (20 × 133 = 2,660) for third-track blocking.

Calculations relating to blocking factor, track capacity, and physical record size are made by the use of reference data for the particular device. The use of reference data is illustrated in the file design example in this chapter.

File Access Methods

Access methods pertain to the sequence in which records will be written or read. The two choices are sequential and random. The sequential access method is simply to start reading or writing at the beginning of the file, and then continue through the file one record after another, in sequence. If a transaction file is being processed against a master file sequentially, both files must have their records in the same sequence. The random access method provides for the reading or writing of unsequenced records, that is, in random order. The advantage of this method is the ability to go directly to the desired record without handling any other records ahead of it. To locate the desired record, its address must be supplied to the operating system before an attempt is made to read the record. The address may be determined either by calculating it with a mathematical algorithm, or formula, or by looking up the address in a table or index generated at the time the file was created. In either case, the address consists of the cylinder number, the track number, and optionally a record number. A *cylinder* is defined as the surface area covered by all read-write heads in one position of the access mechanism. Figure 19.5 depicts a typical magnetic disk drive.

File Organizations

There are many file organizations in use, but we will discuss only three of the most popular: sequential, direct, and indexed sequential.

Sequential files are created by writing records on the storage medium in sequence according to a control field within each logical record; for example, employee records to be recorded alphabetically would be sorted by the name field (control field). The addresses of the records are not recorded as the file is created. Without these addresses, it is not possible to use random access methods. Therefore, the sequential file must always be accessed by starting at the beginning of the file and processing each record in sequence—the order it occurs on the medium. For this reason, transactions should be batched to accumulate a reasonable number of updates for efficient processing. This organization is the only choice for magnetic tape files.

Direct files are created by calculating the address (cylinder, track, and record number, or cylinder and track number) of the record from its control or key field.

Figure 19.5

Magnetic disk unit schematic. A typical magnetic disk data module consists of multiple magnetic disks and read/write heads. When the moveable access arm unit is in one position, each of the read/write heads cover one track of data. All of these covered tracks make up a cylinder of data.

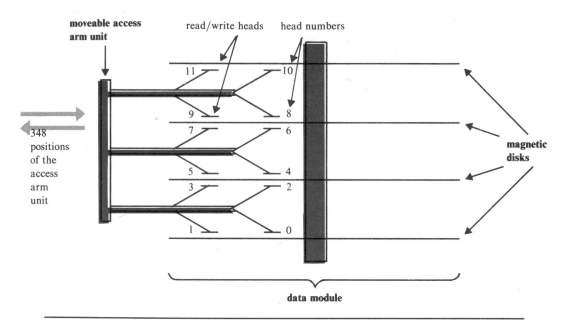

Direct files are always accessed randomly by recalculating the record address each time the record is required; they cannot be accessed sequentially because the records are not physically recorded in any particular sequence. Also, there are probably many addresses throughout the file that are not used, leaving "holes" in the file. In addition, different record keys sometimes produce the same file address. When this occurs, the second record is called a synonym. The synonym record is recorded at the next available location and "chained to" (that is, referred to) the original calculated address by a field indicating the address of the synonym. The only way to reorganize a direct file to eliminate synonyms is to change the algorithm used to calculate the addresses.

Indexed sequential files are created in sequential order just as sequential files are. However, in addition to the file itself, a set of address indexes is created. These indexes may be used to look up the address of any desired record and to access it randomly. If sequential processing is desired, the file may also be accessed sequentially. Thus, the indexed sequential organization is a compromise between a sequential file and a direct file. Figure 19.6 summarizes the advantages and disadvantages of the three file organizations.

Figure 19.6
File organization advantages and disadvantages. There are advantages and disadvantages to all file organizations. Selection should be based on the way in which the file is to be used.

sequential files		
advantages	1.	Efficient use of storage.
	2.	Programming to access records is straightforward.
disadvantages	1.	Transactions must be batched and in file sequence.
	2.	Additions and deletions require rewriting the file.

direct files		
advantages	1.	Can efficiently handle large numbers of unsequenced transactions against a volatile file.
disadvantages	1.	Storage efficiency is reduced by gaps.
	2.	Processing efficiency is reduced by synonyms.
	3.	File may be difficult to reorganize.

indexed sequential files		
advantages	1.	Either sequenced or unsequenced transactions may be routinely processed.
	2.	Programming is relatively straightforward.
	3.	File may be reorganized easily.
disadvantages	1.	Unsequenced transactions tend to reduce processing efficiency.
	2.	A volatile file may have to be re-created frequently.

File Design Examples

Indexed Sequential File Organization

The following example file design is for the ABCO corporation's on-line accounts receivable system (OARS).

Logical record layout. The first step in file design is to determine the record layout for the file. Figure 19.7 depicts a typical worksheet for documenting this record layout. The worksheet has enough room for two records, each with a maximum length of 256 bytes. Each byte of the record is pictured and numbered, and its data format (characteristic) is identified.

The data format of each field must be selected before the field size in bytes can be determined. Figure 19.8 depicts the same worksheets that were used to make a gross storage estimate during the feasibility analysis. They identify the master file record fields and indicate the number of characters in each field (excluding edit symbols in numeric fields). The worksheets are updated to reflect any changes made since the feasibility analysis; they describe the master file record. The analyst indicates on the worksheets the numeric fields to be in packed decimal format and the number of bytes required. In figure 19.8, all fields to be packed are circled, and the packed field sizes are indicated. Figure 19.9 depicts

Figure 19.7

Record layout worksheet. Record layout worksheets are used to document magnetic disk files. The upper portion of the worksheet has enough room to describe two logical records. The bottom portion of the form is used to document file characteristics such as blocking factor, characters per record, and file identification.

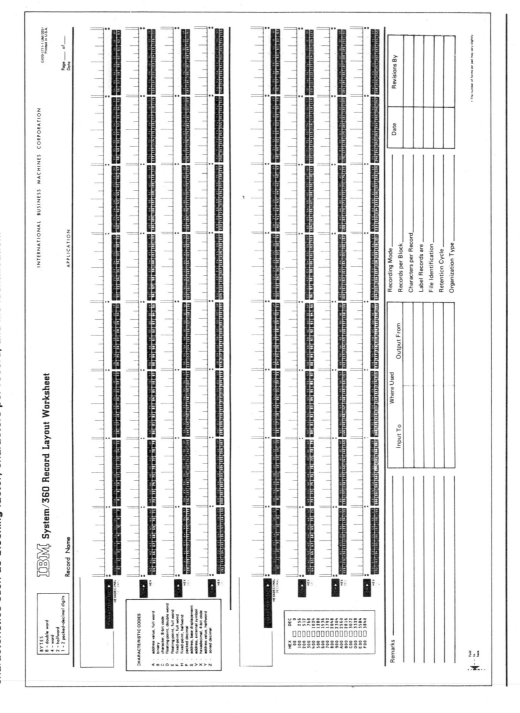

Figure 19.8

Output data element source analysis. The output data element source analysis sheets prepared in the design phase can be used to identify master and transaction file record data elements; they also indicate the number of characters for each data element. The output data element source analysis sheets shown are the same as those in figure 13.8. Of course, the worksheets should be updated to reflect any changes made since the feasibility analysis.

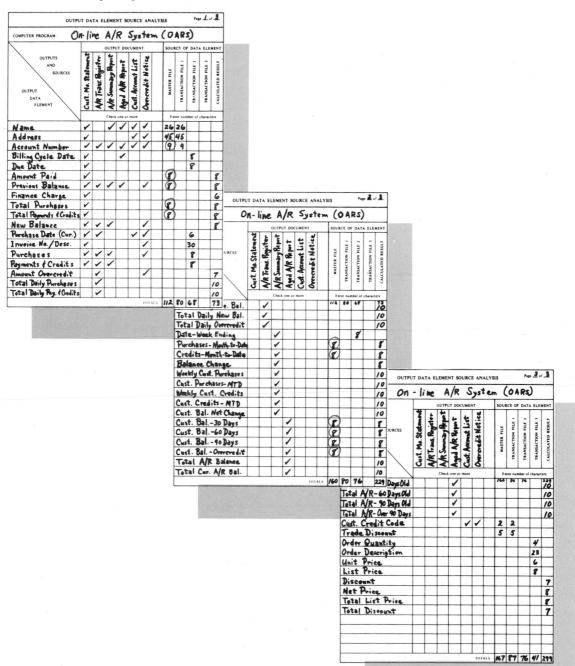

Figure 19.9

OARS record layout worksheet. This disk record layout worksheet was prepared to document the OARS master file and its records. Fields and field sizes were taken from the output data element source analysis sheets of figure 19.8.

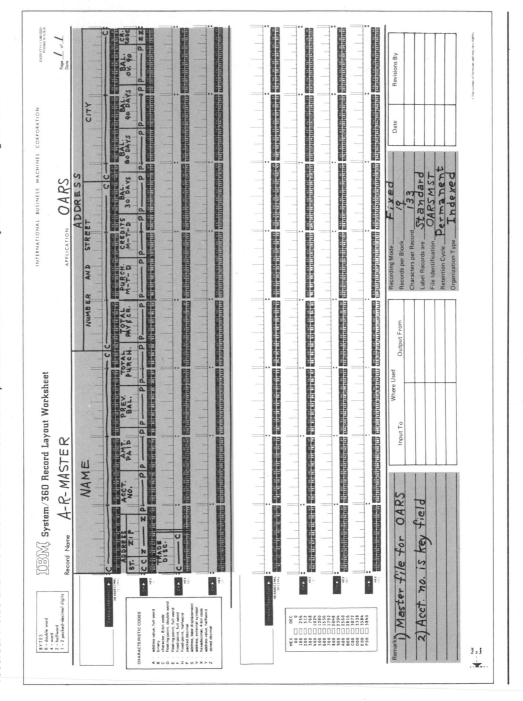

the master file record layout for the OARS project. Each field is identified. Alphabetic or alphanumeric fields have a character (C) characteristic. Numeric fields have either a zoned (Z) or a packed (P) characteristic. Note that the byte count is numbered relative to zero; hence, the record length is 133 bytes even though the last byte in the example record is number 132.

File organization selection. The advantages and disadvantages of sequential, direct, and indexed sequential organizations were listed in figure 19.6. The analyst must evaluate the processing requirements of the systems computer program components identified in the system design activities. High activity files—those with a high percentage of their records processed each run—are very efficient as sequential files. Files that are processed against large numbers of unsequenced transactions or that are volatile (that is, with a large number of records added or deleted) are most efficient as direct files. If the file has both sequential and random processing requirements, an indexed sequential file often is the best choice.

Since magnetic tape can only support sequential files, magnetic disk was the storage medium chosen. The OARS master file organization selection was indexed sequential. The requirements of low activity processing of sales orders and payments combined with the very high activity of customer statement processing, plus the rapid growth projected for the customer accounts, made this the best choice.

Blocking factor selection and file size calculation. The ABCO corporation adopted the standard of third-track blocking for all disk files; in other words, no physical record will be larger than one-third a disk track. This standard was adopted because it was felt that approximately 2,600 bytes (one-third track) per input/output area should be the maximum main storage to be used for this purpose, considering the total main storage available.

There are two sets of reference data, one "with keys" and one "without keys." Recording a copy of the key or control field with the physical record helps to locate records when random access methods are being used. Reference data without keys is used for sequential files. With keys reference data is used with direct files and indexed sequential files. Since the OARS master file organization is indexed sequential, with keys reference data is used.

Figure 19.10, gives the DASD reference data used to select blocking factors and calculate file size for the OARS. Since each physical record can take up to one-third track, there will be three physical records per track. Using figure 19.10, we locate 3 in the records-per-track column. The corresponding entry in the maximum bytes-per-record column indicates a maximum *physical record* size of 2,603 bytes. Figure 19.11 illustrates the remaining calculations required to determine the master file size. The blocking factor is equal to the maximum physical record size (from the reference data table) divided by the logical record size (from the record layout worksheet). Since physical records must be multiples of logical records, disregard any fraction as remainder. The physical record size is equal to the blocking factor times the logical record size plus the key field length. The

Figure 19.10 (upper left)

Typical disk reference data: with keys. This reference data is used to select the blocking factor and to calculate the file size for direct and indexed files.

Figure 19.11 (lower right)

OARS file calculations. The OARS file calculations for the master file size were made using the reference data provided in figure 19.10.

REFERENCE DATA: WITH KEYS

BYTES PER RECORD		RECORDS PER	
MINIMUM	MAXIMUM	TRACK	CYLINDER
4026	8293	1	12
2604	4025	2	24
1892	2603	3	36
1466	1891	4	48
1181	1465	5	60
978	1180	6	72
825	977	7	84
707	824	8	96
612	706	9	108
534	611	10	120
470	533	11	132
415	469	12	144
368	414	13	156
328	367	14	168
292	327	15	180
261	291	16	192
233	260	17	204
208	232	18	
185	207	19	
165	184	20	
146	164	21	
130	145	22	
114	129	23	
100	113	24	
87	99	25	
75	86	26	
63	74	27	
53	62	28	
43	52	29	
34	42	30	
25	33	31	
17	24	32	
10	16	33	
2	9	34	

```
Disk Storage:  With Keys

third-track blocking      =   2,603 bytes maximum

logical record size       =   133 bytes

blocking factor           =   maximum physical record size/logical
                                  record size
                          =   2,603/133
                          =   19.57
                          =   19

physical record size      =   (blocking factor x logical record
                                  size) + key field length
                          =   (19 x 133) + 5
                          =   2,527 + 5
                          =   2,532 bytes

logical records per track =   blocking factor x physical
                                  records per track
                          =   19 x 3
                          =   57

file size in tracks       =   (logical records/load factor)/
                                  logical records per track
                          =   (10,000/.90)57
                          =   11,111/57
                          =   194.92
                          =   194

file size in cylinders    =   file size in tracks/data tracks
                                  per cylinder
                          =   194/9
                          =   21.55
                          =   22
```

key field for the OARS is the customer account number, which is 5 bytes in length. (Note: If the physical record size exceeds the maximum third-track physical record size after the key field length is added, either the blocking factor or the logical record must be reduced in size, and the calculations repeated.)

The number of logical records per track is determined by multiplying the blocking factor by the number of physical records per track (3 for third-track blocking).

The next step is to determine a load factor for the file. The load factor is the percentage of the maximum capacity of the file to be utilized. A load factor of ninety percent was selected for the OARS. This gives a ten percent margin if the projected size of the file is underestimated. The file size, in tracks, is equal to the projected number of logical records divided by the load factor and by the number of logical records per track. Fractional file sizes are rounded upward.

On the ABCO disk drive, each cylinder of storage has 12 tracks. With an indexed sequential file, however, all 12 tracks are not available for data (prime data area). The track covered by head 0 is used as an index (track index), and the last tracks of each cylinder are used as a cylinder overflow area. An overflow area is usually required whenever records are added to the file. Typically, the last two tracks (heads 10–11) are used as the cylinder overflow area. Thus, out of the 12 tracks per cylinder, only 9 are available as the prime data area.

The file size in cylinders is the file size in tracks divided by the number of data tracks per cylinder (9 in this case). Indexed sequential files are required to be multiples of whole cylinders. Thus, the calculated file size in cylinders is always rounded upward. The OARS master file size will be 22 cylinders. There are other indexes (one required, one optional) used with indexed sequential files, but the index sizes are small and do not significantly alter the calculated file size.

Sequential File Organization

If a sequential file organization had been selected for the OARS, the only difference, as illustrated in the following figures, would be in the file size calculations. Figure 19.12 was used to perform the calculations summarized in figure 19.13.

There are four differences in calculations with a sequential file as compared with an indexed sequential file. For the sequential file:

1. The maximum physical record size for third-track blocking is larger.
2. There is no key field length to add to the calculated physical record size.
3. All 12 tracks of each cylinder are available for data.
4. The file size in cylinders need not be rounded up to a whole number of cylinders.

Data Base Management Systems

As the number of computer-based systems within a company increases, larger amounts of data must be stored in support of them. The maintenance and control of a large, complex set of files is a costly and difficult task. Often data elements are recorded in several different files. For instance, employee name and employee number may be recorded in the personnel file, the payroll file, and other employee-related files. These

Figure 19.12 (upper left)
Typical disk reference data: without keys. This reference data is used to select the blocking factor and to calculate the file size for sequential files.

Figure 19.13 (lower right)
Sequential file calculations. These file calculations were made for a sequential OARS master file using the reference data provided in figure 19.12.

REFERENCE DATA: WITHOUT KEYS

BYTES PER RECORD		RECORDS PER	
MINIMUM	MAXIMUM	TRACK	CYLINDER
4101	8368	1	12
2679	4100	2	24
1967	2678	3	36
1541	1966	4	48
1256	1540	5	60
1053	1255	6	72
900	1052	7	84
782	899	8	96
687	781	9	108
609	686	10	120
545	608	11	132
490	544	12	144
443	489	13	156
403	442	14	168
367	402	15	180
336	366	16	192
308	335	17	204
283	307	18	216
260	282	19	228
240	259	20	
221	239	21	
205	220	22	
189	204	23	
175	188	24	
162	174	25	
150	161	26	
138	149	27	
128	137	28	
118	127	29	
109	117	30	
100	108	31	
92	99	32	
85	91	33	
77	84	34	
71	76	35	
64	70	36	
58	63	37	
52	57	38	
47	51	39	
42	46	40	
37	41	41	
32	36	42	
27	31	43	
23	26	44	
19	22	45	
15	18	46	
11	14	47	
8	10	48	
4	7	49	
1	3	50	

```
Disk Storage:  Without Keys

third-track blocking      =   2,678 bytes

logical record size       =   133 bytes

blocking factor           =   maximum physical record size/logical
                                  record size
                          =   2,678/133
                          =   20.33
                          =   20

physical record size      =   blocking factor x logical
                                  record size
                          =   20 x 133
                          =   2,660 bytes

logical records per track =   blocking factor x physical records
                                  per track
                          =   20 x 3
                          =   60

file size in tracks       =   (logical records/load factor)/
                                  logical records per track
                          =   (10,000/.90)/60
                          =   11,111/60
                          =   185.18
                          =   186 tracks

file size in cylinders    =   file size in tracks/data tracks
                                  per cylinder
                          =   186/10
                          =   18.6 cylinders
```

redundant or duplicated data items cause two major problems: (1) they increase the total amount of file storage space needed; and (2) they necessitate multiple updates whenever a change occurs.

A solution for the problem of redundant data is to combine files with common elements into larger, shared files. To eliminate all redundant data, all files could be combined into one large file. However, the creation of large files that are shared by multiple systems introduces two additional problems. First, each program using the file must input and hold all the data in the record, not just the data elements used in its processing. This increases the time required to input or output data and causes a greater amount of main storage to be used for input/output areas. Second, data elements that do not apply to a system may become available to system users. This is especially true for on-line inquiry system (that is, terminals in user locations), where users might see confidential data they are not authorized to use. Considering the public's right to privacy and the several laws that require privacy of confidential data, this must be a major concern for the corporation.

For many companies, the solution to the problems of maintaining and controlling the data base has been the use of a data base management system. A **data base management system (DBMS)** is a software system used to maintain and control large, shared files.

DBMS Functions

Data base management systems have two major functions: (1) to maintain the data base independently from the application programs that use the data; and (2) to provide a measure of data security so that unauthorized users will not have access to the data.

In traditional application programs, all files, records, and data elements used by the program must be described within the program. If the size of the record or any data element changes, all the application programs using the file must be modified to reflect the change. In a shared file, changes may have been made in only one system, but data description changes are required in all programs that use the file.

Data base management systems avoid this problem by removing the data descriptions from the application programs and putting them in the DBMS. This allows the data to be described once, within the DBMS, instead of within every application program. If any changes are made to the data base, only the DBMS description must be altered. Application programs are not affected. The DBMS description of the data elements and the relationships between the data elements in the data base is called the schema.

Since the application program no longer contains data descriptions, it cannot directly access the data base. When an input or output dealing with the data base is required, the application program will "call" for the data from the DBMS. The DBMS will then access the data base and transfer the data to or from the application program. It should be noted that the application program cannot be written in all languages. Each data base management system is designed to interface with one or more host languages. All data base management systems support COBOL as a host language; many support FORTRAN, and a few support RPG and other languages.

Although the application program does not contain data descriptions, it does use a list of data elements that make up the program's logical record. The description of the data elements available to the program is called the subschema. The data elements of the logical record described in the subschema may actually come from several different records in several different files. The DBMS "manufactures" a logical record for the application program to match the subschema. If the contents of the subschema are controlled, control can be exercised also over the data that the user can see. Data elements that do not apply to the user's program or data that the user is not authorized to see will not be provided to the application program. If different subschema are provided to different users, what the user sees is controlled. Users can be limited in what they do with the data. They may be limited to inquiry only, allowed to update data elements, add records to the file, delete records from the file, or any combination of these functions.

An additional security function in most data base management systems is provided by a system of privacy locks or a password system. Before users can access data through a terminal, for example, they are required to enter personal passwords. The password communicates to the DBMS the level of authority the user has in dealing with the data.

DBMS Components

Four major components are common to most data base management systems. The first two are a data description component and a data manipulation component. The use of these components is illustrated in figure 19.14 (DBMS) modules. Application programs call for data through the data description component, which communicates the data requirements to the data manipulation component. All data retrievals, updates, additions, and deletions are accomplished through the data manipulation component. This component then transmits the requested data to the application program, which processes the data and can output the results to the user. These two components are the heart of the DBMS.

A third component is a query language. This is a simplified programming language that allows users to specify the data wanted and the format that will meet user information needs. The query language is easy to use and typically requires only a few key words to create a user output. It is especially valuable in on-line systems where a user can specify the data through the terminal and see the resulting output almost immediately.

The last component is a DBMS utility. It is a series of programs used to create, back up, and restore the data base. The utility programs allow the data processing center to protect itself against possible loss of the data base.

DBMS Architectures

Data base management systems allow the programmer to view the data base as a set of related files. The DBMS architecture describes how the files relate to each other. There are three principle architectures: hierarchical, networked, and relational.

A hierarchical DBMS links files through a superior-subordinate relationship. As an example, the OARS system has a customer file and a file of orders from those customers. The relationship between these two files could be established with the customer

Figure 19.14
Data base management system (DBMS) modules. Data base management systems
consist of two major modules. The data manipulation component makes all data
retrievals, updates, additions, and deletions to the data base. The data description
component communicates the data requirements of the application program to the data
manipulation component.

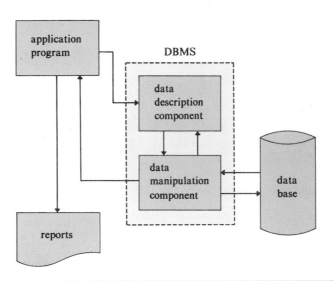

file records being the superior, and the orders for that customer the subordinate. Thus,
if a customer's record is accessed, the DBMS will provide links to all orders for that
customer. In a hierarchical organization the only way to access subordinate records is
through a superior record. A subordinate file may only have one superior file.

The networked architecture is similar to the hierarchical, except that each subordinate record can have multiple superior records. This would allow the access of
customer records from order records, as well as the reverse. In addition, the order file
could be linked to other superior files, such as a file of suppliers.

The relational architecture is the newest of the popular DBMS types. In the hierarchical and networked architectures, the links from one file to another are established within the DBMS. In the relational architecture, the data within the file is used
to associate it with other files. Thus, two files that contain common data element names
are automatically associated with each other. This architecture provides a great deal
of flexibility in that new files may be added to the data base and be immediately linked
to all other files. One disadvantage of the relational architecture has been that performance (or response time) has been slow.

Almost all of the current DBMS are software-oriented. However, some DBMS
suppliers are developing hardware-oriented Data Base Management Systems. These
hardware-oriented systems should be more compact, higher performance systems as
compared to software systems.

It should be noted that there are several suppliers of data base management systems. Each of these systems differs from the others in their approach, capabilities, and/or cost. Converting to a DBMS environment from a nonDBMS environment is a complex task that must be planned carefully. Make sure that the best system is selected for your organization and that you have appropriate personnel to install it. Once converted to any DBMS it is very difficult, if not impossible, to convert to an alternate DBMS.

The Data Base Administrator (DBA)

The data base management system separates data descriptions from the application programs and provides a measure of data security. The **data base administrator** (**DBA**) is the authority that regulates the DBMS by controlling the data base schema and subschema. The DBA also authorizes the use of passwords by users. The job of data base administrator may range from a part-time position in smaller companies to full-time administration with a staff of several assistants in larger companies. The data base administrator's job is to provide services to both data processing personnel and users; this person must be able to work with both groups. The administrator has the ultimate responsibility for the organization and control of the data base.

Summary

The two objectives of file design are effective auxiliary storage and efficient computer programs. Important concepts of file design are data formats and record formats. The basic data formats include (1) EBCDIC characters, (2) packed decimal, (3) fixed point, and (4) floating point. The two record formats are logical and physical. The former is the unit of data operated in by the computer; the latter is the unit of data transferred by one input or output operation. Three common methods of file organization are sequential, direct, and indexed-sequential.

File design proceeds in the following steps:

1. Selection of a data format for each numeric data field—characters, packed-decimal, fixed-point, or floating-point.
2. Selection of the record format—the number of logical records per physical record (blocking factor).
3. Selection of the access method to be used—sequential or random.
4. Selection of the file organization—sequential, direct, or indexed sequential.

After making the above selection decisions, the analyst performs the actual file design, using a record layout worksheet.

In addition, it may be appropriate to consider a data base management system to maintain the several files of data for the organization. If a DBMS is to be selected, it is necessary to evaluate and select a DBMS vendor. There are many choices covering architecture, performance, and cost. The selection must be made only after careful consideration of data base needs.

For Review

For Discussion

1. What are the objectives of file design?
2. In what ways is packed decimal more efficient than zoned decimal?
3. Explain the relationship between a logical record and a physical record.
4. How are record addresses supplied to the computer system to locate records randomly?
5. Which file organization is the simplest? The fastest with unsequenced transactions? Capable of both sequential and random access?
6. What is a DBMS?
7. What are the major functions of a DBMS?
8. What are the functions of the data base administrator (DBA)?

20
Design Phase Report and Review

Preview

At the conclusion of the design phase activities the analyst prepares a report and reviews it with users of the computer-based information systems. The central element of the design phase report is the design specification, which is a "build to" specification. The design phase review is attended by principal users and managers who will be affected by the system; the outcome of the review is a decision whether or not to proceed with the development phase.

Objectives

1. You will become familiar with the content of a design specification.
2. You will learn how to prepare a study phase report.
3. You will understand the purposes of a study phase review.

Key Terms

design phase report   A report prepared at the end of the design phase; it is an extension of the study phase report and summarizes the results of the design phase activities.

design specification   A baseline specification that serves as a "blueprint" for the construction of a computer-based business information system.

design phase review   A review for the dual purpose of presenting results of design phase activities and determining future action.

Design Specification

The *design phase* activities are concluded by three major events. These are (1) completion of the design specification, (2) preparation of the design phase report, and (3) conduct of the design phase review.

The **design specification** is the technical core of the design phase report. This second major baseline document is the "blueprint" for constructing the computer-based business system. It is the communication link between the analyst and the programmers who will be assigned to the project.

Figure 20.1 displays the content of a typical design specification. The design specification is an extension of the performance specification prepared at the conclusion of the study phase. It also is divided into two parts. The first part is an *external*

Figure 20.1
Design specification outline. The design specification is the technical core of the design phase report and an expansion of the performance specification. It has two major sections: the external (to the computer program component) design requirement, and the internal (to the computer program component) design requirement.

design specifications

a. *external design requirement*
  1. flowchart: information-oriented and/or data flow diagram
  2. system output requirements
  3. system input requirements
  4. system interface requirements
  5. system test requirements
  6. equipment specifications
  7. personnel and training requirements

b. *internal design requirements*
  1. computer program component design requirement
    a. flowchart: process-oriented and/or data flow diagram
    b. expanded system flowcharts (process and/or HIPO)
    c. data base requirements
    d. computer program control requirements
    e. computer program test requirements
  2. computer program design requirements (for each program, as required)
    a. detailed system flowcharts
    b. transaction file requirements
    c. control requirements
    d. interface requirements
    e. test requirements
    f. special requirements

design requirement that relates to the interaction between the system and its operating environment. The second part of the design specification is an *internal design requirement*. This part establishes design requirements for the overall computer program component of the business information system and for the individual programs that are the building blocks for this component. The requirements for the computer programs result from the design phase activities, which were described in chapters 16, 17, 18, and 19. The continuing documentation of these activities resulted in the expansion of the *performance specification* into the design specification. This expansion is displayed in figure 20.2. This chart is called a "*gozinto*" chart because it shows how each element of the performance specification goes into ("gozinto") the design specification. The gozinto chart is an excellent illustration of cumulative documentation. Note that the performance specification "descriptions" become design phase "requirements." Not surprisingly, the greatest expansion during the design phase is internal to the computer program component of the business information system.

Design Phase Report

Structure and Content
The structure and content of the *design phase report* are shown in figure 20.3. As shown in this figure, the design phase report has the same five major parts as the study phase report.

Elements of the study phase report, appropriately expanded or modified, are carried forward into the design phase report. For instance, the system scope section is brought forward to refamiliarize reviewers with the project. Of course, the system scope section in the design phase report should identify and explain any changes that occurred during this phase. As mentioned earlier, the performance specification "gozinto" the design specification.

The life-cycle project plan and the cost schedule, which were prepared at the conclusion of the study phase, are updated to show progress in reaching the design phase milestones. Of course, significant departures from or changes in the plan should be noted and explained. A detailed milestone plan and cost schedules are now prepared and presented for the development phase, for which authorization to proceed is being requested. Figure 20.4 lists appropriate milestones. These are described in chapter 21, Development Phase Overview.

Appendices should be included in the design phase report as required. It usually is a good idea to place complicated analyses in appendices. These analyses can be referred to, and the significant results can be presented in the body of the report, without diverting the reader from the "mainstream" message. Other materials, such as tables and charts that support the conclusions and recommendations of the design phase report, should be placed in an appendix unless it is appropriate to present them in the body of the report.

Example Design Phase Report
An example of a design phase report based on the ABCO corporation's on-line accounts receivable system (OARS) appears on the following pages as Exhibit 2.

Figure 20.2
Performance specification—design specification gozinto chart. This gozinto ("goes into") chart illustrates the concept of cumulative documentation. It shows, in detail, how the design phase specification builds upon the performance specification prepared during the study phase.

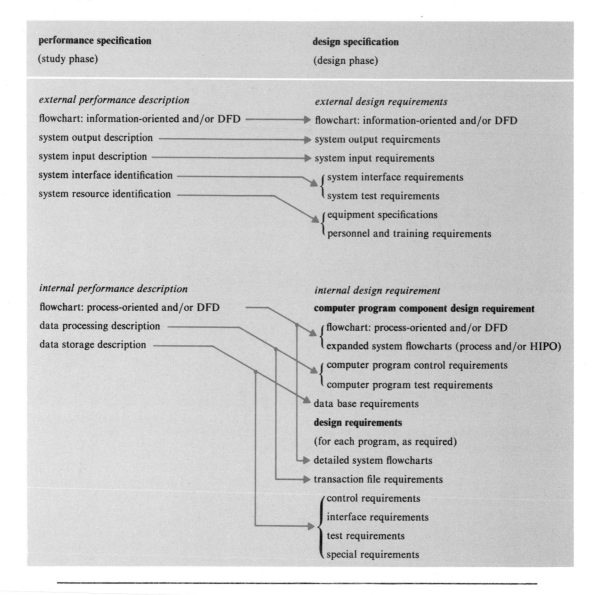

Figure 20.3
Design phase report outline. The design phase report contains the same five major sections as the study phase report: system scope; conclusions and recommendations; design specification; plans and cost schedules; and appendices. However, the content is expanded to include the results of all design phase activities.

design phase report	
i. *system scope*	a. system title
	b. problem statement and purpose
	c. constraints
	d. specific objectives
	e. method of evaluation
ii. *conclusions and recommendations*	a. conclusions
	b. recommendations
iii. *design specification*	a. external design requirement
	b. internal design requirement
iv. *plans and cost schedules*	a. detailed milestones—design phase
	b. major milestones—all phases
	c. detailed milestones—development phase
v. *appendices*	as appropriate

Design Phase Review

The **design phase review** is a particularly critical review. It is a true test of sponsor confidence. Up to this point, the computer-based business system activities were "visible" to the user-sponsor. The principal user was able to follow the study phase efforts that resulted in the preparation of the performance specification and the study phase report. This person can comprehend most of the design tasks that are summarized in the design specification. Now the system is on the verge of moving into the development phase. The user must make a decision about future activities that cannot be visualized clearly or followed in detail. The sponsor is being asked to make the most significant cost commitment to date. Usually, this commitment is for a greatly enlarged project scope, involving many complex development activities over an extended time. For example, programmers will be added to the project, and the results of their activities will not be visible to the user until shortly before the system is scheduled to become operational.

Figure 20.4
Detailed milestones for the development phase. The detailed milestones for the development phase are included in the design phase report in order to provide the user-sponsor with specific information about the activities and associated costs that are to be authorized for the subsequent phase.

development phase—detailed milestones	
implementation plan	test plan
	training plan
	conversion plan
equipment acquisition and installation	
computer program development	computer program design
	coding and debugging
	computer program tests
reference manual preparation	programmer's reference manual
	operator's reference manual
	user's reference manual
personnel training	
system tests	
changeover plan	
system specification	
development phase report	
development phase review	

The analyst must plan the design phase review with great care. The user should be provided with a well-written design phase report in advance of the review, and given the study phase report as a reference document. At the review, an effective presentation is necessary so that the principal user will retain faith in the analyst's ability to continue with the detailed development of the system.

Summary

When the design phase activities are completed, the systems analyst prepares a report. This report, called the design phase report, is the basis for a comprehensive review with the user-sponsor of the computer-related business information system.

The design phase report contains the design specification, which is the second major baseline document. This document is an expansion of the first major baseline document, the performance specification. There are two parts to the design specification, an external design requirement and an internal design requirement. The greatest difference between the performance specification and the design specification is the expansion of an internal performance definition—contained in the study phase report—into a completed internal design requirement for computer programs that are building blocks of the computer program component of the information system.

After the principal user has had an opportunity to study the design phase report a review is held and a decision is made whether to proceed to the development phase. The design phase review is particularly critical, since the principal user must make decisions both about future activities that cannot be followed in detail and a greatly expanded project scope. The user-sponsor's continued confidence in the ability of the systems analyst to manage the project is an important element of the design phase review.

For Review

design phase	"gozinto"	internal design
design specification	**design phase report**	requirement
performance	**design phase review**	external design
specification		requirement

For Discussion

1. What is the purpose of the design phase report?
2. What is the content of the design phase report?
3. Discuss the importance of the design specification, including its relationship to the performance specification.
4. What project plans and schedules are presented at the design phase review?
5. What is the purpose of the design phase review?

Exhibit 2: The Design Phase Report

A case study that follows the life-cycle road map is introduced in chapter 11, Initial Investigation. In the previous edition, the case study resulted in a Modified Accounts Receivable System (MARS), which was an improvement over a predecessor manual system. MARS operated in a batch processing mode and depended upon punched card data preparation. In this edition MARS is the "old" system; the case study updates MARS to an On-line Accounts Receivable System (OARS). A principal goal of OARS is to establish geographically separate profit centers. The evaluated candidate systems are distributed data processing systems, which range from "dumb" terminals to a substantial local data processing capability, and which provide for direct data entry and several communications options.

Exhibits of example study phase, design phase, and development phase reports are included in chapter 14 (Study Phase Report and Review); chapter 20 (Design Phase Report and Review); and chapter 24 (Development Phase Report and Review). The attachment to this chapter is Exhibit 2: The Design Phase Report.

OARS Design Phase Report

I. System Scope
   A. System Title
      On-line Accounts Receivable System (OARS)
   B. Problem Statement and Purpose
      The ABCO corporation's present accounts receivable system is at its maximum capacity of 10,000 accounts. The number of accounts is expected to double to 20,000 accounts over a five-year period. The present system cannot meet this projected growth and satisfy the corporate goal of distributing in-

formation processing resources to regional profit centers. Serious problems already have been encountered in processing the current volume of accounts. Specific problems that have been identified are:

1. Saturation of the capacity of the present computer system, causing difficulties in adding new accounts and obtaining information about the status of existing accounts.
2. Processing delays in preparing customer billing statements because of the batch-oriented design of the current accounts receivable system.
3. Excessive elapsed time between mailing of customer statements and receipt of payments, which creates a high-cost, four-day float.
4. Inadequate control of credit limits.
5. Inability to provide regional centers with timely customer-related information.

Therefore, the purpose of the OARS project is to replace the existing accounts receivable system with one that can eliminate the stated problems and meet ABCO's growth and regional accountability goals.

C. Constraints

The OARS constraints are:

1. Development of the on-line accounts receivable system is to be completed within fourteen months.
2. OARS is to have a growth potential to handle 20,000 customer accounts.
3. OARS is to interface with the existing perpetual inventory system.
4. OARS is to be designed as an on-line system operating in a distributed data processing environment.
5. The design must be compatible with corporate plans to install regional profit centers.

D. Specific Objectives

The specific objectives of OARS are:

1. To establish billing cycles for each region.
2. To mail customer statements no later than one day after the close of a billing cycle.
3. To provide the customer with a billing statement two days after the close of a billing cycle.
4. To speed up collections, reducing the float by 50 percent.
5. To examine customer account balances through on-line inquiry at the time of order entry.

E. Method of Evaluation

After OARS has been operational from sixty to ninety days:

1. A statistical analysis will be made of customer account processing to verify the elapsed time between the close of the billing cycle and the mailing of customer statements.
2. The float time will be measured, and the cost of the float will be calculated at three-month intervals.
3. Periodically, random samples of customer accounts in each region will be audited for accuracy and to validate the effectiveness of on-line inquiry.

4. The validity of OARS transactions that affect the inventory system will be measured by random sampling and physical count.

5. Personal evaluations of the effectiveness of the system will be obtained from its principal users.

II. Conclusions and Recommendations

A. Conclusions

The design phase activities substantiate the results of the study phase activities, and no major changes in the OARS project are required. An evaluation of equipment suitable for the regional minicomputers was completed, and the Excalibur Model V computer system was selected. A letter of intent, subject to confirmation following the design phase review, was placed in order to establish a delivery date that met the development phase schedule.

No special problems were encountered during the design phase, and the OARS project is on schedule. The design phase cost was $39,000, which exceeds the estimated cost of $35,000 by $4000. The extra costs were due to the assignment of an additional person from the accounts receivable department to the project to assist in the minicomputer evaluation and procurement activities.

B. Recommendations

It is recommended the OARS project be approved for the development phase. It is also recommended that the letter of intent be converted into a firm order for the Excalibur V minicomputer system. Also, because of the rapid rate of change in computer technology, it is further recommended that the system be procured by means of a three-year lease, with accruals applied to purchase at the election of ABCO.

III. Design Specification

A. External Design Requirement

1. *Flowcharts* Figure E2.1a is an information-oriented system flowchart for OARS. The accompanying narrative appears as figure E2.1b. Figure E2.2 is the logically equivalent data flow diagram.

2. *System output requirements* The six OARS outputs are:
   a. Customer monthly statement
   b. Accounts receivable transaction register/display
   c. Accounts receivable summary display
   d. Aged accounts receivable report
   e. Customer account list display
   f. Overcredit notification

A print chart and a data element list for an output are presented as figure E2.3.

3. *System input requirements* The three OARS inputs are:
   a. Customer account application
   b. Customer order
   c. Payment/credit

An example of the system inputs is included as figure E2.4.

Figure E2.1a
OARS information-oriented flowchart.

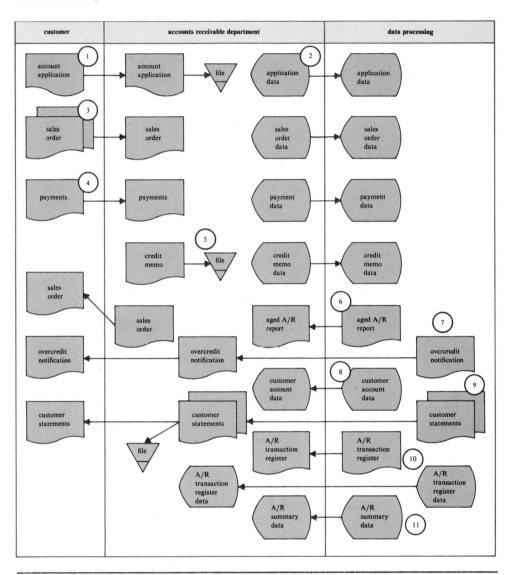

Figure E2.1b
Narrative for OARS information-oriented flowchart.

```
INFORMATION-ORIENTED FLOWCHART NARRATIVE


1.  Customers submit account applications that are processed
    by the accounts receivable department.

2.  If an application is accepted, it is keyed into the
    system.  The application is filed.

3.  Sales orders are entered through on-line order
    processing.  The customer retains a carbon copy.

4.  Payments are sent to the accounts receivable
    department and entered into the system.

5.  Credit memos are generated and the data entered.
    The memo is filed.

6.  An aged A/R report is sent to the accounts receivable
    department from data processing each month.

7.  Data processing sends overcredit notices to the credit
    department whenever a new order would exceed the
    customer's credit limit and the additional credit is
    not approved.  The order and notice are returned to the
    customer.  If the additional credit is approved, the
    order is processed.

8.  Customer account data is displayed on demand.

9.  Customer statements are sent to accounts receivable
    in duplicate.  The original copy is sent to the
    customer; the duplicate is filed.  One-third of the
    statements are produced each ten days of the month,
    that is, on the 1st, 10th, and 20th.

10. A daily A/R transaction register is sent to the accounts
    receivable department as a file copy.  The data is also
    available as a display.

11. The accounts receivable summary data is available as
    a display to the accounts receivable department.
```

4. *System interface requirements*   The OARS must interface with (that is, transfer data to and from) the existing inventory system. Inventory must be on hand for shipment prior to billing the customer, and inventory quantities should be reduced as merchandise is committed for shipment.

5. *System test requirements*   The system tests will be conducted in two stages. The first stage of testing will be run as in an actual operating system but using test input and files.

The second stage will involve the use of live data and a copy of a live file. Initial tests will be with low volume, with the volume gradually increasing as successful tests are completed.

All tests will utilize user personnel under the supervision of the test team.

Figure E2.2
OARS data flow diagram.

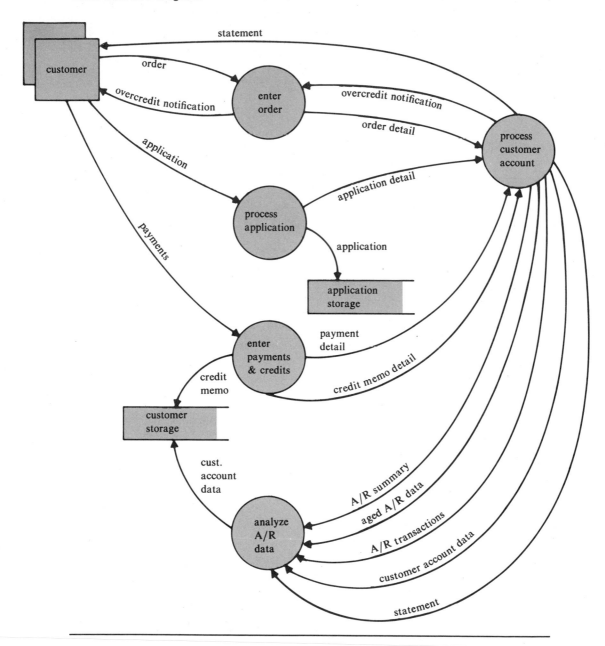

Figure E2.3
OARS output design requirement: overcredit notification.

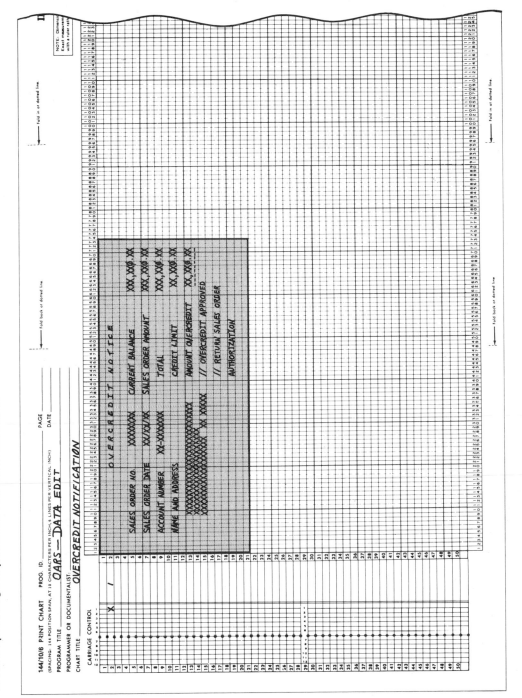

Figure E2.3
continued

OUTPUT SPECIFICATION

TITLE: Overcredit Notification

LAYOUT:

Overcredit Notification

Sales Order No. _____
Sales Order Date _____
Account Number _____
Name and Address
_____
_____
_____

Current Balance _____
Sales Order Amount _____
Total _____
Credit Limit _____
Amount Overcredit [_____]

☐ Overcredit Approved
☐ Return Sales Order

_____
AUTHORIZATION

FREQUENCY: Daily
SIZE: 1 page
QUANTITY: 80 max.
COPIES: 2

DISTRIBUTION: 1. Accts. Receivable Dept.

COMMENTS:

DATA ELEMENT LIST		

TITLE: Overcredit Notification

DESCRIPTION	FORMAT	SIZE
Sales Order Number	XXXXXXXX	8 characters
Sales Order Date	XX/XX/XX	8 "
Account No: Region No. – Sequence No.	XX-XXXXXXX	10 "
Name		26 "
Address		
Street		20 "
City		18 "
State		2 "
Zip Code		5 "
Current Balance	XXX,XXX.XX	10 "
Sales Order Amount	XXX,XXX.XX	10 "
Total	XXX,XXX.XX	10 "
Credit Limit	XX,XXX.XX	9 "
Amount Overcredit	XX,XXX.XX	9 "

Figure E2.4
OARS input design requirement: customer account application.

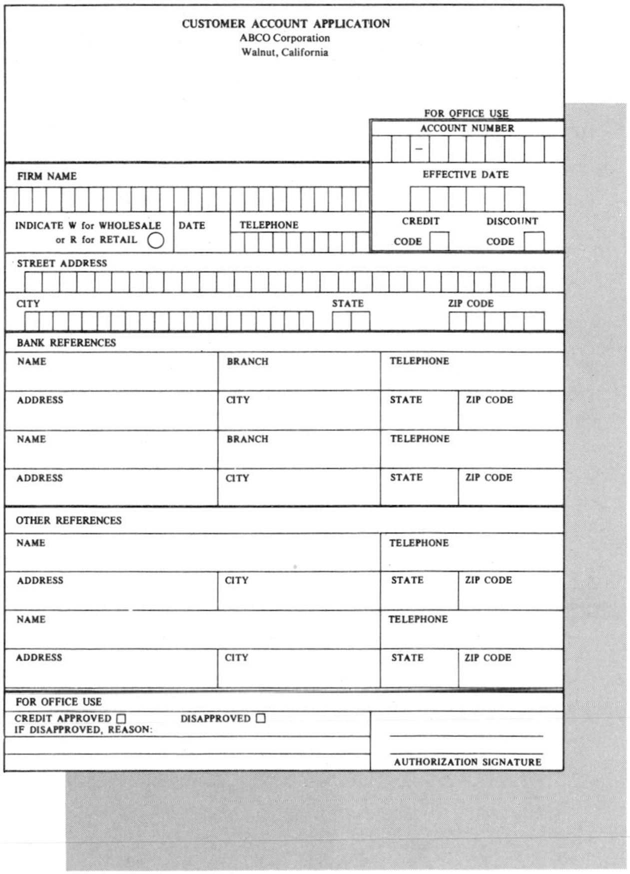

CUSTOMER ACCOUNT APPLICATION
ABCO Corporation
Walnut, California

FOR OFFICE USE
ACCOUNT NUMBER
EFFECTIVE DATE
CREDIT CODE
DISCOUNT CODE

FIRM NAME
INDICATE W for WHOLESALE or R for RETAIL
DATE
TELEPHONE
STREET ADDRESS
CITY
STATE
ZIP CODE

BANK REFERENCES
NAME | BRANCH | TELEPHONE
ADDRESS | CITY | STATE | ZIP CODE
NAME | BRANCH | TELEPHONE
ADDRESS | CITY | STATE | ZIP CODE

OTHER REFERENCES
NAME | TELEPHONE
ADDRESS | CITY | STATE | ZIP CODE
NAME | TELEPHONE
ADDRESS | CITY | STATE | ZIP CODE

FOR OFFICE USE
CREDIT APPROVED ☐ DISAPPROVED ☐
IF DISAPPROVED, REASON:
AUTHORIZATION SIGNATURE

6. *Equipment Specifications* The current central-site computer system will be augmented by five Excalibur V minicomputer systems, one for each region. Each regional system will have the following configuration:
   a. CPU with 128K positions of main memory
   b. One 600 line:-per-minute printer
   c. One 130 megabyte magnetic disk drive
   d. One console keyboard/printer
   e. Two CRT keyboard/display terminals, expandable to eight stations
7. *Personnel and training requirements* One additional programmer will be required prior to start of the development phase. Staff will have to be selected from the regional centers for training in the use of the On-line Accounts Receivable System (OARS).

Figure E2.5a
OARS process-oriented flowchart and narrative.

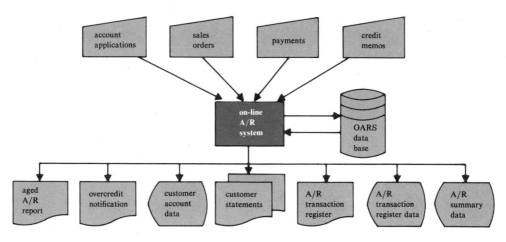

```
SYSTEM FLOWCHART NARRATIVE
```

```
1.  Customer account applications, payments, credit memos,
    and sales orders are the four major system inputs.

2.  The A/R program inputs the transaction data and reads/
    updates the master file on magnetic disk.

3.  The seven major outputs of the system are the aged A/R
    report, overcredit notification, customer account data
    display, customer statements, A/R transaction register,
    A/R transaction register data display, A/R summary data
    display.
```

B. Internal Design Requirement
   1. Computer program component design requirements
      a. Flowchart. This system flowchart is shown in figure E2.5a. Figure E2.5b is the logically equivalent data flow diagram from which E2.5a was derived.
      b. Expanded flowcharts. Figure E2.6a depicts the OARS expanded system flowchart. Figure E2.6b is the accompanying narrative. The expanded data flow diagram, used to develop to figure E2.6a, appears as figure E2.6c. Figure E2.6d shows the HIPO charts for OARS.

Figure E2.5b
OARS data flow diagram—processing detail.

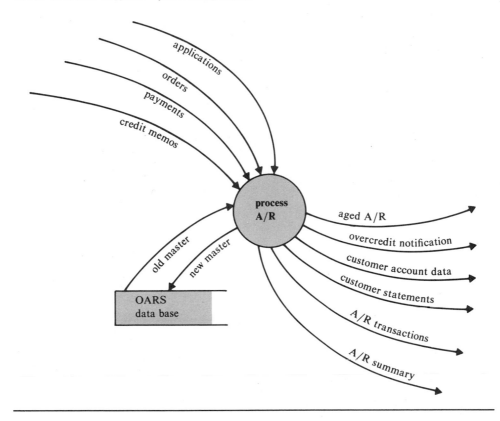

c. Data base requirements. The OARS master files will be indexed sequential files. Each logical record will be 133 bytes in length. Figure E2.7 is a disk record layout. The blocking factor is to be 19.

d. Computer program conrol requirements. The OARS computer programs will provide the following controls:
   (1) Edit all input data for validity
   (2) Detect overcredit conditions prior to processing invoices
   (3) Detect short or out-of-stock conditions prior to processing invoices

e. Computer program component test requirements. The computer program component is to be tested by the test team, top-down, as program modules are developed. Testing will continue until the entire computer program component is tested. Tests will include both valid and invalid data.

Figure E2.6a
OARS expanded process-oriented flowchart.

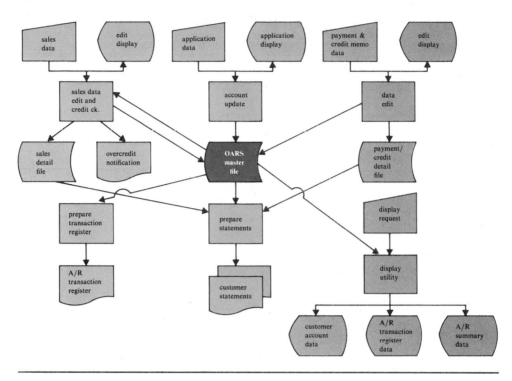

2. *Computer program design requirements*—Data Edit Program
   a. Detailed process-oriented flowchart. As an example, the customer credit program module is presented as figure E2.8.
   b. Transaction file requirements. The transaction file requirements for OARS are shown in figure E2.9. This figure shows the CRT display for the customer application.
   c. Control requirements. The data edit program must edit all input data for validity and detect overcredit conditions. If an overcredit condition exists, an overcredit notification is to be produced. Records containing invalid data are to be noted on an edit error report.
   d. Interface requirements. The data edit program output, on disk, must interface with the invoice program that is part of the perpetual inventory system.
   e. Test requirements. The programmer is responsible for initial testing of a program module in accordance with procedures approved by a programming supervisor. When satisfied that a module is performing according to design specifications, the programmer is to return it, along with all test data and results, to the test team.

Figure E2.6b
Narrative for OARS expanded process-oriented flowchart.

```
OARS EXPANDED PROCESS-ORIENTED FLOWCHART NARRATIVE


Sales data edit and credit check:

   The sales data is entered via terminal keyboard.  A
numeric edit is performed.  Bad data discovered during the
edit is displayed to the terminal operator.  If the data is
valid, a check is performed to determine whether or not the
customer will exceed their credit limit.  An overcredit
notification is produced if the limit is exceeded and the
data is recorded in the sales detail file.

Account update:

   Application data is entered via terminal keyboard.  The
OARS master file is updated to reflect the new account.

Payment/credit memo data edit:

   The payment and credit memo data is entered via terminal
keyboard.  A check is made to verify account numbers and
valid numeric data.  Good data updates the master file and
is recorded in the payment/credit detail file.

Prepare transaction register:

   At the end of each month, a transaction register is printed
using data from the OARS master file.

Prepare customer statements:

Display utility:

   Requests to display customer account data, A/R transaction
register data, or A/R summary data are made through any
terminal with proper password clearance.  All data comes from
the OARS master file.
```

Note: This example design specification is not complete. The data edit program is shown as one example of program design requirements. A complete computer program component specification would contain a description of all program modules.

IV. Plans and Cost Schedules
A. Detailed Milestones—Design Phase

Figure E2.10 depicts the detailed schedule for the nearly completed OARS design phase. The design phase is on schedule.

The design phase costs ran higher than estimated by approximately $4,000. Figure E2.11 shows the design phase costs.

Figure E2.6c
OARS expanded data flow diagram.

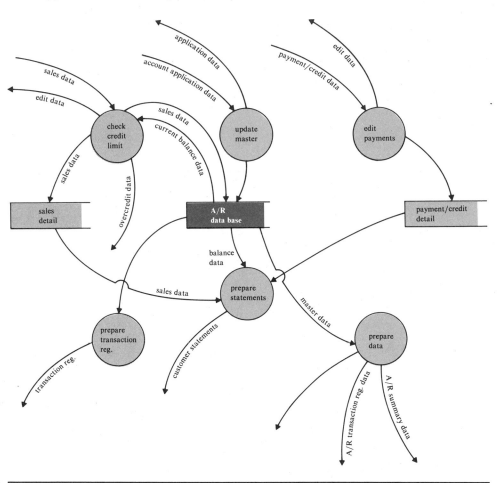

B. Major Milestones—All Phases

Figure E2.12 is a schedule for the entire OARS project. The project is on schedule. The thirty-week development phase is scheduled for completion on 11/11/xy.

The estimated cumulative cost for the entire project is graphed in figure E2.13. The total cost is estimated to be $179,000.

C. Detailed Milestones—Development Phase

Figure E2.14 presents the detailed projections for the Development Phase over the thirty-week period.

Figure E2.15 is the accompanying cumulative cost estimate for the development phase. The total development phase cost is estimated to be $127,500.

Figure E2.6d
HIPO charts for OARS.

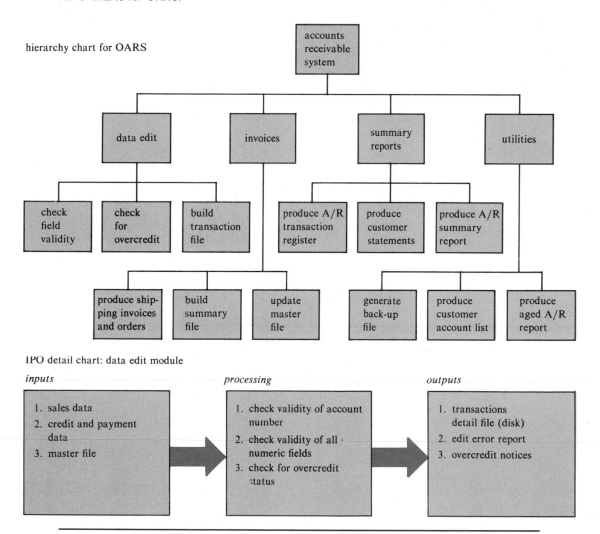

hierarchy chart for OARS

IPO detail chart: data edit module

V. Appendices
   Note: The appendices include all supporting data for the report. For the purpose
         of this exhibit, such detail is not included. The following is a list of typical
         appendices.
         Project directive
         Study of alternatives for computer and computer source selection
         Details of development phase costs and schedules

Figure E2.7
OARS disk record layout.

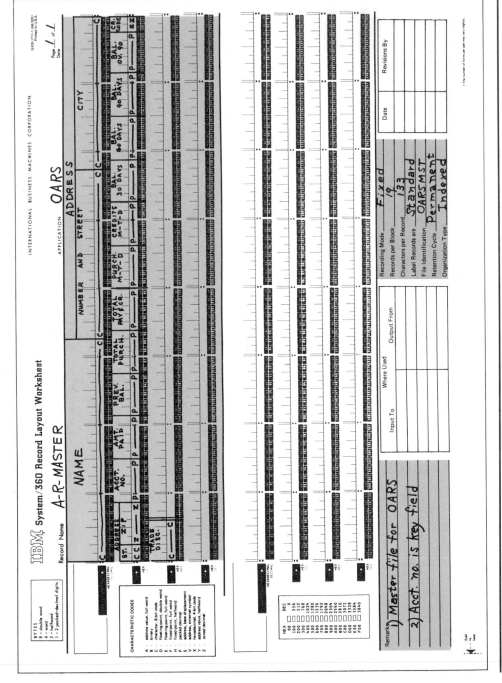

Figure E2.8
Data edit program module.

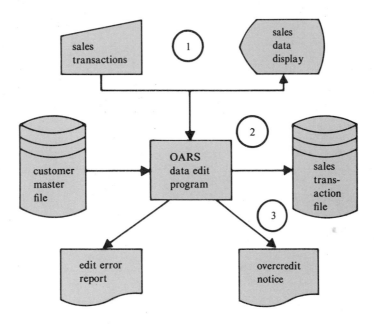

data edit programs-OARS

1. Sales data is entered through CRT terminals.

2. All fields are checked for validity. The new balance is calculated and compared to the customer's credit limit (master file).

3. Valid sales transactions that are not overcredit are recorded in the sales transaction file. Invalid data generates an error report entry. Overcredit conditions cause an overcredit notice to be printed.

Figure E2.9
OARS transaction file requirements: customer application.

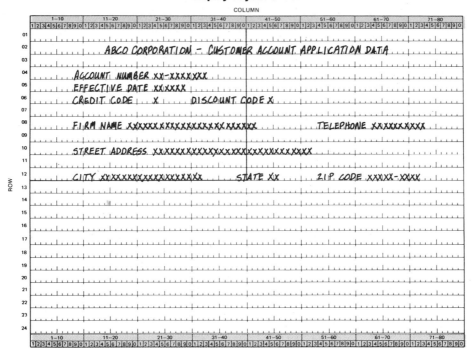

Figure E2.10
Project plan and status report—design phase.

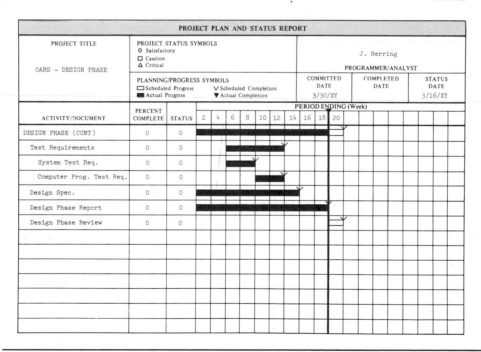

Figure E2.11
Project cost report—design phase.

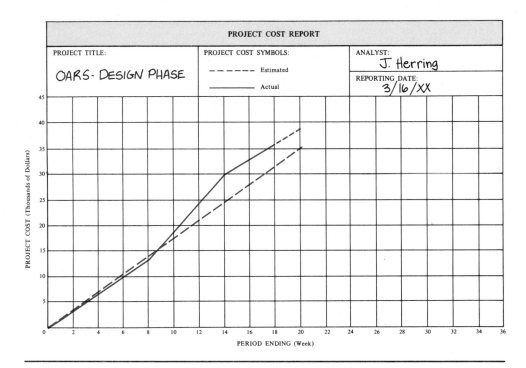

Figure E2.12
Project plan and status report—OARS major milestones.

PROJECT PLAN AND STATUS REPORT																				

PROJECT TITLE
OARS - MAJOR MILESTONES

PROJECT STATUS SYMBOLS
O Satisfactory
□ Caution
△ Critical

PLANNING/PROGRESS SYMBOLS
□ Scheduled Progress   V Scheduled Completion
■ Actual Progress   ▼ Actual Completion

J. Herring
PROGRAMMER/ANALYST

COMMITTED DATE 11/17/XY   COMPLETED DATE   STATUS DATE 3/16/XY

PERIOD ENDING (Week)

ACTIVITY/DOCUMENT	PERCENT COMPLETE	STATUS	2	4	6	8	10	12	14	16	18	20	22	24	26	28	30			
STUDY PHASE	95	O																		
Initial Investigation	100	O																		
Performance Spec.	100	O																		
Study Phase Report	100	O																		
Study Phase Review	0	O																		
DESIGN PHASE																				
Allocation of Functions	0	O																		
Computer Prog. Functions	0	O																		
Test Requirements	0	O																		
Design Spec.	0	O																		
Design Phase Report	0	O																		
Design Phase Review	0	O																		

PROJECT PLAN AND STATUS REPORT																				

PROJECT TITLE
OARS - MAJOR MILESTONES

PROJECT STATUS SYMBOLS
O Satisfactory
□ Caution
△ Critical

PLANNING/PROGRESS SYMBOLS
□ Scheduled Progress   V Scheduled Completion
■ Actual Progress   ▼ Actual Completion

J. Herring
PROGRAMMER/ANALYST

COMMITTED DATE 11/17/XY   COMPLETED DATE   STATUS DATE 3/16/XY

PERIOD ENDING (Week)

ACTIVITY/DOCUMENT	PERCENT COMPLETE	STATUS	32	34	36	38	40	42	44	46	48	50	52	54	56	58	60			
DEVELOPMENT PHASE	0	O																		
Implementation Plan	0	O																		
Equipment Acquisition	0	O																		
Computer Program Dev.	0	O																		
Personnel Training	0	O																		
System Tests	0	O																		
Changeover Plan	0	O																		
System Spec.	0	O																		
Dev. Phase Report	0	O																		
Dev. Phase Review	0	O																		

Figure E2.13
Project cost report—total project.

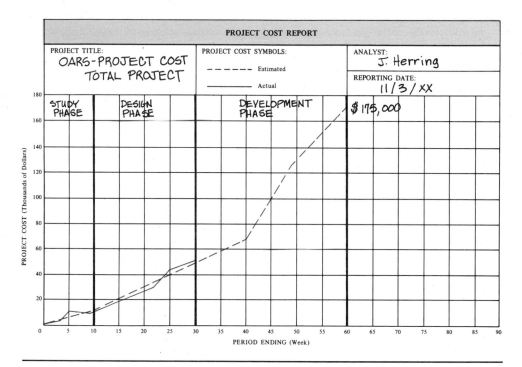

Figure E2.14
Project plan and status report—development phase.

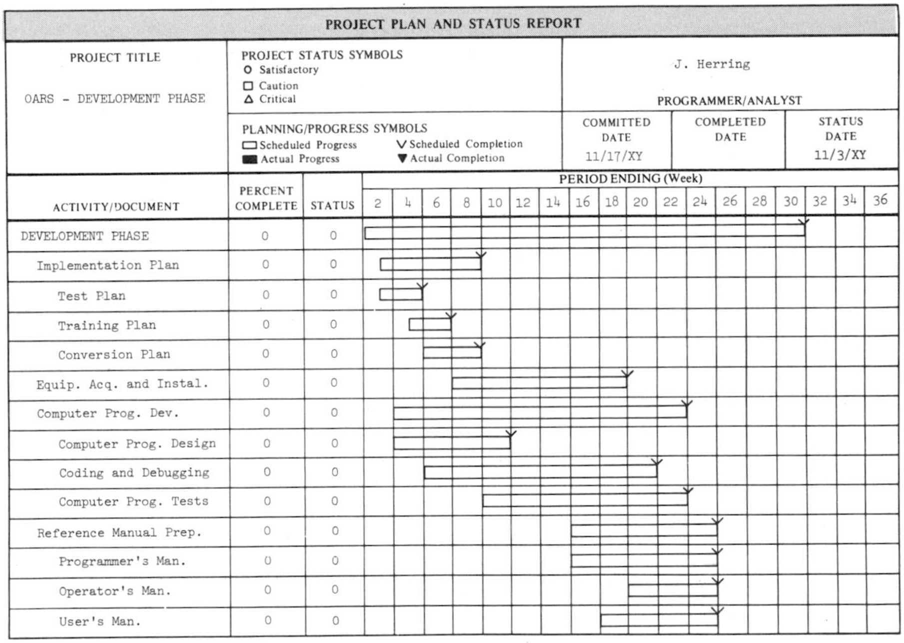

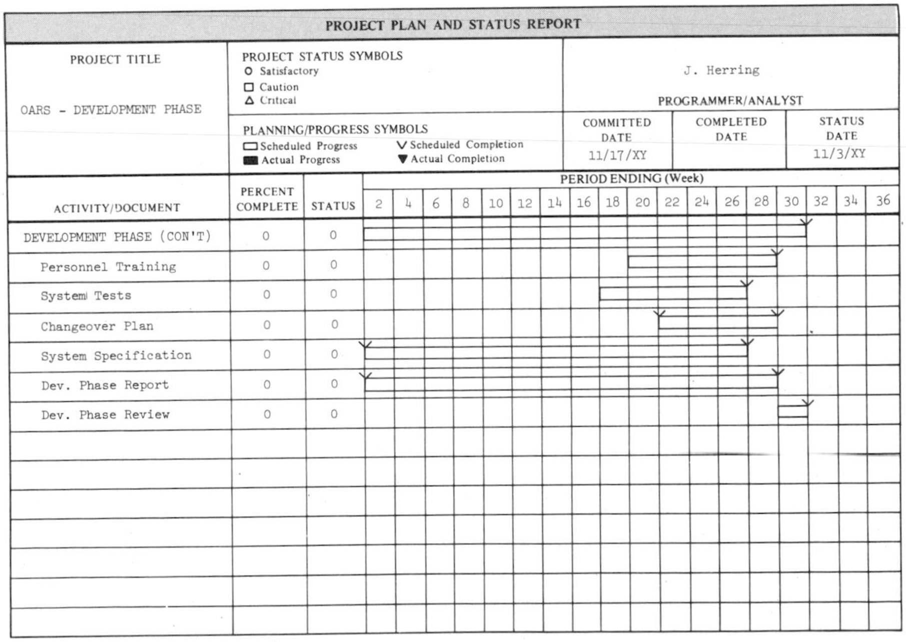

Figure E2.15
Project cost—development phase.

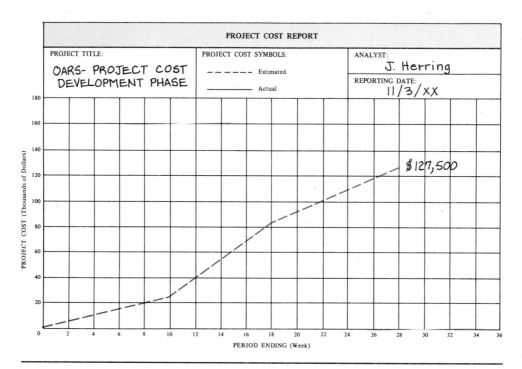

The Development Phase

Unit 5

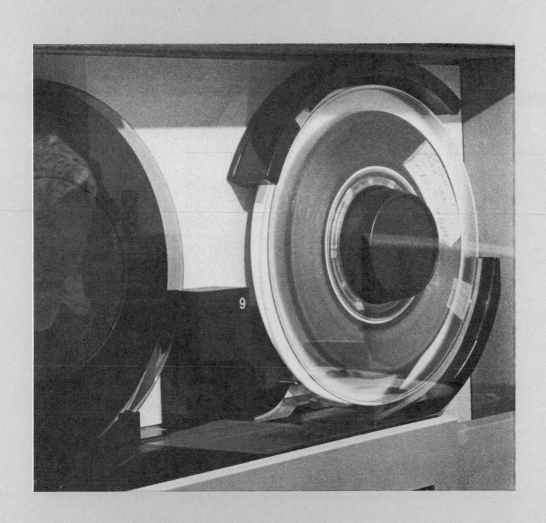

21
Development Phase
Overview

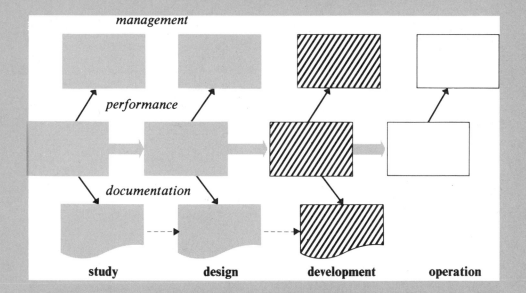

Preview

In the development phase the computer-related business information system is constructed from the specification prepared in the design phase. A principal activity of the development phase is coding and testing the computer programs that make up the computer program component of the overall system. Other important activities include implementation planning, equipment acquisition, and system testing. The development phase concludes with a development phase report and a user review.

Objectives

1. You will become familiar with the major development phase activities.
2. You will acquire a perspective of the development phase that will serve as a reference and a guide as you study the specific topics in the chapters in this unit.

Key Terms

development phase   The life-cycle phase in which the system is constructed according to the design phase specification.

system specification   A baseline specification that contains all of the essential system documentation; it is a complete technical specification.

development phase report   A report prepared at the end of the development phase; it is an extension of the design phase report and summarizes the results of the development phase activities.

development phase review   A review held with the user organization at the conclusion of the development phase to determine whether or not to enter the operation phase.

Development Phase Organization

The **development phase** is the third of the four life-cycle phases. In the development phase the computer-based business system is developed to conform to the design specification prepared in the preceding phase. The *design specification,* a "build to" specification, evolves into the *system specification,* an "as built" specification. The largest project expenditures occur during this third phase. Additional personnel, such as analyst/programmers, programmers, and technical writers, are assigned to the project. Additional dollars are committed for the use of computer facilities. The increased expenditure rate usually continues for a relatively long period. It is not uncommon for the development phase to be two or three times as long as the combined study and design phases. Hence, the project management techniques introduced during the study and design phases should be expanded to correspond to the enlarged project scope. Project plan, status report, and project cost report schedules should be prepared for the principal development phase activities. The level of detail should correspond to the complexity and scope of the work to be done.

The principal activities performed during the development phase can be divided into two major related sequences. These are (1) external (to the computer program component) system development; and (2) internal (to the computer program component) system development. The primary external system development activities are implementation planning; preparation of manuals and personnel training; and equipment acquisition and installation. The principal internal system development activities are computer program development and performance testing. An overview of the major development phase activities is provided in the section that follows.

Development Phase Activities

Figure 21.1 displays the relationship between the principal development phase activities. Each of the activities shown in figure 21.1 is discussed briefly:

1. *Implementation planning*   After the initiation of the development phase is approved, implementation planning begins. Essential parts of the *implementation plan* are:
    a. A plan for testing the computer program component, both as the integrated assembly of its individual programs and as an element of the overall business system.
    b. A plan for training the personnel who are to be associated with the new system. This includes persons who will provide inputs to, receive outputs from, and operate or maintain the new system.
    c. A conversion plan. This plan provides for the conversion of procedures, programs, and files preparatory to actual changeover from the old system to the new one. The *conversion plan* also includes a preliminary plan for the changeover.
2. *Computer program design*   Computer program design is begun parallel with the implementation planning effort. As necessary, system flowcharts are expanded to show additional detail for the computer program components. The

Figure 21.1
Development phase activity flowchart. The principal activity sequences of the
development phase relate to (1) implementation planning, including conversion to the new
system; and (2) computer program design and system testing.

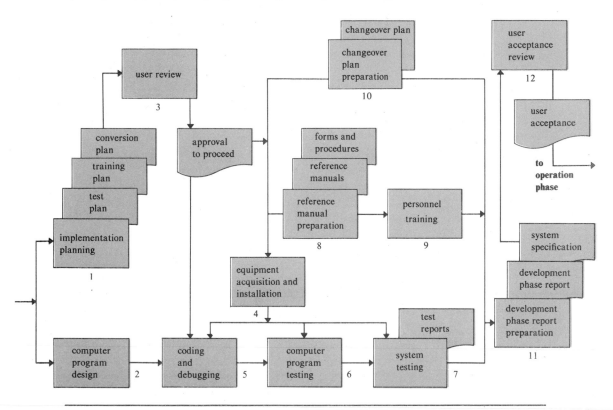

complete data base is developed. Input and output files are identified and
computer program logic flowcharts prepared for each computer program
component.

3. *User review*   Reviews are held with the principal user throughout the
development phase. The first review block shown in figure 21.1 is indicative of
an interim *development phase review.* This type of review normally is not held
to reevaluate the decision that initiated the development phase, but rather to
keep the user informed of general project progress and to secure cooperation in
areas in which the sponsor can be of assistance to the project. As illustrated, a
review of test plans, training plans, and conversion plans is essential because
user personnel are directly involved in the implementation activities. The user's
concurrence with the implementation plan reaffirms support, which is
documented by a written approval to proceed. As shown in figure 21.1, the
approval to proceed also applies to the ongoing computer program design and
development activities.

4. *Equipment acquisition and installation*   In the design phase special hardware required to support the system may have been identified. If not ordered during the design phase, this equipment is ordered at this time, and delivered, installed, and tested. Often all hardware components need not arrive at the same time because needs vary as the computer programs develop from coding through testing. Therefore, an appropriate schedule is established and maintained for the acquisition of hardware items.

5. *Coding and debugging*   Each of the computer programs that make up the computer program component of the overall system is coded and debugged. This means that each is compiled without error and successfully executes its program logic, using data supplied by the programmer.

6. *Computer program testing*   The computer programs are tested in a planned, top-down sequence which includes structured walk-throughs. The testing continues until the programs can be assembled as a component that can be tested as a unit. The analyst supplies data for testing the programs.

7. *System testing*   System tests are performed to verify that the computer-based business system has met its design objectives. The system includes the computer program component as one of its major elements. The user is responsible for supplying the input data and for participating in the evaluation of the system test results. System test reports are prepared to validate system performance.

8. *Reference manual preparation*   Appropriate reference manuals for the various individuals who will work with the new computer-based information system must be prepared. These reference documents are based upon the system specification. The three principal manuals are for programmers, operators, and users.

   Forms and procedures are important elements of the reference manuals. Procedures are written, and appropriate forms are designed. The forms are prepared in-house or ordered from a manufacturer of forms.

9. *Personnel training*   Operating, programming, and user personnel are trained using the reference manuals, forms, and procedures as training aids. The training schedule is closely coordinated with the schedule for completing the development phase. All essential training must be completed prior to the user acceptance review, which occurs at the end of the development phase.

10. *Changeover plan preparation*   The preliminary changeover plan, which was an element of the conversion plan, is updated. Changeover from the old to the new system takes place at the beginning of the operation phase. The *changeover plan* specifies the method of changeover, giving a detailed schedule of activities to be performed and identifying the responsibilities of all personnel involved in these activities.

11. *Development phase report preparation*   At the conclusion of the development phase the development phase report is prepared, documenting the development of the system in accordance with requirements specified in the design phase report. This report contains a summary of all of the pertinent activities undertaken during the development phase. The development phase report includes a *system specification*—the third major baseline document—which evolves from the performance and the design specifications. The system specification contains the complete technical specification for the computer-

related business information system and its components. It contains, for instance, detailed flowcharts, data base specifications, and computer program listings. The system specification contains all of the essential system documentation; it is the baseline reference for the preparation of manuals and training aids.

12. *User acceptance review*   At the conclusion of the development phase the computer-based business system is reviewed by the management of the user organization. Representatives of the information service organization and other affected organizations participate in this review. The principal documents upon which the *acceptance review* is based are the design phase report, the development phase report, test reports, and the changeover plan.

After the conclusion of a successful acceptance review the user organization issues a written memorandum of acceptance and the system enters the operation phase of its life cycle.

Summary

The development phase is the third of the four life-cycle phases. This is the phase in which the computer-related business information system is developed to conform to the requirements prepared in the design phase. The information system project is expanded to include additional personnel, such as programmers. The principal activities performed during the development phase relate to implementation planning, including preparation of a changeover plan; equipment acquisition; computer program design, debugging, and testing; personnel training; and system tests.

User reviews are held throughout the development phase as needed. The development phase concludes with a formal review for the purpose of deciding whether or not to accept the system and proceed into the operation phase. A development phase report is prepared prior to this review. This report contains the system specification, which is the complete technical specification for the computer-related business information system. This specification contains the essential system documentation and is the basis for the preparation of reference manuals and training aids.

For Review

system specification   conversion plan   computer program
design specification   changeover plan   acceptance review
development phase   coding   **development phase**
report   debugging   **development phase**
implementation plan   **review**

For Discussion

1. Define development phase.
2. What occurs during implementation?
3. Define and distinguish between conversion and changeover.
4. What is the purpose of the development phase report? The development phase review?
5. What is the system specification? How does it relate to documentation?

22
Preparing for Implementation

Preview

Implementation preparation is an activity that continues throughout the development phase. It is the process of bringing a developed system into operational use and turning it over to the user. An implementation plan is needed to schedule and manage the implementation activities. The implementation plan provides for test plans, training plans, equipment installation, and a plan for converting from the old to the new system.

Objectives

1. You will become familiar with the major implementation activities.
2. You will learn the importance of plans for testing, training, installing equipment, and converting to the new system.
3. You will understand the usefulness of network techniques, such as PERT, for managing complex projects.

Key Terms

implementation   The process of bringing a developed system into operational use and turning it over to the user.

top-down computer program development   A structured technique that starts with a general description of the system and expands into successively greater levels of detail.

implementation plan   A plan for implementing a system that includes test plans, training plans, an equipment acquisition plan, and a conversion plan.

conversion   The process of performing all the operations that directly result in the turnover of the new system to the user.

changeover   The process of changing over from the old to the new system; the transition from the development phase to the operation phase.

Implementation Planning

The Implementation Process

Implementation is the process of bringing a developed system into operational use and turning it over to the user. Implementation activities extend from planning through conversion from the old system to the new. At the beginning of the development phase a preliminary implementation plan is created to schedule and manage the many different activities that must be integrated into the plan. The implementation plan is updated throughout the development phase, culminating in a changeover plan for the operation phase.

The Implementation Plan

A common implementation management technique is to assign the responsibility for each element of the implementation plan to a team. The head of each team is selected from the organization best qualified to perform the specific implementation task. For example, a user-manager would head the conversion team, and the data processing manager would head the equipment installation team.

The major elements of the **implementation plan** are test plans, training plans, an equipment installation plan, and a conversion plan. We will discuss each of these elements in later sections of this chapter. In a final section, we will describe a critical path network technique that can be used to manage the diverse implementation activities.

Test Plans

The implementation of a computer-based system requires that test data be prepared and that the system and its elements be tested in a planned, structured manner. The computer program component is a major subsystem of the computer-based information system, and particular attention should be given to the testing of this system element as it is developed.

There are two methods of planning for the development and testing of computer programs. These are the traditional "bottom-up" method and the contemporary "top-down" method, which has reduced the severity of problems associated with the traditional method. The latter is the preferred method.

Bottom-up Computer Program Development

The traditional method for scheduling and managing the tests of computer programs is to develop a hierarchical structure within which the lowest level programs are tested individually and then combined into higher level modules, which are tested next. This process, which sometimes is called "string testing," is illustrated in figure 22.1, an example of *bottom-up computer program development*.

A typical development and testing sequence (from 1 to 11) is shown in this figure. Modules that have been coded from the bottom-up and those that are not yet coded are shown. Eventually, all the modules will be strung together at successively higher levels to form the complete computer program.

There have been many difficulties with this traditional method of developing computer programs. Often special programs, called "driver" programs, have to be

Figure 22.1
Traditional bottom-up computer program development. The bottom-up method for
developing computer programs is based upon proceeding from lower-level modules to
higher-level, more complex ones. Failure of all of the modules to mesh at the highest
level often produced cost overruns and inabilities to meet schedules.

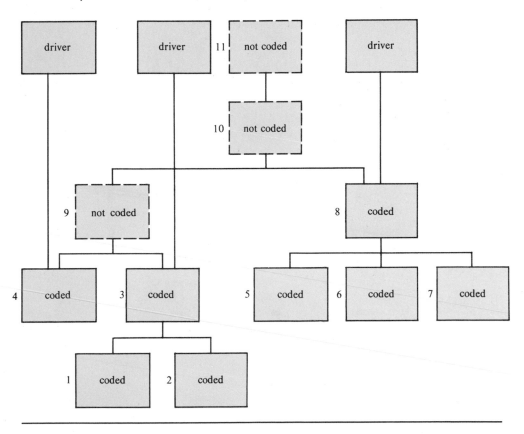

written to test the higher-level modules as they are created. For example, in figure
22.1, modules 3, 4, and 8, which are the highest-level coded modules at the stage of
development depicted, must be tested with driver programs that supply calling and
control instructions not yet available from other modules in the computer program
development hierarchy.

In addition, interfaces between modules must be developed, and the modules must
be integrated successfully to create a complete and functional computer program. Failure of all the components to mesh at the end of the project has caused serious errors.
Changes made at this time, high in the level of the system hierarchy, could cause much
of the lower-level development and testing to be redone, causing overruns in cost and
failures to meet schedules. Because of these problems, the top-down method for developing computer programs has become a part of the life-cycle method.

Figure 22.2
Contemporary top-down computer program development. The top-down method for computer program development is based upon proceeding from less-detailed, higher-level modules to more detailed, lower-level ones. This approach reduces system testing and problems related to final system integration.

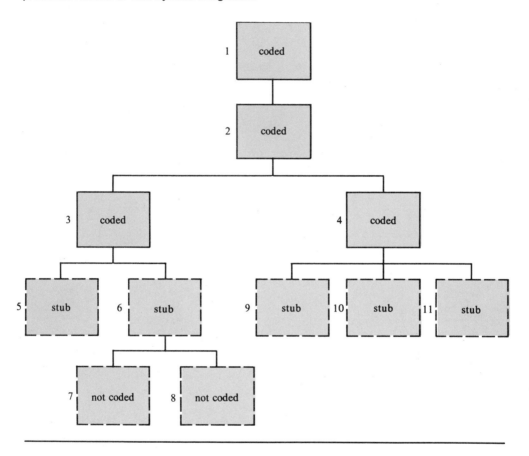

Top-down Computer Program Development

The **top-down computer program development** and testing approach is a structured technique that starts with a general description of the system and expands into successively greater levels of detail.

The top-down approach to computer program development and testing is shown in figure 22.2. This structured technique for computer program development is a logical extension of the top-down, hierarchical approach to system design characteristic of *HIPO charts* (chapter 16).

As the typical development sequence (from 1 to 11) in figure 22.2 shows, modules are developed downward from nucleus at the top of the computer program hierarchy. Driver programs are not necessary. Instead, modules that display a message acknowledging receipt of higher-level program control are used. These modules are called *stubs;* their use also is illustrated in figure 22.2.

Figure 22.3
Comparative machine usage for top-down and bottom-up computer program development.
With the top-down method for computer program development the initial investment in
machine usage is higher. The total machine usage, however, is less.

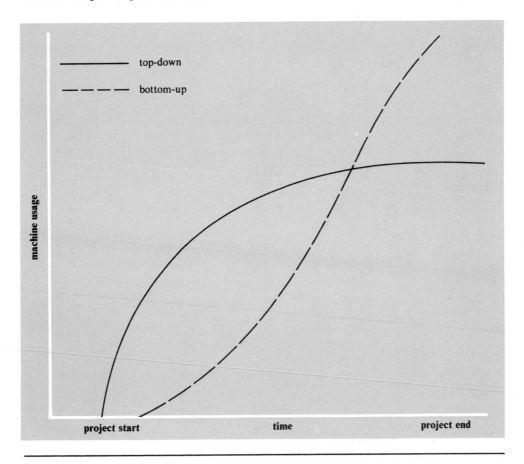

A major advantage of top-down structured testing and development is that the
computer program continues to operate as stubs are removed and modules added.
Managers thus have continuous control over the computer program development process, and the problems that can arise from an overall integration effort are minimized.
An important aspect of this advantage is illustrated in figure 22.3. An end-of-project
integration effort is eliminated, and total machine usage for testing is reduced.[1] Early
testing can verify the correctness of the higher-level computer program routines responsible for the major logic and overall program control, and can increase the reliability of the final product.

---

1. Joan C. Hughes and Jay I. Michtom, *A Structured Approach to Programming* (Englewood Cliffs:
Prentice-Hall, Inc., 1977).

Figure 22.4

Annotated test plan format. A plan is necessary for testing the programs that make up the computer program component of a computer-related system. The scope of the plan, the data to be collected and identified, and special procedures are included.

(subsystem/system) test plan

scope

1. *name*	A name or number that identifies the test
2. *purpose*	**Why:** the specific objectives of the test, including identification of the computer program components involved in the test
3. *location*	**Where** the test is to be performed
4. *schedule*	**When** the test is to be performed
5. *responsibilities*	**Who:** the individuals involved in the test and their specific duties
6. *general procedures*	**What:** a general overview of the test inputs, events, and anticipated results

data collection and evaluation

1. *data to be used*	**What:** a detailed description of live or simulated input data to be used in the test
2. *data to be collected*	**What:** a detailed description of the data to be obtained as test results
3. *method of data recording*	**How** the data is to be recorded, for example, listings, punched cards, and so forth
4. *method of data evaluation*	**How** the test results are to be analyzed

special procedures

What procedures are unique to this test, for example, equipment operating procedures, operator intervention procedures, abnormal condition procedures.

Formal Test Planning

Plans must be made for the formal evaluation of the computer program as it is developed, as well as for internal technical reviews, such as *structured walk-throughs*. The results of these tests should become part of the cumulative development phase documentation.

A standardized test plan document is useful. A general format for a test plan is shown and annotated in figure 22.4. The comments relate to the questions of "why, where, when, who, what, and how" that a test plan should answer. The format illustrated is suitable for both computer subsystem and overall system tests. An entry identifying the type of test is inserted at the top of the test plan document.

At the conclusion of each test a written report is prepared to record the results; a standardized test report document relating to the test document can be used. Figure 22.5 outlines a general test report format and describes the entries to be made on the document.

Figure 22.5
Annotated test report format. A written record of test results is required. This report defines the tests performed and describes the results of the tests.

(subsystem/system) test report	
scope	
1. *name*	the name shown on the corresponding test plan
2. *purpose*	the purpose stated on the corresponding test plan
3. *references*	identification of the corresponding test plan and other pertinent documents, such as previous test results
description of results	
1. *test methods*	how the test was performed
2. *objectives met*	identification of specific test accomplishments
3. *problem areas*	discussion of problems encountered
4. *recommendations*	specific actions to be taken, for example, accept test results, perform additional tests, revise coding

Final system tests are performed after the subsystem tests have been completed. Their purpose is to exercise the entire system, including the computer program subsystem, under "live" environmental conditions. The main objective of system tests is to subject the computer-based business system to all foreseeable operating conditions. The user should specify and conduct these tests; system test responsibility should not be delegated to programmers or even to systems analysts.

Systems tests become progressively more complex. Initial tests may involve selected samples of input data and a small test file; later, pilot tests (called "history processing") can be performed using complete files and a large number of past data transactions.

The persons who will be involved in the "live" operation of the system should prepare the system test data for computer processing. If the user is involved appropriately in the system level tests, progress can be made toward two supplementary goals: user training and user acceptance. User training is vital to the successful conversion to the new system. User participation in earlier top-down tests of the computer program subsystem and in structured walk-throughs is a powerful training technique. The final system tests, if sufficiently thorough, can serve as acceptance tests, which can assure the user that the system is ready for operation.

Training

Training: An Overview

Training plans are an important element of the implementation plan. Their purpose is to ensure that all the personnel who are to be associated with the computer-based business system possess the necessary knowledge and skills. Operating, programming, and user personnel are trained using *reference manuals* as training aids. The training schedule is coordinated with the schedule for completing the development phase, for all essential training must be completed prior to the user acceptance review at the end of this phase. However, training should not be completed too far in advance of need, or the personnel are liable to lose interest or forget their training.

Training programs begin with the selection of appropriate participants and the preparation of different types of training programs for programmers, operators, and user personnel. User personnel are those persons who will prepare inputs, follow procedures, and use outputs. Additional training programs must be conducted for management level personnel, not only to familiarize managers with the new system, but also to obtain their active support and cooperation during implementation.

Before training programs can be initiated, materials must be prepared. The basic training resources are reference manuals appropriate to the needs and interests of each type of trainee—programmer, operator, or user. These reference manuals are based largely upon the *system specification*. This baseline document usually is not in a final format before the end of the development phase, at which time it is included in the development phase report. However, because the design specification is added to continuously throughout the development phase, an interim system specification exists. Drafts of the reference manuals can be prepared for the initial training programs from the interim specification. The reference manuals are prepared in their final formats at the end of the development phase, and they are available for training purposes throughout the life of the system.

In the next section we will describe the training programs established to meet the needs of programmers, operators, and users, outlining the general content of the training manuals.

Programmer Training

Programmers are assigned to the computer-based business system project at the beginning of the development phase. The programmers assigned to develop the computer program modules are indoctrinated by the analysts and programmer/analysts who prepared the design specification. They help to create the programmer's reference manual, which is used to train other programmers assigned to the system throughout its operational life.

The programmer's reference manual is the most comprehensive of the reference manuals. It informs an experienced programmer, unfamiliar with the system, about all of the aspects of the computer program. The manual should enable this person to (1) understand existing program components; (2) modify existing program components; and (3) write new program components. Figure 22.6 is a guide to typical contents of such a manual. The external specification relates to elements of the system external to the computer program subsystem; the internal specification relates to elements internal to the computer program subsystem.

Figure 22.6

Programmer's reference manual. The programmer's reference manual is a comprehensive guidebook, containing both internal (to the computer program component), and external (to the computer program component) system specifications.

guide to programmer's reference manual

1. *title* — name of computer-based system
2. *purpose* — general description of system and its major objectives
3. *external specification*
   - a. information-oriented flowchart and/or data flow diagram
   - b. output, input, and interface descriptions
   - c. system level test input-output samples
   - d. equipment specification
4. *internal specification*
   - a. computer program component:
     1) process-oriented flowcharts
     2) expanded system flowcharts
     3) data base specification
     4) control specifications
     5) test input-output samples
   - b. computer programs:
     1) detailed system flowcharts
     2) logic flowcharts
     3) transaction file specifications
     4) control specifications
     5) interface specifications
     6) listings
     7) test input-output samples

Operator Training

If new equipment is to be installed, operator training is completed in conjunction with its installation and checkout. If new equipment is not required for the computer-based system, operators still must become familiar with the operational requirements of the new system. Different kinds of operational personnel may be involved in the operation of the system, such as computer operators, console operators, and data entry operators. Therefore, more than one type of manual may be required. Each manual should acquaint the operators with the system and its purpose and provide a ready reference to specific duties and step-by-step operating instructions. These instructions should cover both normal and abnormal situations. They should be written in a style that is easily understood by the intended users of the manual; detailed discussions of complex and lengthy system descriptions should not be included. A guide to the typical contents of a reference manual is shown in figure 22.7.

Training programs for operators are scheduled to coincide with the needs of the computer-based business system as it is developed, tested, and approaches operational status. Users, analysts, and programmers may participate in the training of operators.

Figure 22.7

Operator's reference manual. An operator's reference manual contains all the information an operator needs to perform specific tasks, including input, processing, and output descriptions and instructions for handling errors and exceptions.

---

guide to operator's reference manual

1. *title* — name of computer-based system

2. *purpose* — general description of system

3. *operating procedures* — as appropriate to specific operational duties:

    a. operator inputs: a complete description of all inputs, including:
       1) purpose and use
       2) title of input
       3) input and media
       4) limitations
       5) format and content

    b. operator outputs: a complete description of all outputs, including:
       1) purpose and use
       2) title of output
       3) output media
       4) format and content

    c. file summary: a complete description of all files, including:
       1) file identification
       2) medium
       3) type: master, transaction, and so forth

    d. error and exception handling: procedures for handling hardware and software error conditions

---

User Training

For the system to begin operation, a sufficient number of users must be trained before the end of the development phase. Thereafter, additional personnel are trained, and training continues throughout the operational life of the system. As with operator's reference manuals, usually more than one type of user's reference manual is required. Each one should be self-contained and should explain the system in terms of the user's specific needs. The text should be factual, concise, specific, and clearly worded and illustrated. Sentences should be simple and direct; discussions of theory and detailed technical matters should be avoided. The manual should provide users with a general overview of the system. However, primary emphasis is on the specific steps to be followed, the results to be expected, and the corrective actions to be taken when such results are not obtained. Figure 22.8 lists the typical contents of a user's reference manual.

    Training sessions for user personnel usually involve larger numbers of people than do operator or programmer training programs. These sessions should be planned to meet the needs of each type of user. Normally, several sessions should be scheduled

Figure 22.8
User's reference manual. A user's reference manual contains all the information needed to train a user of the computer-related information system in input, output, and operating procedures.

guide to user's reference manual	
1. *title*	name of computer-based system
2. *purpose*	general description of system and its major objectives, including an information-oriented flowchart and/or data flow diagram
3. *user procedures*	as appropriate to each user:

      a. instructions for input preparation:

        1) title
        2) description
        3) purpose and use
        4) media
        5) limitations
        6) format and content
        7) relation to outputs

      b. instructions for output use:

        1) title
        2) description
        3) purpose and use
        4) media
        5) limitations
        6) format and content
        7) relation to inputs

      c. operating instructions: procedures for operating equipment with which the user must be familiar

for all trainees so that they fully understand the new system and have an opportunity to familiarize themselves with the handling of documents and equipment.

The training team should be certain that sufficient user personnel are thoroughly trained and prepared to support the new system at the time of its implementation. Individuals who are not willing to cooperate or who cannot follow procedures can cause great difficulties at the time of changeover from the old system to the new.

Management Orientation

The life-cycle process for the development of computer-based business systems automatically includes numerous reviews to keep user management informed of and committed to the support of the project. However, before changeover to a new system, it is important to augment the scheduled reviews with a series of management presentations. The purpose of these presentations is to inform all managers affected by the new system and to solicit their support during its implementation. These presentations, which should be made by the senior personnel involved in the development of the system, should include these subjects:

1. Review of system objectives, costs, and benefits.
2. Organizational and procedural changes associated with the new system.
3. Responsibilities of the organizations that report to the management attendees.

It is important that the implementation team's responsibilities be understood and that each organization involved in the implementation be assigned a constructive role during the critical changeover period.

Equipment Installation

Earlier in the life cycle of a computer-based system, fundamental equipment decisions were made. During the study phase, for example, alternative configurations were evaluated and decisions made about using available in-house equipment and obtaining new computer components or systems. Early in the design phase, when functions were allocated between manual, hardware, and software tasks, a final process of equipment evaluation and vendor selection took place. If new equipment was needed, it was placed on order. The development phase implementation plan must include all activities related to the installation and check-out of equipment scheduled to be delivered at various times throughout the development phase.

The principal equipment-related activities that must be implemented are (1) site preparation, (2) equipment installation, and (3) hardware and software check-out.

Equipment vendors can provide the specifications for equipment installation. They usually work with the project's equipment installation team in planning for adequate space, power, and light, and a suitable environment (for example, temperature, humidity, dust control, and safety measures). After a suitable site has been completed, the computer equipment can be installed. Although equipment normally is installed by the manufacturer, the implementation team should advise and assist. Participation enables the team to aid in the installation and, more importantly, to become familiar with the equipment.

Usually manufacturers will check out the hardware and the software they supply. The implementation team also should perform its own check-out tests, using application-oriented test programs.

Conversion

Conversion: An Overview
Conversion is the process of performing all of the operations that result directly in the turnover of the new system to the user. Conversion has two parts:

1. The creation of a conversion plan at the start of the development phase and the implementation of this plan throughout the development phase.
2. The creation of a system changeover plan at the end of the development phase and the implementation of the plan at the beginning of the operation phase.

We shall discuss the conversion plan and its implementation in this chapter. We shall also discuss the changeover plan, the implementation of which is described in chapter 26, Changeover and Routine Operation.

Conversion Activities (Development Phase)

A conversion plan is prepared at the start of the development phase. Its principal elements concern procedures, program, and file conversion.

Procedures Conversion   Often a new system will incorporate many of the old system's forms and procedures, but some of these may require modification to fit into the new system. Also, the new system may interface with a network of other systems; this, too, may cause some modification of procedures. The procedures that require change must be identified, and the changes explained during training of personnel.

Program Conversion   The new computer-based system may include some computer programs that are part of an existing system. A conversion problem may arise if new equipment is installed, if the inputs and outputs of existing programs change, or if the existing programs are not efficient in their new environment. Even if new equipment is not involved, all the existing programs must be reevaluated. Reprogramming should be considered when programs are poorly documented, heavily patched, or not efficient enough. For instance, many small programs might be replaced by a single program that performs a repetitive function more effectively.

System interfaces with other computer programs also must be examined. Programming modifications may be required to enable the new system to supply or receive data through these interfaces.

File Conversion   File conversion can be the most time-consuming and expensive step in the entire project. The magnitude of this task often is underestimated. For example, if many thousands of customer account records are to be stored on a magnetic disk instead of kept in filing cabinets—possibly located in different parts of the company—the conversion effort could be extensive. Existing files must be converted into a format acceptable to the computer program and equipment. Duplicate files must be consolidated and errors corrected before changeover to the new system starts. Otherwise a series of data errors may plague users of the new system for a considerable time after its implementation.

File conversion activities include many basic systems analysis activities, such as fact finding and analysis, forms design, procedure writing, and computer program design. We can divide file conversion into a sequence of three major activities. These are (1) collection of file conversion data, (2) conversion of files, and (3) testing of converted files.

In many circumstances file conversion data must be collected from a variety of sources. Some data may already be in machine-readable form; however, it often is necessary to create new data to supplement that which is already filed in some form. Forms may have to be designed and procedures written to transfer data from an existing file to a new file. Often intermediate files are created. For example, the data to be transferred may be entered into specially designed input documents and then punched into cards for verification before being written onto a disk or tape.

Verifying data going into the new files is an important and often laborious task. All too often a high percentage of "current" data is incomplete or in error. Discrepancies are common, for instance, between data stored in two redundant files that are to be consolidated into a shared data base. Before consolidation can take place, it must be determined which (if either) file is correct. Verification usually requires the extensive assistance of user personnel. The analyst must remember that the new file manager is to be a computer, which will not be as flexible as the human file manager it may replace. Humans often can detect and ignore "garbage." The computer usually cannot.

Computer programs are required to perform the actual file conversion. These programs must be written and checked out before they are needed. They must sort data, validate data, and create the file in the new format. After the files have been converted, the conversion team must check their accuracy. Even if the original files have been "purified" before the data stored on them is entered into the new system, errors may be introduced during conversion. All file data should be printed out and verified. This, again, requires the assistance of user personnel. Special file correction forms may have to be designed, and several conversion runs may be required if a large data base is being assembled. Involving the user in file conversion activities is healthy because it tends to build user confidence in the new data base and in the computer-based system.

Changeover Plan (Operation Phase)

The activities described in the preceding sections were the elements of a conversion plan that is implemented throughout the development phase. By the time of the acceptance review at the end of the development phase, forms and procedures have been prepared and used, computer programs have been written and tested, and old files have been converted to new files. The next step in the conversion process is the actual changeover from the old system to the new, which takes place at the beginning of the operation phase.

A changeover plan that identifies and schedules all changeover activities should be available at the acceptance review. It should specify the method of changeover and identify the roles and responsibilities of all personnel. The three general methods of **changeover** from the old system to the new system are parallel operation, immediate replacement, or phased changeover.

In *parallel operation,* data is processed by both the old and new systems. In theory, this method offers many advantages. Users have maximum flexibility because they do not have to begin using the new system until they are certain it is producing acceptable outputs. They know they can always revert to the old system in the event of disaster. In practice, unfortunately, there are several reasons why a "pure" parallel operation method of changeover seldom is possible:

1. The new system is "different" from the old system. It probably has been designed to perform functions and produce outputs that were not available with the old system.
2. Parallel processing may be too time-consuming or expensive, particularly if personnel are not available to operate both systems. This is particularly true if the volume of work is large.

3. Determining which system is in error can be difficult. People tend to be biased toward the familiar; in this case, toward the old system.
4. Parallel processing tends to delay adoption of the new system. People tend to cling to a "security blanket," thus prolonging the problems the new system was designed to solve.

Immediate replacement—requiring immediate use of the new system—is a risky alternative. The outputs of the new system may be compared with the "last" outputs of the old system to identify errors. Correcting errors usually creates crisis situations during the early stages of immediate replacement. The circumstances under which immediate replacement usually occurs are those in which:

1. A high percentage of outputs are new.
2. The system is not so critical that failure is a disaster.
3. No type of parallel processing is possible.
4. The user exerts "pressure" for use of the system outputs.
5. An alternate, or fallback, system is available.

Phased changeover is a compromise between parallel operation and immediate replacement; it is recommended over the other two methods. In this method users process some percentage, perhaps 10 percent of their normal volume of transactions, through the new system, with the remainder processed through the old system. Thus, users can become familiar with the operation of the new system, and the task of correcting errors is manageable with existing resources. After the users are assured that normal transactions are being processed correctly, more complex transactions can be introduced. The volume of data handled by the new system can then be increased until the old system is phased out. The success of phase changeover is enhanced if the changeover plan is properly scheduled for sequence and timing, for introducing elements of the new system, and for terminating corresponding parts of the old system.

Whatever changeover method is selected, problems will arise. We will discuss the actual changeover in chapter 26, Changeover and Routine Operation.

An initial operating schedule also should be prepared. Its purpose is to demonstrate how the new computer program can perform its functions in a timely fashion in the "real world." The schedule should identify groups of computer program components that must be run without interruption on a regular basis. These groups are called run modules. Such a schedule is prepared in conjunction with the operations staff of the data processing department.

Implementation Management

The Implementation Committee

We have discussed the four major elements of the implementation plan: test plans, training plans, an equipment installation plan, and a conversion plan. Each of these elements may be complex, involving the combined efforts of a large number of people, many of whom are unfamiliar with the new system. At the beginning of this chapter we suggested that one technique for managing the entire implementation task was to establish implementation teams, each one to be responsible for a major element of the implementation plan. However, for the team approach to be effective, there must be a central reporting point. Often the implementation teams will report to a coordination committee. The membership of such a committee should include the heads of the implementation teams, the primary users of the system, and representatives from the information service organization. The senior systems analyst is responsible for forming this committee and for selecting its members. In doing so, the analyst should consult with the principal user of the new system.

The responsibility for heading the *implementation committee* belongs to the principal user, not the systems analyst. This user already has a major financial commitment to the success of the system. With the principal user as head of the implementation committee, his or her emotional commitment and readiness to accept the system as it approaches operational status will increase. There are also other advantages:

1. User leadership will demonstrate that the system is "real" and not just an "exercise."
2. The user, as a "line" manager, can bring additional authority to the project. The user can take direct action to resolve and to prevent problems.
3. Because of involvement in implementation management, the user will be prepared to conduct a knowledgeable acceptance review at the end of the development phase. The user will be predisposed to accept a system that he or she feels is ready for operational status.
4. The user's involvement in and commitment to the success of the system will carry over into the operation phase. This will help the analyst during the critical changeover period.

Although the systems analyst does not head the implementation committee, the analyst is an important member. Often the systems analyst functions as the committee secretary, providing the planning and scheduling skills needed to manage the many implementation tasks. Because these tasks can be defined and are similar for almost all computer-based systems, it is appropriate to consider using critical path networks, such as were mentioned in chapter 7, Charting Techniques. The use of these network techniques to manage the implementation tasks is discussed in the following section.

Network Techniques: PERT

As introduced in chapter 7, critical path networks are project management charting techniques that use a graphical format to depict the relationships between tasks and schedules. As an example, we shall consider the application of one of these techniques, *PERT* (Program Evaluation Review Technique), to managing the implementation of a computer-based business system.

PERT is a management planning and analysis tool that uses a graphical display, called a *network,* to show relationships between tasks that must be performed to accomplish an objective. PERT is a means of creating a "master plan" for the control of complex projects. Developed by the United States Navy, Lockheed Aircraft Corporation, and the consulting firm of Booz, Allen, and Hamilton for use on the Polaris submarine program, PERT has been applied widely to both civil and military projects.

PERT is a management tool. It provides a manager with an orderly approach to planning. By forcing the manager to construct a network, PERT points out relationships between tasks that might be otherwise overlooked. It also brings about coordination of effort, since it requires that participants in a project communicate with each other in order to establish and review the network. In short, PERT is a technique that helps managers answer questions such as these:

1. What work is to be done?
2. How will the work be done?
3. When is the work to be done?
4. What management actions can be taken?

We will develop the basic knowledge required to use PERT as we show how it can help provide answers to the above questions.

To answer the question "What is to be done?" we must specify objectives and develop a plan identifying the tasks to be completed to achieve these objectives. In PERT the plan is represented by a network like that in figure 22.9, which displays related activities and events. A network is a graphical representation of related activities and events. An *activity* is the application of time and resources to achieve an objective. It is measured in units of time, usually weeks, and is represented on the PERT network by an arrow. The arrows labeled *a, b,* and *c* in figure 22.9 represent activities. These activities are similar to horizontal bars on a Gantt chart; however, they differ in that elapsed time is not necessarily proportional to the length of the arrow.

An *event* is a point in time at which an activity begins or ends. It is represented on a PERT network by a circle. Thus, in figure 22.9, event 2 represents the end of activity *a* and the start of activity *b*. That is, each internal activity has a predecessor and a successor event. Events are similar to milestones on a Gantt chart, but the relationships between events are expressed much more explicitly than are those between milestones. Typically, events are identified by phrases such as "training manuals prepared," "training completed," and "equipment installed."

Figure 22.9
Elementary PERT network. A critical path network, such as PERT, is a management planning and analysis tool. The building blocks of a PERT network are activities and events.

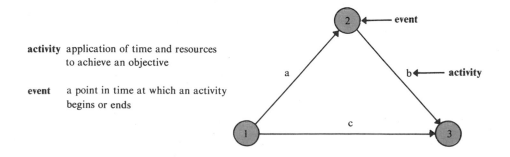

activity application of time and resources to achieve an objective

event a point in time at which an activity begins or ends

The PERT network helps to answer the question "How will the work be done?" by displaying the sequences in which activities must occur if specified events are to be reached. For example, the network of figure 22.9 tells us that three separate jobs, *a, b,* and *c,* must be performed in order to achieve the end objective denoted by event 3. Further, it identifies two independent paths along which activities must be completed. Along one path, activities *a* and *b* must be performed in sequence. Along the other path, activity *c* must be performed concurrently with activities *a* and *b.*

PERT networks also help to answer the question "When is the work to be done?" This is accomplished by estimating an expected time for each activity. The expected activity time is based upon three estimates: optimistic, pessimistic, and most likely. It is calculated according to the following formula:

$$t_e = \frac{O + 4M + P}{6}$$

where

    t_e = Expected activity time (in weeks)
    O = Optimistic estimate (how long the activity would take if everything went well)
    M = Most likely estimate (the normal time the activity should take)
    P = Pessimistic estimate (how long the activity could take under adverse conditions)

The above formula is the essential difference between a similar network technique, CPM (Critical Path Method), and PERT. CPM uses only one estimate to obtain a value for t_e. The time to reach any event along a network path can be calculated as the sum of the activity times along the path. However, since more than one path may lead to an event, it is necessary to select the largest sum of activity times, that is, the longest path, as the determining time. This time is defined as the expected event time, T_E. Thus:

$$T_E = \text{sum of all expected activity times } (t_e\text{'s}) \text{ along} $$
$$\text{the longest path leading to an event.}$$

The longest T_E is the time needed to proceed by the longest path from the first to the last event in the network. This is the minimum amount of time that must be scheduled for the project represented by the PERT network. The path along which the longest T_E is measured is called the *critical path.* Slippage along the critical path can cause the scheduled completion date, which usually corresponds to the longest T_E, to be missed.

There is time to spare along all other paths in the network leading from the first event to the last event (unless there are multiple critical paths). This time to spare is called slack (*s*). Slack is calculated for each event by subtracting the T_E for that event from the latest allowable time, T_L, which is the latest time that an event can be reached without causing any path on which the event lies to exceed the critical path. Thus:

$$s = T_L - T_E$$

Figure 22.10
Example of PERT calculations. PERT network calculations are based upon optimistic, most likely, and pessimistic estimates of activity times. Because there is no slack along the path S-3-F, this path is the critical path.

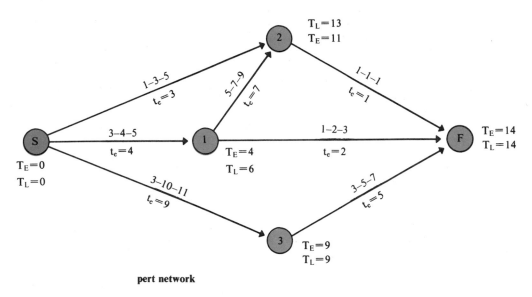

pert network

event	predecessor events	expected event time T_E	latest allowable time T_L	slack $s = (T_L - T_E)$
S	—	0	0	0
1	S	4	6	2
2	S, 1	11	13	2
3	S	9	9	0
F	1, 2, 3	14	14	0

table of pert calculations

We will use the network previously presented as figure 7.19 in chapter 7 to illustrate the basic calculations performed to obtain values for t_e, T_E, T_L, and s. This network is redrawn as the first part of figure 22.10. The values of the optimistic, most likely, and pessimistic expected activity times are shown above each activity line. For example, for the activity extending from event S (start) to event 2:

$$t_e = \frac{1 + 4(3) + 5}{6} = \frac{18}{6} = 3 \text{ weeks}$$

The second part of figure 22.10 is a useful format for a table for calculations. We will illustrate its use as we proceed with our example. The procedure for the use of the table is as follows:

1. All events are entered in the event column in sequence, from first to last.
2. The numbers of all events that immediately precede each event are entered in the predecessor events column. Note that an event may have more than one predecessor.
3. The values of the expected event time, T_E, are calculated and entered in the expected event time column. All T_E's are calculated by starting with the first event and continuing until the last event is reached. The successive values are cumulative and are calculated from the following relationship:

$$T_E = T_E \text{ (predecessor)} + t_e \text{ (activity)}.$$

Thus:

T_E (event S) $= 0$
T_E (event 1) $= 4$
T_E (event 2) $= 4 + 7 = 11$, or 3
Event 2 has two predecessor events. Hence, there are two possible values for T_E. We use the larger, in this case, 11.
T_E (event 3) $= 9$
T_E (event F) $= 4 + 2 = 6$, or $11 + 1 = 12$, or $9 + 5 = 14$
We use the largest, which is 14. To illustrate the results, we have entered the values for all T_E's on the PERT network.

4. The values for the latest allowable time, T_L, are calculated by subtracting t_e from the T_L for its successor event. T_L is calculated by working from the last event toward the first event. The relationship for calculating T_L is:

$$T_L = T_L \text{ (successor)} - t_e \text{ (activity)}.$$

Whenever there are multiple paths leading back to an event, there is more than one possible value for T_L. The smallest value is used. Thus:

T_L (event F) $= T_E$ (event F) $= 14$
T_L (event 3) $= 14 - 5 = 9$
T_L (event 2) $= 14 - 1 = 13$
T_L (event 1) $= 13 - 7 = 6$, or $14 - 2 = 12$, and we use 6.
T_L (event S) $= 13 - 3 = 10$, $6 - 4 = 2$, or $9 - 9 = 0$, and we use 0.

Again, for illustrative purposes, we also have entered the values of T_L on the PERT network. The reason why there is no slack along paths that contain events S, 3, and F is that each of these events appears on the critical path, which is S-3-F.

It is important to note that slack applies to an entire path and not to each event on the path. Also, the value of slack is not always obvious. For example, it would be incorrect to assume that there is a slack of 10 weeks along path S-2-F by observing that the expected times from event S to event 2 and from event 2 to event F are 3 weeks and 1 week, respectively. Subtracting 4 weeks from the critical path time of 14 weeks would, in this case, result in an erroneous result. Event 2 cannot be completed in 3 weeks. It requires the prior completion of event 1. The path S-1-2 (which is the T_E for event 2) is 11 weeks long. The identification of slack time by inspection is very difficult for complex networks. In such cases, the use of a computer is warranted.

We now can consider the question: "What management actions can be taken?" The PERT network lends itself to "exception" reporting. This means that the manager need focus attention only on those activities and events that are not proceeding according to schedule. The PERT network can be expanded to provide more detailed coverage in areas requiring management attention. Also, conventional Gantt-type charts may be prepared from the PERT networks.

PERT provides the manager with an early warning of possible difficulties. The manager has many ways of reacting to problems if made aware of them with sufficient time for action. For example, the manager may:

1. Add new resources along a path with zero or negative slack. (Negative slack occurs when the slippage is such that the path length exceeds that of the critical path.)
2. Trade off resources by shifting them from less critical to more critical activities.
3. Extend the scheduled completion time.

The manager may utilize PERT networks to answer "What if"? questions. This is an effective technique for exploring the implications of alternative actions when complex PERT networks are maintained on a computer. The PERT network becomes a model that can be used to simulate the effect of changes of allocations of time and other resources.

PERT Network for System Implementation

To further illustrate the value of PERT, we will consider, as an example, a PERT network for implementing a computer-based system.

Figure 22.11 is a PERT network that displays the major implementation activities we have discussed in this chapter. We will use this network to illustrate some of the management options made possible by PERT. The events are numbered and described in the table accompanying the network. The expected activity times (t_e) have been calculated and appear beneath each activity line. The second part of figure 22.11, the table of PERT calculations for this network, presents the results of the significant calculations. Examination of this table leads to the following observations.

1. The critical path is the path along which the computer program development takes place (S-4-5-6-10-11-12-F). The length of this path is 43 weeks. The next longest path is that along which the interim and final system specifications are prepared (S-2-10-11-12-F). This path has a slack of 4 weeks. The manager may be able to reduce the longest path by several weeks by diverting resources from other paths and applying them to computer program development and testing.

Figure 22.11

Implementation network. PERT networks are useful in managing complex tasks, such as the implementation of a computer-based information system. The complexity of the network is due to the number of activities that must be completed. In this example, the critical path is S-4-5-6-10-11-12-F.

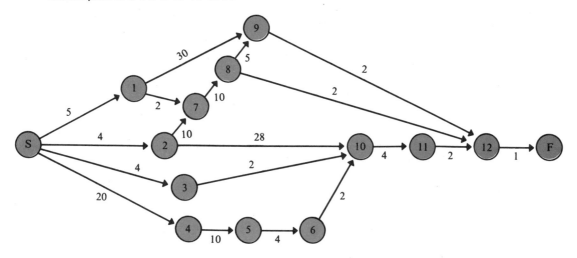

event	description
S	start of development phase
1	implementation plan prepared
2	interim system specification prepared
3	equipment installed
4	computer programming and testing completed
5	system tests completed
6	test reports prepared

event	description
7	training manuals prepared
8	completion of training
9	changeover plan prepared
10	final system specification completed
11	development phase report completed
12	acceptance review completed
13	approval to proceed received

event	predecessor events	T_E	T_L	S
S	—	0	0	0
1	S	5	10	5
2	S	4	8	4
3	S	4	18	14
4	S, 3	20	20	0
5	4	30	30	0
6	5	34	34	0

event	predecessor events	T_E	T_L	S
7	1, 2	14	25	11
8	7	24	35	11
9	1, 8	35	40	5
10	2, 3, 6	36	36	0
11	10	40	40	0
12	8, 9, 11	42	42	0
F	12	43	43	0

critical path   S-4-5-6-10-11-12-F

TABLE OF PERT CALCULATIONS FOR IMPLEMENTATION NETWORK

2. The path along which training occurs (S-2-7-8-12-F) has a slack of 11 weeks. This may present both a problem and an opportunity. The problem is that the expected event time, T_E, for event 8, completion of training, is 24 weeks. Personnel training will be completed too soon, and people may forget their training. The opportunity is the possibility of rescheduling the training program and using the available resources to accelerate progress along the critical path and along other paths with little slack.

3. Along the critical path (S-4-5-6-10-11-12-F) the longest T_E is associated with event 4, complete system tests. The manager may wish to review all program development and test plans. They may already be represented by subordinate PERT networks. If they are not, the manager may request that such networks be constructed to aid in the review.

The above network example illustrates some of the possible uses of PERT as a management tool. For actual management control of a project of moderate size, the network would be expanded to at least one additional level of detail. In practice the manager would probably want to use a computer to maintain and analyze the detailed PERT network. PERT computer programs and other software tools are available to aid in the design, construction, and implementation of computer-based systems.

Summary

Implementation is the process of bringing a developed system into operational use and turning it over to the user. Often it is effective to form implementation teams. An implementation plan is necessary; its major elements include test plans, training plans, an equipment installation plan, and a conversion plan. The test plan provides for the preparation of test data and for testing the system in a planned, structured manner. Top-down computer program development, as contrasted with bottom-up computer program development, is a contemporary technique for testing computer program modules as they are developed. Test reports document the performance of the system and its subsystems.

Training plans are necessary to insure that all persons who are associated with the computer-related information system have the necessary knowledge and skills. Reference manuals are prepared to assist in the training of programmers, operators, and user personnel. An important activity related to training is management orientation.

Equipment implementation is an activity that is completed during the development phase. The principal equipment-related activities are: site preparation, equipment installation, and hardware and software check-out.

Conversion is the process of initiating and performing all of the physical operations that result directly in the turnover of the new system to the user. There are two parts to conversion: (1) a conversion plan, which is implemented throughout the development phase, and (2) a changeover plan for moving the system from the development phase into the operation phase. The conversion plan includes procedures conversion, program conversion, and file conversion. The changeover plan also specifies the method of changeover from the old to the new system. Choices of changeover methods include parallel operation, immediate replacement, or phased changeover. The latter usually is the recommended method.

An implementation management team, which includes the head of each implementation team, primary users of the system, and members of the information service organization, oversees the implementation preparation activities. Implementation activities are complex. PERT is a management planning analysis tool that is often used in such situations. PERT (Program Evaluation Review Technique) is a network technique that relates activities—which are the application of time and resources to achieve objectives—to events, which are the points in time at which activities begin or end. PERT helps managers to foresee potential difficulties and to reallocate resources to alleviate critical problems.

For Review

implementation	HIPO chart	parallel operation
bottom-up computer program development	structured walk-through	immediate replacement
top-down computer program development	system specification	phased changeover
	reference manual	PERT
stub	**conversion**	network
implementation plan	**changeover**	activity
	implementation committee	event
		critical path

For Discussion

1. Distinguish between implementation, conversion, and changeover.
2. Define and describe bottom-up and top-down development and testing of computer programs.
3. What are the advantages of top-down development and testing of computer programs?
4. Discuss the importance of testing prior to changeover.
5. What are the principal reference manuals? Describe the content of each.
6. What are the values of including management orientation sessions in a training program?
7. Describe and distinguish between the three general changeover methods.
8. What is an implementation team? What is the implementation committee? What roles should the principal user and the systems analyst play?
9. Describe PERT and discuss its value as a management tool.

23
Computer Program Development

Preview

It is the responsibility of programmers to write the computer programs that make up the computer program component of the overall business information system, but it is the systems analyst who provides the programmers with the program requirements. These requirements include a brief written statement of the problem, a copy of the detailed process-oriented system flowchart, and copies of the input, output, and file design specifications. These requirements are provided to the programmer, along with the documentation developed in the design phase.

Objectives

1. You will understand the steps that are used by programmers to develop computer programs and to ensure that they function properly.
2. You will see how the programmer uses the flowcharts, input design, output design, and file design results of the design phase activities to develop a computer program.

Key Terms

algorithm   A set of rules or instructions used to accomplish a task.

coding   The process of writing instructions in a programming language, such as COBOL, PL/I, or RPG.

debugging   The process of testing a computer program for errors and correcting any errors found.

Steps in Computer Program Development

The steps in the development of each of the computer programs that make up the computer program component of a system are (1) define the function of the program; (2) plan the logic of the program; (3) code the program; (4) test and debug the program; and (5) complete the documentation.

Although the programmer is responsible for writing the computer program, the systems analyst must communicate the computer program requirements to the programmer. The function of each program was defined for the programmer when functions were allocated during system design. A detailed system *flowchart* is prepared for each program from the expanded *process-oriented system flowchart* created during the design phase. This flowchart and its narrative define the function of each program.

In *program* planning, the logic to be used to solve the problem is developed. Algorithms, computer program logic flowcharts, and *HIPO charts* are useful tools for program planning. **Algorithms** are sets of rules or instructions used to accomplish tasks. They may be stated as formulas, decision tables, or narratives. The program logic flowchart and algorithms that result from program planning are retained and become part of the project documentation.

The next step, writing, or **coding,** a program, is the actual writing of computer instructions. These instructions will be translated to machine code and followed by the computer; they should follow the steps of the program logic plan.

Several *programming languages,* particularly COBOL, PL/I, and RPG, are commonly used to solve business problems. Each language has its advantages and disadvantages. Most computer installations have a standard language used by their programmers. Programmers usually are not given a choice of languages unless some special circumstances exist.

Testing and debugging a program involve (1) translating the coded program into machine language, a process called *compilation;* and (2) testing the translated program with sample data and checking the result. If the results of testing are not correct, the program is said to have "bugs." **Debugging** is the process of correcting computer programs to obtain correct results.

As emphasized in chapter 22, testing must be planned and structured to reduce the chance that errors will be overlooked.

The last step is to complete the documentation for the program. The documentation must include a statement of the purpose of the program (from step 1), a description of the solution logic (step 2), a listing of the program instructions (step 3), and sample outputs from the completed programs (step 4). Information provided to the programmer by the analyst, such as descriptions of program inputs, outputs, and files, should be included. Instructions to operators explaining how the program is to be used must be written before the program documentation is complete.

Program Coding and Debugging Example

The data edit program of the OARS (On-Line Accounts Receivable System) provides an example of computer program development.

The problem definition is provided for the programmer in the form of a detailed process-oriented system flowchart and narrative.

Figure 23.1

Process-oriented system flowchart—OARS data edit program. This process-oriented system flowchart was derived from the OARS expanded flowchart drawn early in the design phase (figure 16.1a).

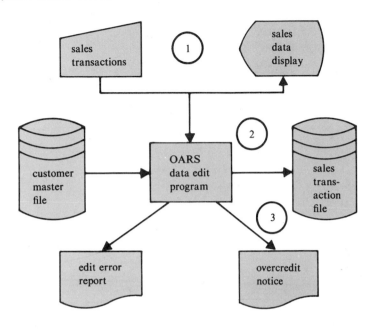

data edit programs-OARS

1. Sales data is entered through CRT terminals.

2. All fields are checked for validity. The new balance is calculated and compared to the customer's credit limit (master file).

3. Valid sales transactions that are not overcredit are recorded in the sales transaction file. Invalid data generates an error report entry. Overcredit conditions cause an overcredit notice to be printed.

Figure 23.1 illustrates the required detail. The flowchart is derived from the OARS expanded flowchart, figure 16.1 (chapter 16).

Before the programmer can prepare the computer program, additional details of the input, storage, and output must be provided.

Input layouts were developed during input design (chapter 18), print charts and screen layouts during output design (chapter 17), and file descriptions during file design (chapter 19). Figure 23.2 is a print chart from output design.

Figure 23.2

Overcredit notification print chart. Output descriptions created during the design phase are used to communicate system output requirements to the programmer.

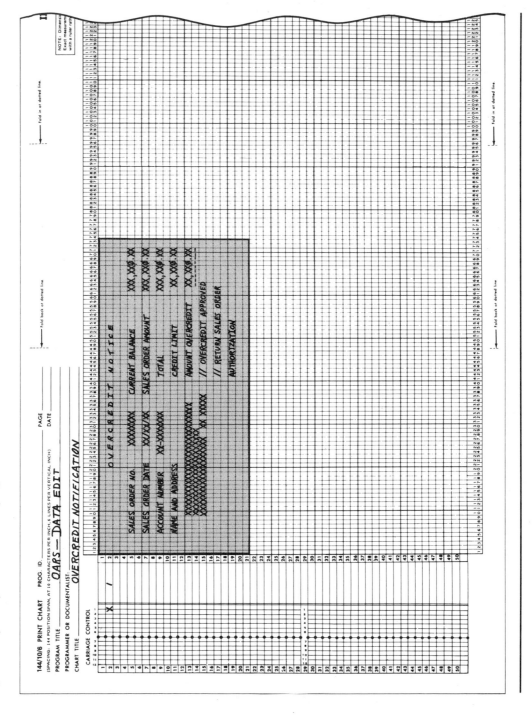

Figure 23.3

Overcredit decision tree. Decision trees can be used to communicate required processing logic. This example communicates the logic used to determine whether or not an overcredit notification is required.

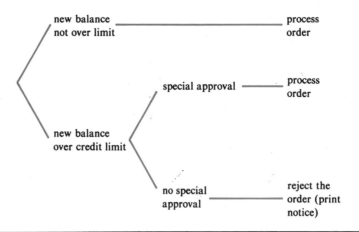

To aid the programmer, algorithms to express the rules or logic of solving the problem, as well as *decision tables* or *decision trees,* may be provided. The programmer then develops a computer program logic flowchart and/or a logic description in *pseudocode.*

Figure 23.3 is a decision table that depicts the logic used to determine the requirement for an overcredit notice. Figure 23.4a shows one page of a computer program logic flowchart. Figure 23.4b shows the equivalent plan using pseudocode.

After completing the logic plan the programmer can code the program logic.

Figure 23.5 depicts the coding of the segment in figure 23.4. It uses COBOL (COmmon Business Oriented Language) coding.

Each program should be tested by the programmer until all errors are found and corrected. Each output must match the planned output in the print charts and display layouts.

Figure 23.6 illustrates the overcredit notice produced by the data edit program. Note that the output corresponds to the print chart of figure 23.2.

After testing and debugging each program, the programmer turns over all documentation to the analyst so that the program can replace its "stub" in the overall top-down procedure for the design, development, and testing of the computer program component of the computer-based business information system.

442

Figure 23.4a

OARS computer program flowchart segment. The programmer will use flowcharts or other planning techniques to design the computer program logic. This example includes the logic of the decision tree shown in figure 23.3.

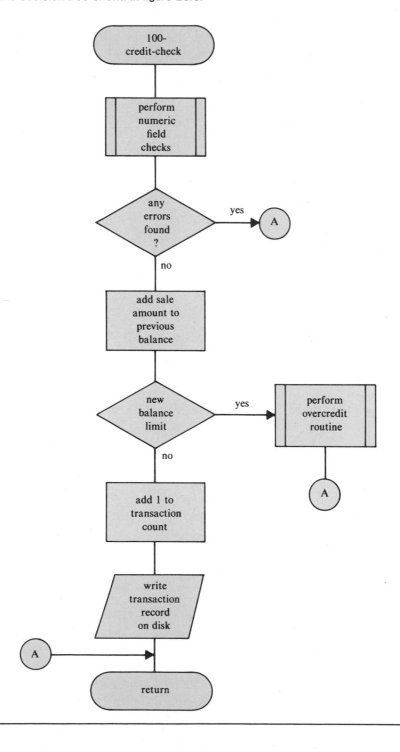

Figure 23.4b
OARS pseudocode segment. Pseudocode is a common alternative to flowcharting for planning program logic. This pseudocode example conveys logic that is equivalent to the flowchart of figure 23.4a.

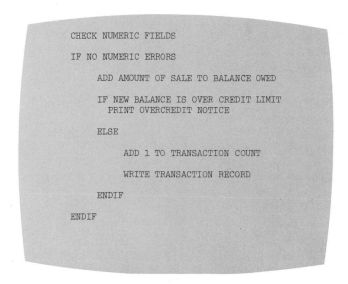

```
CHECK NUMERIC FIELDS

IF NO NUMERIC ERRORS

        ADD AMOUNT OF SALE TO BALANCE OWED

        IF NEW BALANCE IS OVER CREDIT LIMIT
           PRINT OVERCREDIT NOTICE

        ELSE

                ADD 1 TO TRANSACTION COUNT

                WRITE TRANSACTION RECORD

        ENDIF

ENDIF
```

Figure 23.5
OARS data edit coding sample. This coding sample, using COBOL, was written following the logic plan depicted in figures 23.4a and 23.4b.

```
080010************************************************************
080120*                                                          *
080030*        THIS ROUTINE IS PERFORMED FOR CREDIT LIMIT        *
080040*        CHECKS.  IT ALSO CHECKS ALL NUMERIC FIELDS        *
080050*        TO VERIFY THAT THEY CONTAIN VALID NUMERIC DATA.   *
080060*                                                          *
080070************************************************************
080080
080090 100-CREDIT-LIMIT-ROUTINE.
080100     PERFORM 1001-NUMERIC-FIELD-CHECK.
080110     IF NUMERIC-ERRORS IS EQUAL TO 'NO'
080120         ADD SALE-AMOUNT TO ACCOUNT-BALANCE
080130         IF ACCOUNT-BALANCE IS GREATER THAN CREDIT-LIMIT
080140             PERFORM 1002-OVERCREDIT-ROUTINE
080150         ELSE
080160             ADD 1 TO TRANSACTION-COUNT
080170             WRITE TRANSACTION-RECORD.
```

Figure 23.6
OARS data edit sample output. This overcredit notification is an example of an actual output produced by the OARS system. It matches the output design for the overcredit notification shown in figure 23.2.

```
              O V E R C R E D I T    N O T I F I C A T I O N

   Sales Order No.   00069721      Current Balance        2,320.50

   Sales Order Date  10/15/84      Sales Order Amount     1,516.02

   Account Number    02-0000367        Total              3,836.52

   Name and Address                 Credit Limit          3,500.00

      ABC Corporation              Amount Overcredit         336.52
      2234 Washington Blvd.                              ----------
      Big City          CA 91789   / / Overcredit Approved

                                   / / Return Sales Order

                                   Authorization
```

Summary

It is the responsibility of programmers to write the programs required for the computer program component of a business information system. These programs, however, must be written according to the specifications provided by the analyst. The specifications provided usually include the design of the inputs, outputs, and files, along with a copy of the detailed process-oriented system flowchart.

Five steps in the development of a computer program are: (1) define the problem to be solved, (2) plan the logic for a solution to the problem, (3) write the program following the logic plan, (4) test and debug the program, and (5) complete the documentation for the program. When these steps are completed the program and its documentation are ready to be included in the system tests.

For Review

algorithm
coding
debugging
decision table

flowchart
HIPO charts
program
decision tree
compilation

process-oriented
  system flowchart
programming languages
pseudocode

For Discussion

1. What are the five steps in computer program development?
2. What is program planning?
3. What is an algorithm? Give an example.
4. Who is responsible for testing a computer program? For subsystem testing?
5. What is pseudocode? Its relationship to structured English?
6. What are responsibilities of the systems analyst for computer program development?

24
Development Phase
Report and Review

Preview

At the conclusion of the development phase the analyst prepares a report and reviews it with the principal user of the computer-based business information system. The central element of the development phase report is the system specification, which is an "as built" specification. The development phase review is attended by all of the users and managers who will be affected by the system. This is a particularly critical review, since a successful outcome initiates the changeover from a "project" to an "operational" system.

Objectives

1. You will become familiar with the content of a system specification.
2. You will learn how to prepare a development phase report.
3. You will understand the purposes of a development phase review.

Key Terms

development phase report   A report prepared at the end of the development phase; it is an extension of the design phase report and summarizes the results of the development phase activities.

system specification   A baseline specification that contains all of the essential system documentation; it is a complete technical specification.

development phase review   A review held with the user organization at the conclusion of the development phase to determine whether or not to enter the operation phase.

System Specification

The development phase is concluded by the completion of three major milestone activities. These are:

1. Preparation of the system specification
2. Preparation of the development phase report
3. Conduct of the development phase review

The **development phase report** is built around the system specification—the third and final major *baseline document*. The system specification evolves from the design specification, just as that baseline specification evolved from the performance specification. Whereas the *design specification* is a "build to" specification, the system specification is an "as built" specification. It is the major reference document for all personnel who will use, maintain, or operate the computer-based business system.

Figure 24.1 illustrates the content of a typical system specification. Like the performance and design specifications, the **system specification** is divided into two parts. The first part is an external specification relating to the interaction of the information system with its environment; the second part is an *internal system specification* that completely documents the computer program component of the system. The system specification is the result of documenting the development phase activities previously described in chapters 22 and 23.

Figure 24.2 is an extension of figure 20.2. It is a gozinto chart that traces the growth of the baseline documentation from the performance specification to the system specification.

Development Phase Report

Structure and Content

Figure 24.3 displays the structure and content of the development phase report, which is similar to the study phase and design phase reports. It has the same five major divisions, and is a continuation of the cumulative documentation performed in the predecessor phases. When completed, the development phase report completely documents the internal system design.

Unless changes to the performance specification, the design specification, or both, occur during the development phase, the system scope section is not altered. If there have been any changes in scope, these should be discussed fully.

The conclusions and recommendations relate to the next life-cycle phase, the operation phase, and focus on the next major decision. This is the decision to change over from the existing system to the new system. Important inputs to the changeover decision are (1) the completed system specification, (2) satisfactory system test reports, (3) the availability of trained personnel, and (4) a changeover plan.

In preparing conclusions and recommendations, the analyst must take into consideration the environment in which the computer-based business information system will be maintained and operated after complete conversion. Preliminary operational schedules, prepared as part of the *changeover plan,* should be used to demonstrate that the new system will mesh with the schedules for ongoing data processing jobs.

Figure 24.1
System specification outline. The system specification is the technical core of the development phase report and an expansion of the design specification. It has two major sections: the external (to the computer program component) system specification, and the internal (to the computer program component) specification.

system specification	
a. *external system specification*	1. flowchart: information-oriented and/or data flow diagram
	2. system output specification
	3. system input specification
	4. system interface specification
	5. system test specification
	a. test data
	b. test results-samples
	6. equipment specification
	7. personnel specification and training procedures
	a. user's reference manual
	b. programmer's reference manual
	c. operator's reference manual
b. *internal system specification*	1. computer program component specification
	a. flowchart: process-oriented and/or data flow diagram
	b. expanded system flowcharts and HIPO charts
	c. data base specification
	d. computer program control specification
	e. computer program test specification
	(1) test data
	(2) test results-samples
	2. computer program specification (for each program, as required)
	a. detailed system flowcharts
	b. computer program logic flowchart
	c. transaction file specifications
	d. control specifications
	e. interface specifications
	f. computer program listings
	g. test specifications
	(1) test data
	(2) input-output samples
	h. special specifications

Figure 24.2

Performance specification-design specification-system specification gozinto chart. This gozinto ("goes into") chart is an extension of figure 20.2. It shows how each specification builds upon a predecessor specification, with the system specification becoming the final and most complete specification.

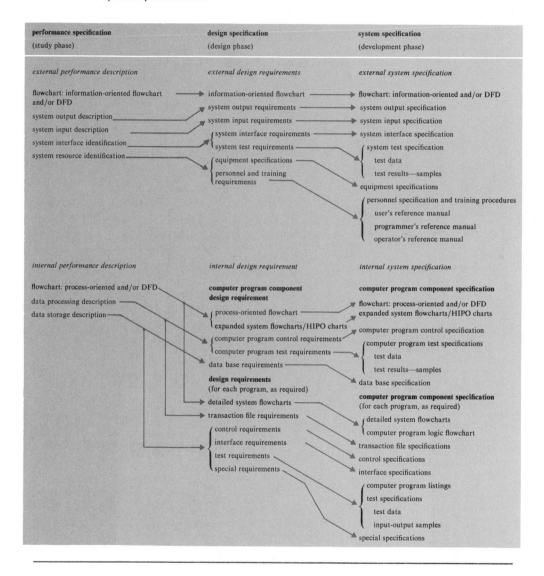

Figure 24.3
Development phase report. The development phase report contains the same five
sections as the study phase and design phase reports: system scope; conclusions and
recommendations; system specification; plans and cost schedules; and appendices.
However, the content is expanded to include results of the development phase activities.

development phase report	
i. *system scope*	a. system title
	b. problem statement and purpose
	c. constraints
	d. specific objectives
	e. method of evaluation
ii. *conclusions and recommendations*	a. conclusions
	b. recommendations
iii. *system specification*	a. external system specification
	b. internal system specification
iv. *plans and cost schedules*	a. progress plans
	1. detailed milestones—development phase
	2. major milestones—all phases
	3. changeover plan
	4. operational plan
	b. cost schedules
	1. project cost—development phase
	2. project cost—major milestones
	3. operation phase—recurring costs
v. *appendices*	

The system specification follows the conclusions and recommendations in the development phase report. Major elements of the system specification were outlined in figure 24.1. Referring to the figure, we note that the reference manuals for programmers, operators, and users appear as part of the *external system specification.* These manuals were discussed previously (chapter 22) as important training tools. Because much of the content of these documents is based on other parts of the system specification, and because they are stand-alone manuals that are used for training, the entries shown under the external specification may refer the reader to appendices in which these manuals are contained. Figure 24.4 identifies the elements of the report on which the content of each reference manual is based.

The systems analyst updates the major milestones project plan and the cost schedules to display the completion of the development phase. The analyst identifies and describes significant changes that have occurred since the design phase review, and also prepares cost estimates for operating and maintaining the new system elements.

The appendices that appeared in the design phase report are reviewed, modified, and incremented as necessary.

Example Development Phase Report

An example of a development phase report appears on the following pages as Exhibit 3, OARS development phase report.

Development Phase Review

A successful **development phase review** marks the beginning of the transition from a "project" to an "operational" system. The immediate consequence of an approval to proceed is the initiation of the changeover activity. Therefore, it is critically important that all the factors that will ensure an effective transition from the old system to the new system be considered. All organizations involved in or significantly affected by the pending conversion should be alerted and be invited to participate in the review: the development phase report should be distributed to them beforehand; the study phase and design phase reports, prepared at the conclusion of earlier phases, are provided as reference documents. In addition, the review group should have had an opportunity to read the changeover plan and the test reports. A commitment to the changeover activities should be obtained from each major participant before a decision is made to embark on the operation phase.

Figure 24.4

Relationship between development phase report elements and reference manuals. The development phase report is the source for the principal reference manuals. This figure shows the extent to which the programmer's, operator's, and user's reference manuals are based upon information in the development phase report.

DEVELOPMENT PHASE REPORT ELEMENT	Programmer's Reference Manual	Operator's Reference Manual	User's Reference Manual
SYSTEM SCOPE			
Title	✓	✓	✓
Problem Statement and General Objective	✓	✓	✓
SYSTEM SPECIFICATION—EXTERNAL			
Information-Oriented Flowchart	✓	✓	✓
System Output Specification	✓	✓	✓
System Input Specification	✓	✓	✓
System Interface Specification	✓	✓	✓
System Test Specification	✓	✓	
Equipment Specification	✓	✓	
SYSTEM SPECIFICATION—INTERNAL			
Computer Program Component			
Process-Oriented Flowcharts	✓	✓	✓
Expanded System Flowcharts	✓	✓	
Data Base Specification	✓		
Control Specifications	✓	✓	
Test Specifications	✓	✓	
Computer Programs			
Detailed System Flowchart	✓	✓	
Logic Flowcharts	✓		
Transaction File Specifications	✓	✓	
Control Specifications	✓	✓	
Interface Specifications	✓	✓	
Listings	✓		
Test Specifications	✓	✓	
Special Specifications	✓	✓	

Summary

Upon completion of the development phase activities, the systems analyst prepares a design phase report. This report contains the system specification, which is the complete technical specification for the computer-based business information system. The system specification is the expansion of the design specification into the final baseline specification. The system specification contains two parts: an external system specification and an internal system specification. The external specification relates to the interaction of the information system with its environment. For example, it contains the training procedures and user manuals. The internal specification provides the complete documentation for the computer program component of the information system.

After the principal user has had an opportunity to study the development phase report a review is held and a decision is made whether or not to enter the operation phase. A decision to proceed marks the transition from a "project" to an "operational" system. Therefore, the changeover plan is an important element of the development phase review, and a commitment to the changeover activities must be obtained from all major users of the system prior to initiation of the operation phase.

For Review

development phase report   baseline document   internal system specification

system specification   **development phase review**   external system specification

design specification   changeover plan

For Discussion

1. What is the purpose of the development phase report?
2. What is the content of the development phase report?
3. Discuss the importance of the system specification, including its relationship to the performance and design specifications.
4. What project plans and schedules are presented at the development phase review?
5. What is the purpose of the development phase review?

Exhibit 3: The Development Phase Report

A case study that follows the life-cycle road map is introduced in chapter 11, Initial Investigation. In the previous edition, the case study resulted in a Modified Accounts Receivable System (MARS), which was an improvement over a predecessor manual system. MARS operated in a batch processing mode and depended upon punched card data preparation. In this edition MARS is the "old" system; the case study updates MARS to an On-line Accounts Receivable System (OARS). A principal goal of OARS is to establish geographically separate profit centers. The evaluated candidate systems are distributed data processing systems, which range from "dumb" terminals to a substantial local data processing capability, and which provide for direct data entry and several communications options.

Exhibits of example study phase, design phase, and development phase reports are included in chapter 14 (Study Phase Report and Review); chapter 20 (Design Phase Report and Review); and chapter 24 (Development Phase Report and Review). The attachment to this chapter is Exhibit 3: The Development Phase Report.

OARS Development Phase Report

  I. System Scope
    A. System Title
       On-line Accounts Receivable System (OARS)
    B. Problem Statement and Purpose
          The ABCO corporation's present accounts receivable system is at its maximum capacity of 10,000 accounts. The number of accounts is expected

to double to 20,000 accounts over a five-year period. The present system cannot meet this projected growth and satisfy the corporate goal of distributing information processing resources to regional profit centers. Serious problems already have been encountered in processing the current volume of accounts. Specific problems that have been identified are:

1. Saturation of the capacity of the present computer system, causing difficulties in adding new accounts and obtaining information about the status of existing accounts.
2. Processing delays in preparing customer billing statements because of the batch-oriented design of the current accounts receivable system.
3. Excessive elapsed time between mailing of customer statements and receipt of payments, which creates a high-cost, four-day float.
4. Inadequate control of credit limits.
5. Inability to provide regional centers with timely customer-related information.

Therefore, the purpose of the OARS project is to replace the existing accounts receivable system with one that can eliminate the stated problems enumerated above and meet ABCO's growth and regional accountability goals.

C. Constraints

The OARS constraints are:

1. Development of the on-line accounts receivable system is to be completed within fourteen months.
2. OARS is to have a growth potential to handle 20,000 customer accounts.
3. OARS is to interface with the existing perpetual inventory system.
4. OARS is to be designed as an on-line system operating in a distributed data processing environment.
5. The design must be compatible with corporate plans to install regional profit centers.

D. Specific Objectives

The specific objectives of OARS are:

1. To establish billing cycles for each region.
2. To mail customer statements no later than one day after the close of a billing cycle.
3. To provide the customer with a billing statement two days after the close of a billing cycle.
4. To speed up collections, reducing the float by 50 percent.
5. To examine customer account balances through on-line inquiry at the time of order entry.

E. Method of Evaluation

After OARS has been operational from sixty to ninety days:

1. A statistical analysis will be made of customer account processing to verify the elapsed time between the close of the billing cycle and the mailing of customer statements.

2. The float time will be measured, and the cost of the float will be calculated at three-month intervals.
3. Periodically, random samples of customer accounts in each region will be audited for accuracy and to validate the effectiveness of on-line inquiry.
4. The validity of OARS transactions that affect the inventory system will be measured by random sampling and physical count.
5. Personal evaluations of the effectiveness of the system will be obtained from its principal users.

II. Conclusions and Recommendations
   A. Conclusions

   The On-line Accounts Receivable System (OARS) is ready for operation. The system tests have been successfully completed; all required personnel have completed training; and the changeover plan has been reviewed and approved. All phases were completed on schedule and within budget.
   B. Recommendations

   It is therefore recommended that the OARS project be accepted for changeover.

III. System Specification
   A. External System Specification
      1. *Flowcharts*   Figure E3.1a is an information-oriented system flowchart for OARS. The accompanying narrative appears as figure E3.1b. Figure E3.2 is the logically equivalent data flow diagram.
      2. *System output specification*   The six OARS outputs are:
         a. Customer monthly statement
         b. Accounts receivable transaction register/display
         c. Accounts receivable summary display
         d. Aged accounts receivable report
         e. Customer account list display
         f. Overcredit notification
      Print charts and data element lists for each output are presented such as the overcredit notification shown in figure E3.3.
      3. *System input specification*   The three OARS inputs are:
         a. Customer order
         b. Billing notice
         c. Payment/credit
      An example of the system inputs is included as figure E3.4.

Figure E3.1a
OARS information-oriented flowchart.

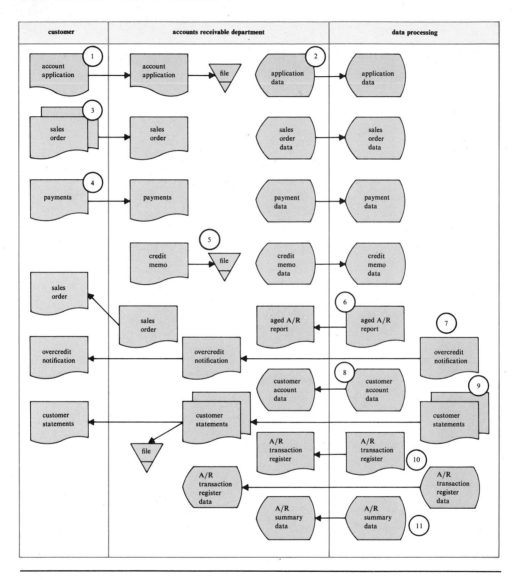

Figure E3.1b
Narrative for OARS information-oriented flowchart.

```
                INFORMATION-ORIENTED FLOWCHART NARRATIVE


      1.  Customers submit account applications that are processed
          by the accounts receivable department.

      2.  If an application is accepted, it is keyed into the
          system.  The application is filed.

      3.  Sales orders are entered through on-line order
          processing.  The customer retains a carbon copy.

      4.  Payments are sent to the accounts receivable
          department and entered into the system.

      5.  Credit memos are generated and the data entered.
          The memo is filed.

      6.  An aged A/R report is sent to the accounts receivable
          department from data processing each month.

      7.  Data processing sends overcredit notices to the credit
          department whenever a new order would exceed the
          customer's credit limit and the additional credit is
          not approved.  The order and notice are returned to the
          customer.  If the additional credit is approved, the
          order is processed.

      8.  Customer account data is displayed on demand.

      9.  Customer statements are sent to accounts receivable
          in duplicate.  The original copy is sent to the
          customer; the duplicate is filed.  One-third of the
          statements are produced each ten days of the month,
          that is, on the 1st, 10th, and 20th.

     10.  A daily A/R transaction register is sent to the accounts
          receivable department as a file copy.  The data is also
          available as a display.

     11.  The accounts receivable summary data is available as
          a display to the accounts receivable department.
```

4. *System interface specification*   The OARS must interface with (that is, transfer data to and from) the existing inventory system. Inventory must be on hand for shipment prior to billing the customer, and the inventory quantities should be reduced as merchandise is committed for shipment.
5. *System test specification*   The system was tested according to the OARS test plan. No major problems were detected. A copy of the test plan, test data, and test results can be found in the appendix.

Figure E3.2
OARS data flow diagram.

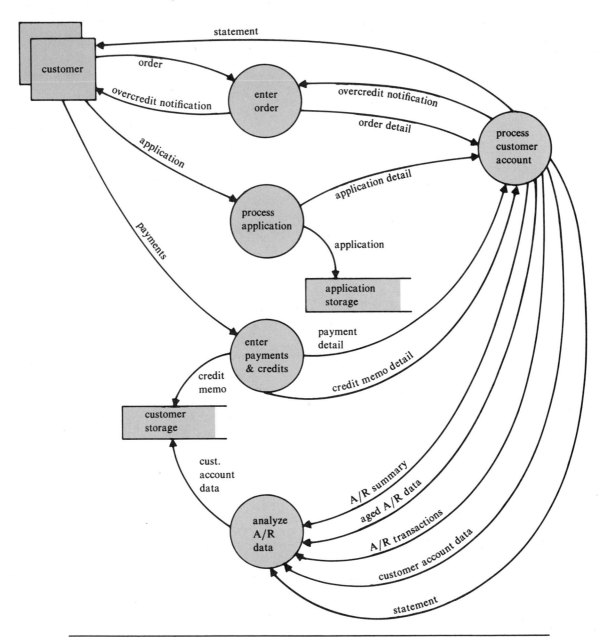

Figure E3.3
OARS output design requirement: overcredit notification.

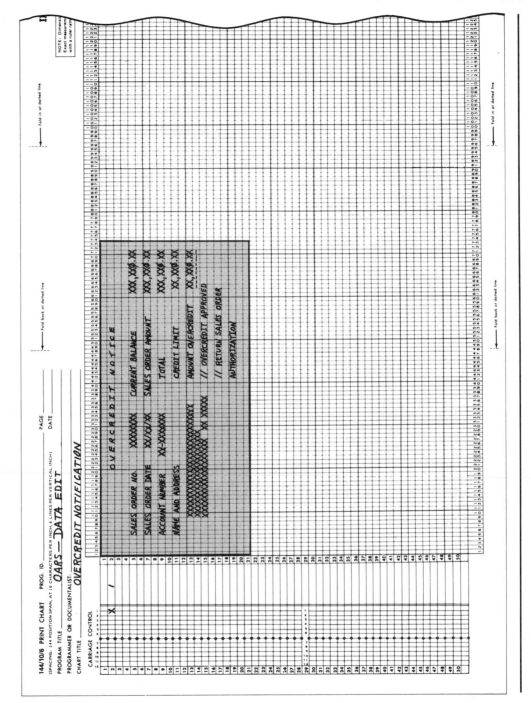

Figure E3.3
continued

OUTPUT SPECIFICATION

TITLE: Overcredit Notification

LAYOUT:

Overcredit Notification

Sales Order No. _____
Sales Order Date _____
Account Number _____
Name and Address _____

_____
_____

Current Balance _____
Sales Order Amount _____
Total _____
Credit Limit _____
Amount Overcredit [_____]

☐ Overcredit Approved
☐ Return Sales Order

AUTHORIZATION

FREQUENCY: Daily
SIZE: 1 page
QUANTITY: 80 max.
COPIES: 2

DISTRIBUTION: 1. Accts. Receivable Dept.

COMMENTS:

DATA ELEMENT LIST		

TITLE: Overcredit Notification

DESCRIPTION	FORMAT	SIZE
Sales Order Number	XXXXXXXX	8 characters
Sales Order Date	XX/XX/XX	8 "
Account No.: Region No.—Sequence No.	XX-XXXXXXX	10 "
Name		26 "
Address		
Street		20 "
City		18 "
State		2 "
Zip Code		5 "
Current Balance	XXX,XXX.XX	10 "
Sales Order Amount	XXX,XXX.XX	10 "
Total	XXX,XXX.XX	10 "
Credit Limit	XX,XXX.XX	9 "
Amount Overcredit	XX,XXX.XX	9 "

Figure E3.4
OARS input design requirement: customer account application.

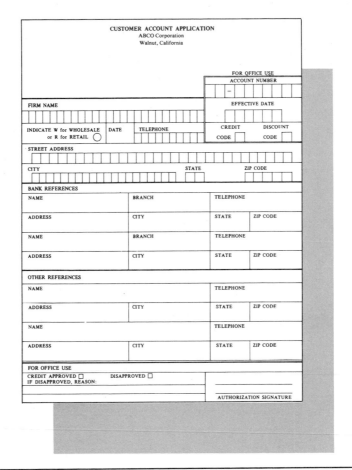

6. *Equipment Specifications*   The current central-site computer system will be augmented by five Excalibur V minicomputer systems, one for each region. Each regional system will have the following configuration:
   a. CPU with 128K positions of main memory
   b. One 600 line:-per-minute printer
   c. One 130 megabyte magnetic disk drive
   d. One console keyboard/printer
   e. Two CRT keyboard/display terminals, expandable to eight stations
7. *Personnel specification and training procedures*   The user's reference manual, programmer's reference manual, and the operator's reference manual are attached to this system specification as appendices.

Figure E3.5a
OARS process-oriented flowchart and narrative.

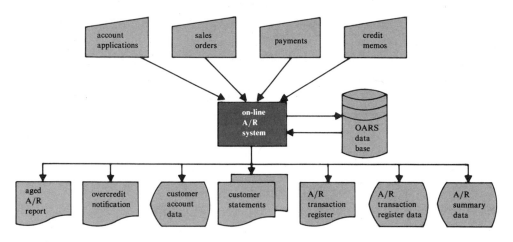

```
SYSTEM FLOWCHART NARRATIVE
```

1. Customer account applications, payments, credit memos, and sales orders are the four major system inputs.

2. The A/R program inputs the transaction data and reads/ updates the master file on magnetic disk.

3. The seven major outputs of the system are the aged A/R report, overcredit notification, customer account data display, customer statements, A/R transaction register, A/R transaction register data display, A/R summary data display.

B. Internal System Specification
   1. *Computer program component specification*
      a. Flowcharts. The process-oriented system flowchart is shown in figure E3.5a. Figure E3.5b is the logically equivalent data flow diagram from which figure E3.5a was derived.
      b. Expanded flowcharts. Figure E3.6a depicts the OARS expanded system flowchart. Figure E3.6b is the accompanying narrative. The expanded data flow diagram, used to develop to figure E3.6a, appears as figure E3.6c. Figure E3.6d shows the HIPO charts for OARS.

Figure E3.5b
OARS data flow diagram.

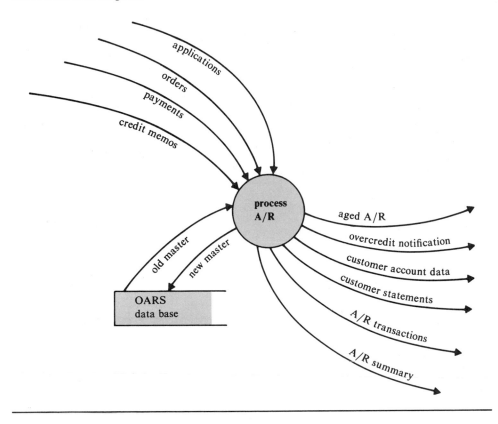

c. Data base requirements. The OARS master files will be indexed sequential files. Each logical record will be 133 bytes in length. Figure E3.7 is a disk record layout. The blocking factor is to be 19.

d. Computer program control requirements. The OARS computer programs will provide the following controls:

(1) Edit all input data for validity.

(2) Detect overcredit conditions prior to processing invoices.

(3) Detect short or out-of-stock conditions prior to processing invoices.

e. Computer program component test specification. The OARS test plan, test data, and test results are included in the report appendix. All tests were in accordance with the test plan.

Figure E3.6a
OARS expanded system flowchart—processing detail.

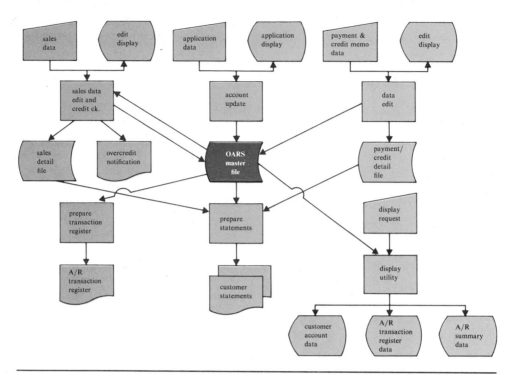

2. *Computer program specification—data edit*
   a. The process-oriented system flowchart, is presented as figure E3.8.
   b. Computer program logic flowchart. Figure E3.9 illustrates the computer logic flowchart for the data edit program.
   c. Transaction file specifications. An example of the transaction file requirements for OARS is shown in figure E3.10.
   d. Control specifications. The data edit program must edit all input data for validity and detect overcredit conditions. If an overcredit condition exists, an overcredit notice is to be produced. Records containing invalid data are to be noted on an edit error report.
   e. Interface specifications. The data edit program output, on disk, must interface with the invoice program that is part of the perpetual inventory system.
   f. Computer program listings. The data edit program source listing is included as figure E3.11.
   g. Test specification. The test plan, test data, and the test input-output samples are included in the report appendix.
      Note: Only the data edit program module is shown in this example.

Figure E3.6b
Narrative for OARS expanded process-oriented flowchart.

```
OARS EXPANDED PROCESS-ORIENTED FLOWCHART NARRATIVE


Sales data edit and credit check:

   The sales data is entered via terminal keyboard.  A
numeric edit is performed.  Bad data discovered during the
edit is displayed to the terminal operator.  If the data is
valid, a check is performed to determine whether or not the
customer will exceed their credit limit.  An overcredit
notification is produced if the limit is exceeded and the
data is recorded in the sales detail file.

Account update:

   Application data is entered via terminal keyboard.  The
OARS master file is updated to reflect the new account.

Payment/credit memo data edit:

   The payment and credit memo data is entered via terminal
keyboard.  A check is made to verify account numbers and
valid numeric data.  Good data updates the master file and
is recorded in the payment/credit detail file.

Prepare transaction register:

   At the end of each month, a transaction register is printed
using data from the OARS master file.

Prepare customer statements:

Display utility:

   Requests to display customer account data, A/R transaction
register data, or A/R summary data are made through any
terminal with proper password clearance.  All data comes from
the OARS master file.
```

IV. Plans and Cost Schedules
   A. Detailed Milestones—Development Phase
        The project progress for the OARS development phase is depicted in figure E3.12. The development phase was completed on schedule with the acceptance review scheduled for 11/10/xy. The costs for the development phase are shown in figure E3.13.
   B. Major Milestones—All Phases
        The progress plan and status report for the OARS project is included as figure E3.14. Figure E3.15 depicts the total OARS project costs.
   C. Changeover
        The approved changeover plan appears in the report appendix.
   D. Operational—Recurring
        The weekly recurring costs per account for OARS are depicted in figure E3.16.

Figure E3.6c
OARS expanded data flow diagram.

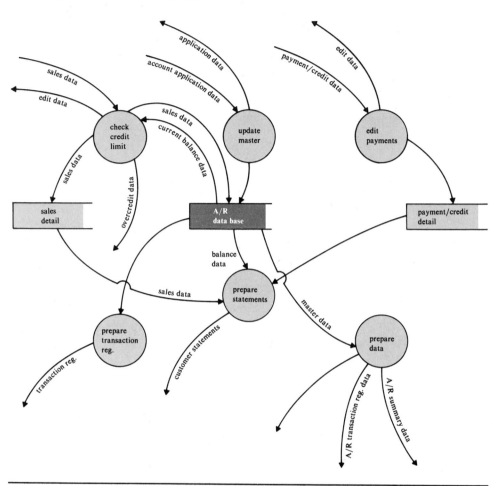

V. Appendices
   Note: The appendices include all supporting data for the report. For the purpose
         of this exhibit, such detail is not included. The following is a list of typical
         appendices.
         Project directive
         System test plan
         System test data and results
         User, programmer, and operator manuals
         Computer program component test plan

Figure E3.6d
HIPO charts for OARS.

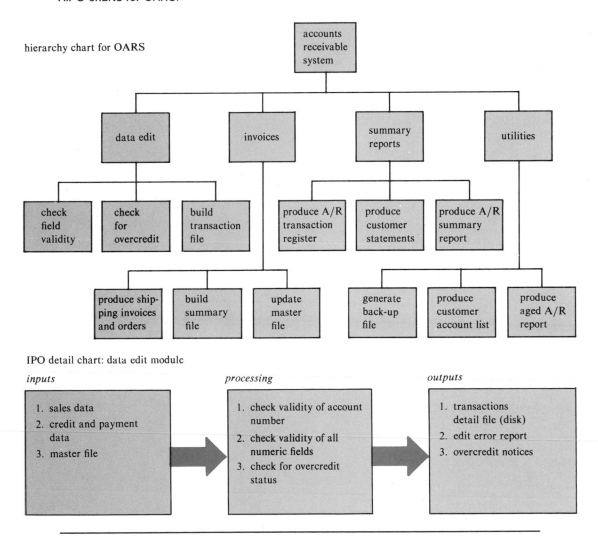

hierarchy chart for OARS

IPO detail chart: data edit module

Computer program component test data and results
Program module test plans
Program module test data and results
System changeover plan
System operational plan

Figure E3.7
OARS disk record layout.

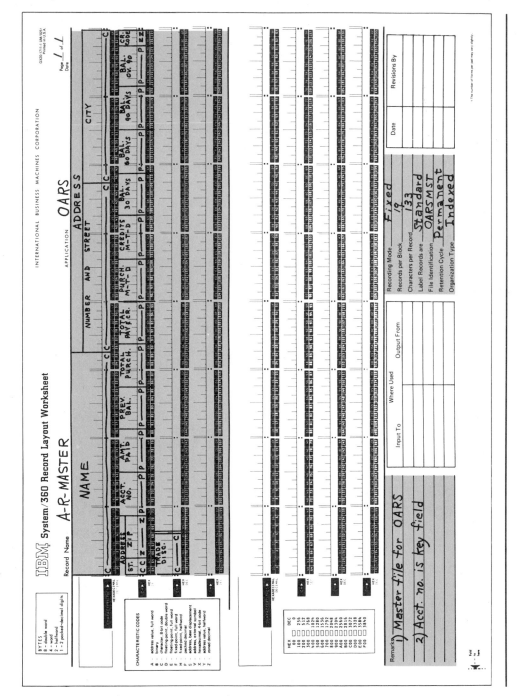

Figure E3.8
Data edit program module: process-oriented flowchart.

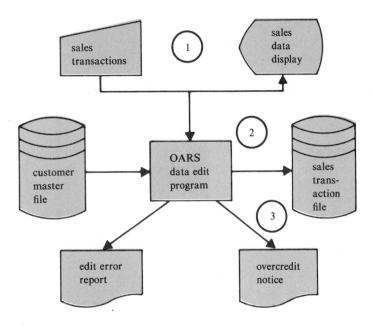

data edit programs-OARS

1. Sales data is entered through CRT terminals.

2. All fields are checked for validity. The new balance is calculated and compared to the customer's credit limit (master file).

3. Valid sales transactions that are not overcredit are recorded in the sales transaction file. Invalid data generates an error report entry. Overcredit conditions cause an overcredit notice to be printed.

Figure E3.9
Data edit program: computer logic flowchart.

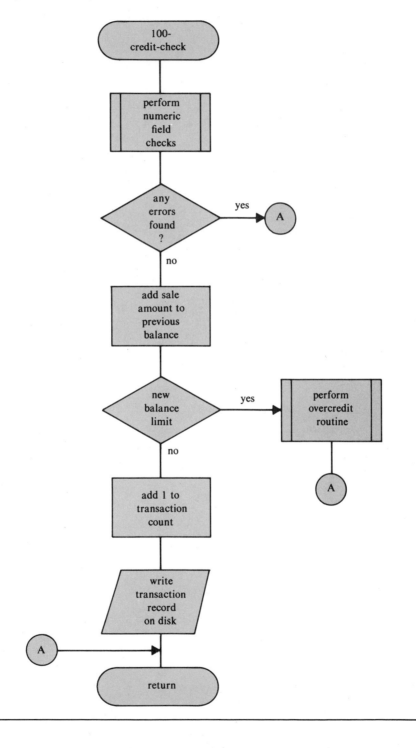

Figure E3.10
OARS transaction file requirements: customer application.

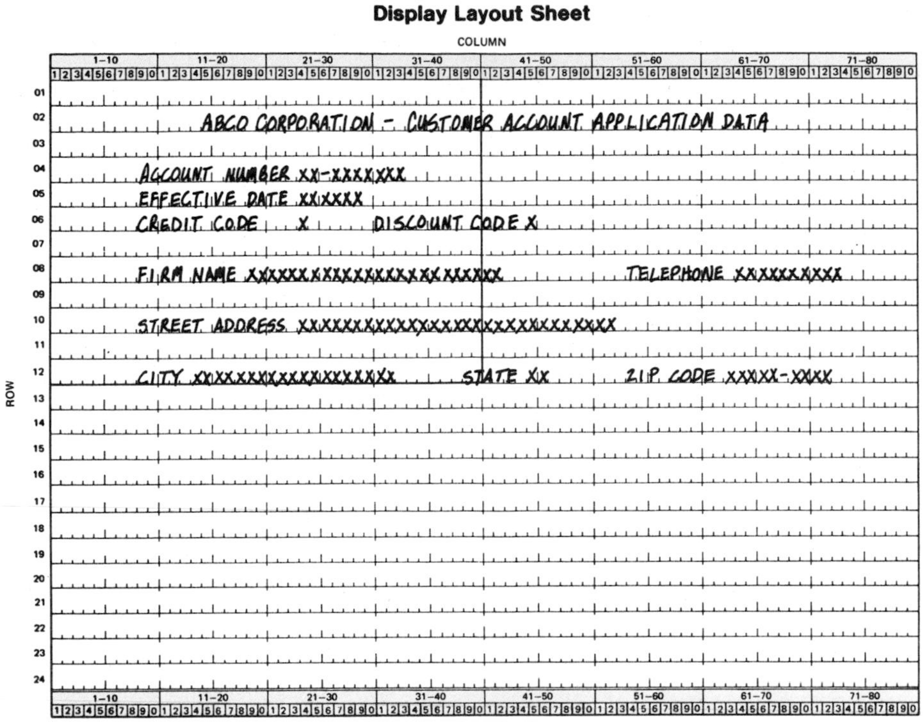

Display Layout Sheet

Figure E3.11
Data edit program: source listing.

```
080010***********************************************************
080120*                                                         *
080030*        THIS ROUTINE IS PERFORMED FOR CREDIT LIMIT        *
080040*        CHECKS.  IT ALSO CHECKS ALL NUMERIC FIELDS        *
080050*        TO VERIFY THAT THEY CONTAIN VALID NUMERIC DATA.   *
080060*                                                         *
080070***********************************************************
080080
080090 100-CREDIT-LIMIT-ROUTINE.
080100     PERFORM 1001-NUMERIC-FIELD-CHECK.
080110     IF NUMERIC-ERRORS IS EQUAL TO 'NO'
080120        ADD SALE-AMOUNT TO ACCOUNT-BALANCE
080130        IF ACCOUNT-BALANCE IS GREATER THAN CREDIT-LIMIT
080140            PERFORM 1002-OVERCREDIT-ROUTINE
080150        ELSE
080160            ADD 1 TO TRANSACTION-COUNT
080170            WRITE TRANSACTION-RECORD.
```

Figure E3.12
Project plan and status report—development phase.

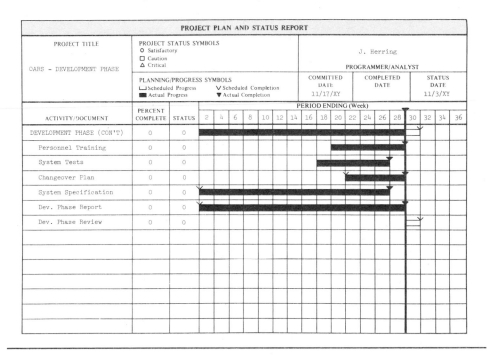

Figure E3.13
Project cost—development phase.

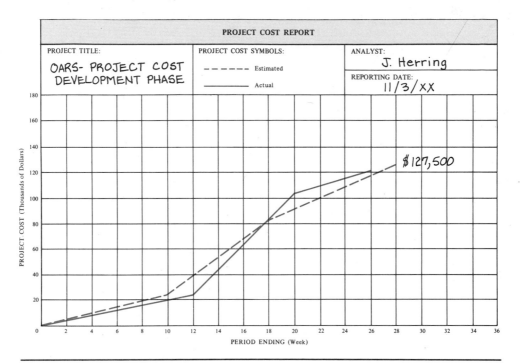

Figure E3.14
Project plan and status report—OARS major milestones.

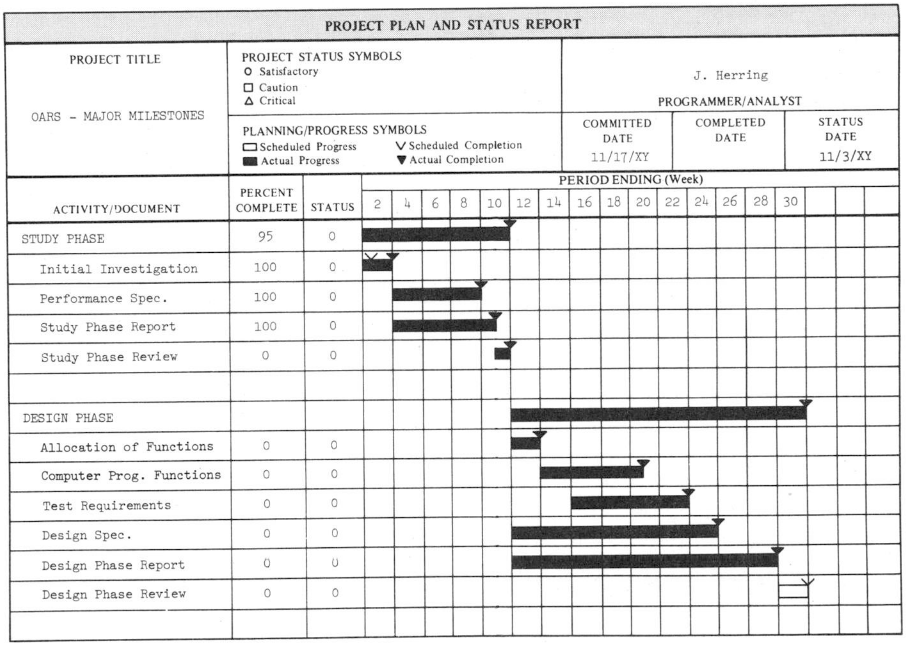

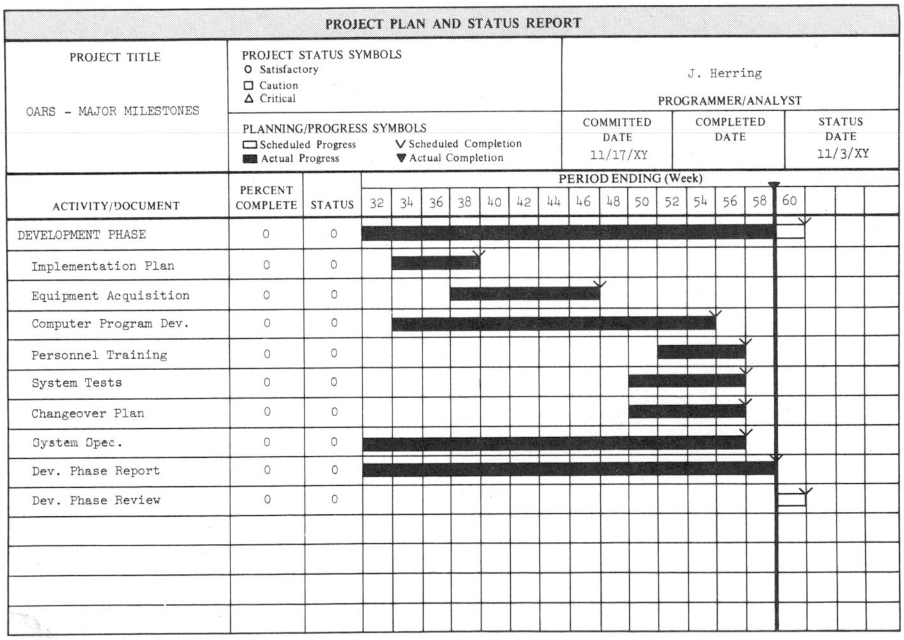

Figure E3.15
Project cost report—total project.

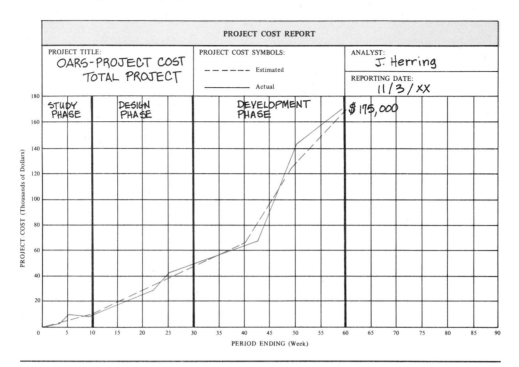

Figure E3.16
Weekly operating costs per account.

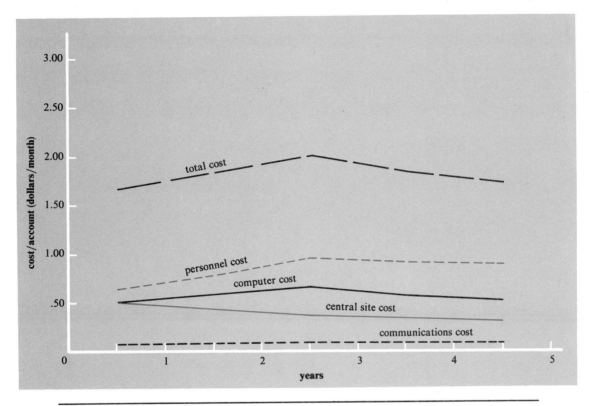

The Operation Phase
Unit 6

25
Operation Phase
Overview

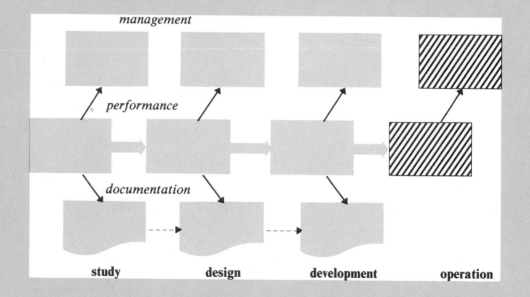

Preview

In the operation phase, the system that was designed and developed in the preceding phases is operated by its users and maintained as an on-going system. The operation phase has four distinct stages: system changeover, routine operation, system performance evaluation, and system change.

Learning Objectives

1. You will become familiar with the operation phase activity flowchart.
2. You will be able to define and distinguish the four stages of the operation phase.

Key Terms

operation phase   The phase involving changeover from the old to the new system, where the system is operated, evaluated, and changed as necessary.

system changeover   The period of transition from the old to the new system.

Operation Phase Stages

The **operation phase** follows the development phase. Usually it is the longest of the life-cycle phases and is characterized by four distinct stages. Initially, the new system must be introduced into the business activity mainstream. This stage is called changeover. The changeover transition period may take weeks or even months. After it is completed, the system enters the operation and routine maintenance stage. Early in this stage an evaluation should be made based on performance measurements that determine whether the specific benefits claimed for the system have been achieved. Finally, the new system, like all operational systems, must be able to accommodate change. Change is perhaps the most important stage in the life of a computer-based business system. Whether or not change can be managed is the final measure of the success or failure of the entire system effort.

The principal activities and documents that characterize the stages of the operation phase are described briefly in the following section.

Operation Phase Activities

The major activities of this phase are shown pictorially in the flowchart of figure 25.1. Each of these activities is summarized as follows:

1. *System Changeover*   Normally a period of transition is required to change from an old system to a new one. If all the development phase implementation activities have been performed adequately, the necessary manuals and documentation for the new system are available. There is a nucleus of trained personnel (user, programming, and operations) to assume responsibility for the new system. However, it is critically important for the project team to remain heavily involved and in control during changeover. Changeover usually is a one-way process; it must result in a system that is operationally acceptable. No matter how completely changeover activities are planned, numerous unforeseen incidents and problems will arise. **System changeover** is the most critical period in the entire life cycle of the computer-based system. Positive support by all user organizations is essential.
2. *Routine Operation*   At the conclusion of the changeover process the system is considered to be operational. The user organization and other operating personnel assume their respective responsibilities, and procedures are established for change control. Except for routine surveillance and participation in subsequent change activities, the systems analysts' responsibilities for the project are reduced. They and other members of the project team become available to assume other assignments.
3. *System Performance Evaluation*   After the computer-based business system has been operational for a reasonable period, its performance is formally evaluated. The results of the evaluation are documented in an evaluation report, which should be presented to a management review board, typically called a performance review board. Although the information service organization

Figure 25.1
Operation phase activity flowchart. Principal activity sequences of the operation phase relate to changeover from the old system to the new; routine operation; and system change, which usually results from system interaction with the business environment.

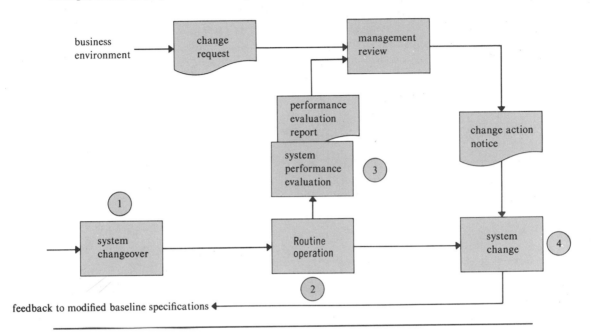

should be represented, the performance review board should be mainly user-oriented. The board should be headed by the principal user of the system.

4. *System Change*   The modern business system environment is dynamic, subject to many internal and external influences. As shown in figure 25.1, the business environment may trigger a change request, which is then reviewed by management. This process may range from a brief analysis of the requested change to an extensive investigation. This investigation may cause a return to an early point in the life cycle. How far back in the life cycle the investigation might reach would depend upon the original baseline specification affected by the change. The investigation could cause a return to the study phase, in which case the resulting new design and development activities might yield a greatly modified system.

At the conclusion of the review and analysis of the requested change, the responsible management organization issues a change action notice. The actual change action is then taken. The potential impact of the change action is shown in figure 25.1 by the arrow that indicates feedback to the modified baseline specifications.

Summary

The operation phase is the last of the four life-cycle phases. It is also the longest of the life-cycle phases, since it includes not only changeover from the old system to the new, but also routine operation, performance evaluation, and change. Routine operation of the system becomes the responsibility of its users. Periodically, the system is examined by a performance review board. If the system must be modified as a result of changing business needs, the review board may request that the system be changed. The change process could vary anywhere from a brief effort to a total reentry into the life cycle as far back as the study phase.

For Review

operation phase   routine operation   system change
system changeover   system performance
                             evaluation

For Discussion

1. Define operation phase.
2. What are the four stages of the operation phase?
3. Why is the operation phase usually the longest of the four life-cycle phases?
4. Give some examples of events that could cause extensive changes in an operational business information system. How far back in the life-cycle would each extend?

26
Changeover and Routine Operation

Preview

System changeover is the first stage of the operation phase. Changeover is the period of transition from the old system to the new one. After the new system has settled it is turned over to the user organization where it enters a stage called routine operation. During routine operation the computer-based business information system is maintained by the information services organization. The trend of the 1980s and 1990s is for this organization to report to a vice-president of information services. This corporate reporting level is necessary for effective information resource management.

Routine operation is challenging because information systems of the Maturing Era will be characterized by the merging of the data processing, automated office, and communications technologies, and by widespread applications of distributed data processing. Success or failure of routine operations will depend upon the usability of the system and the information that it provides. Therefore, it is important that information be made available in a timely manner and that good customer relations be maintained throughout the operational life of the computer-related information system.

Objectives

1. You will understand the difficulties of changeover and the challenges of routine operation.
2. You will become familiar with the functions performed by an information services organization.
3. You will learn the importance of standards, timeliness, and good customer relations.

Key Terms

system changeover   The period of transition from the old system to the new system.

routine operation   The stage in the operation phase that follows system changeover.

standards   The rules under which personnel in an information services organization work.

response time   The elapsed time between the release of input data by a user and the receipt of computer output.

usability   The value of an information system as perceived by its users.

Changeover

The Changeover "Crisis"

The acceptance review, held at the conclusion of the development phase, assured the principal user that the new computer-based business system was ready for operation. At that review the user was presented with evidence that system-level tests had verified the performance of the system; with assurances that personnel had been trained; and with a detailed plan for **system changeover.** The principal user then decided to proceed to the first stage of the operation phase, changeover from the old system to the new one. It might seem, then, that there are no further obstacles to prevent successful implementation and that the project team can be released for new assignments. Nothing could be farther from the truth. Changeover, which is the period of transition from the old system to the new system, is the most critical and problem-beset period in the life of most computer-based systems. It is the period that tends to prove the truth of the so-called "Murphy's Law": "Anything that can go wrong will go wrong."

The reasons for *changeover crisis* are these:

1. Implementation planning, however complete, cannot possibly take into account all the real-life situations that can occur.
2. The more complete the implementation planning, the fewer the unforeseen problems. However, no matter how complete rehearsals have been, it is unrealistic to expect perfect harmony during the initial system performance. For this to occur, all the computer programs must function without error; people must not make mistakes; equipment must not malfunction; files must not contain residual elements of contaminated data; and, above all, everyone should be pulling for the success of the system. It is unlikely that the orchestration will be this perfect when the system goes "on stage" for the first time.
3. All changeover methods contain risks. In chapter 22, Preparing for Implementation, we discussed three changeover methods: parallel operation, immediate replacement, and phased changeover. All introduce new tasks into an actual operational environment. Therefore, mistakes and problems are to be expected.
4. Changeover is an emotional activity. Change suddenly becomes a reality. Things will be different. This realization is sufficient to create tensions that cause and amplify mistakes.

If we were to prepare a graph to illustrate the frequency of the occurrence of crises during a typical changeover, it might look like figure 26.1. Initially, everything appears to go wrong at once. Then, after a period of time, the crisis environment begins to improve. The frequency and the magnitude of the crises tend to become less. Finally, a relatively tranquil state is reached. This usually occurs from sixty to ninety days after changeover begins.

Figure 26.1
Changeover crisis frequency. Changeover from an old to a new system is a critical period beset by frequent crises. With effective implementation planning and strong user support, crises tend to abate with time.

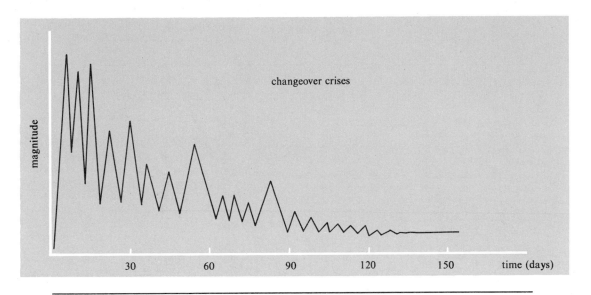

Changeover Activities

The project team must remain intact during changeover. The team contains members of user organizations who must be available on a full-time basis. Here are some practical guidelines for the team's activities during turnover.

1. Compare new outputs with old outputs as much as possible. If outputs differ only in format, it usually is possible to verify content.
2. Check inputs and outputs to be certain that they conform to specifications.
3. Follow up immediately on all errors. Correct the manual or machine processes causing the errors.
4. Keep a log. Use a *changeover action log* to record actions, responsibilities, assignments, and completion dates. Figure 26.2 is an example.
5. Solve all problems promptly. Seek the cooperation of the user groups in resolving problems and in deciding on immediate corrective measures.
6. Defer any refinements or changes in the system until changeover has been completed.
7. Never expect problems that have not been solved before changeover begins to be solved during changeover. Unsolved problems only tend to multiply.

Changeover is the time at which the project leader must make every effort to keep team spirits high. The project must maintain its momentum. The situation is similar to a football team that has the ball on the two-yard line: the entire team must

Figure 26.2

Changeover action log. An action log is a useful management tool to use during changeover. This log lists in numbered sequence the actions needed, responsibilities for actions, and dates of action assignment and completion.

	CHANGEOVER ACTION LOG			
ACTION NUMBER	ACTION DESCRIPTION	RESPONSIBILITY	DATE ASSIGNED	DATE COMPLETED

be psyched up to carry the ball over the goal. At this time the earned support and confidence of user groups will pay off. Even a poor system can be made to work with user support. Without it, a good system will fail. Systems are never more vulnerable to attack than during changeover.

If allowed to stay on the sidelines, some users (who are perhaps a bit fearful of the system anyway) will derive satisfaction from the "failures of the computer." It appears to be human nature to mock the failure of a machine, even if the machine is only performing as instructed. To keep them involved, the systems analyst should assign specific responsibilities to all users. They should be made responsible for the prompt solution of systems problems that arise in their areas of operation.

The project leader should strive to sell "success." The leader must keep the team success-oriented by emphasizing the obstacles overcome, rather than those yet to be encountered. Success will breed more success. The crisis environment will begin to abate, and the system will approach a level of performance that justifies its turnover to the user.

User Turnover

As we noted when discussing figure 26.1, crises tend to be reduced to manageable proportions by about sixty to ninety days after the initiation of changeover. The analyst can then consider turning over the system to its users. At this time, all the error conditions noted during changeover will have been corrected, and the system will have gone through several cycles of successful operation. It then becomes the responsibility of the user to operate the system. The project team can be disbanded and its members can return to their own organizations. Although the information services organization will retain some responsibilities for the system, these will be of a maintenance nature. An analyst will be assigned as a liaison with the user, to participate in resolving system problems. The manner in which maintenance and changes to the system are handled is discussed in chapter 27, Performance Evaluation and Change Management.

The data processing department assumes full responsibility for data processing activities. *User turnover* initiates a new stage in the operation phase. This stage is called routine operation.

Routine Operation

Organizing for Data Processing

We concluded our discussion of changeover by stating that after the system had been turned over to the user, its routine operation became the responsibility of the data processing department. In a large data processing department routine day-to-day processing is anything but "routine." The operations environment itself is crisis-laden. The typical environment is one in which a large number of jobs, some scheduled and some unscheduled, must be processed amid changing priorities and daily emergencies.

Before discussing **routine operations** in greater detail, let us review the organizational location of data processing within the company, as well as its internal organization. As we have discussed in earlier chapters, the location of the systems and data processing functions is changing. Figure 26.3 illustrates the trend in locations for the data processing organization:

1. *dp–1*   Data processing reports to the financial function and provides programming and machine operation support. The systems developed are largely financial, and few services are provided to other organizations. This reporting relationship was common throughout the *Growing Era* and the early *Refining Era,* when the principal data processing applications were financial.

2. *dp–2*   Data processing reports to a general administrator who also has other functional responsibilities. Support services, essentially programming and machine operation, are provided to all functional groups, for example, finance, marketing, and product development. Systems analysis responsibility is fragmented among users. Applications are limited in scope. The data processing manager usually does not have sufficient authority to initiate or to implement systems that extend beyond organizational boundaries. This reporting relationship was not uncommon in the Refining Era.

3. *dp–3*   Data processing reports to a top executive whose principal responsibility is the development of information systems. These systems often are corporate-wide. A strong systems analysis capability exists in the information services

Figure 26.3
Trends in location of data processing organization. The trend in location of the data processing group is toward placement in an information services organization that reports directly to the president of the corporation. This trend is a result of the increasing importance of information resource management and the pervasiveness of computer-related information systems.

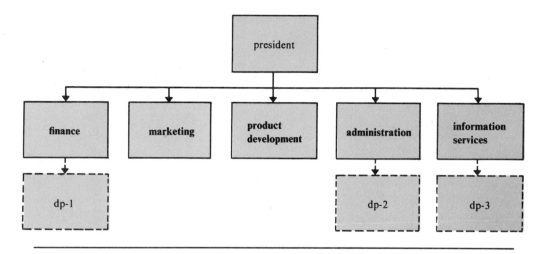

organization, which is concerned with the merging data processing, automated office, and communicative technologies. Innovation is encouraged and is made possible by the status of the organization.

The observable trend in the location of the data processing organization is from *dp–1* toward *dp–3*. We shall limit our discussion to a *dp–3*-type organization, which is characteristic of the *Maturing Era*.

Even in an environment in which the data processing and systems analysis functions report to a senior information services executive, there are many organizational possibilities. For example:

1. Programming and systems analysis may be separate or combined.
2. System analysis may be organized by functions or by projects.

Again, the observable trend is toward an information services organization in which programming is integrated with systems analysis and the overall organization is mixed. That is, some activities are organized by function, and others are organized by project. Most commonly, new systems applications are handled on a project basis, and other operations are performed according to function. This flexibility provides for effective information resource management.

Figure 26.4 is derived from figure 3.6; it illustrates a typical *Maturing Era* organization structure. The director of corporate systems and the director of data processing report to the vice-president of information services. The corporate systems

Figure 26.4
Contemporary information services organization. In a typical contemporary information services organization the two principal functions—both reporting to a vice-president—are systems and data processing. The former includes all systems and applications programming activities. The latter includes data processing operations and also communications and reprographics.

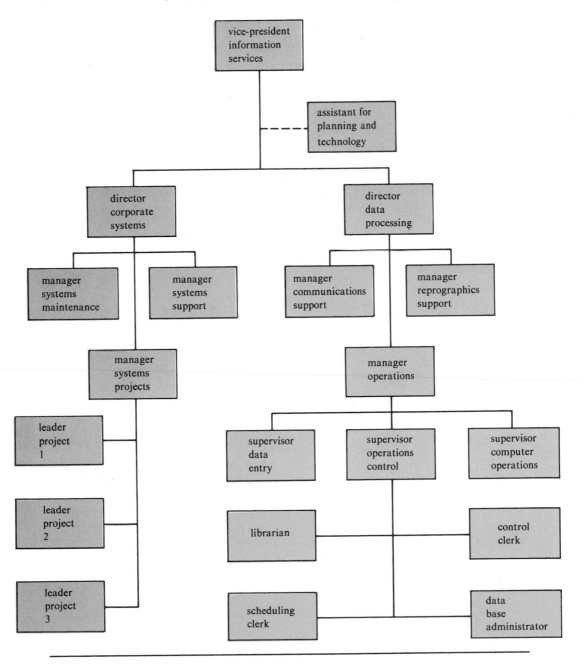

activity includes both systems analysis and programming. The organization structure is mixed. Systems projects are set up for each new application and, as such, are carried through to turnover to the user. Thereafter, the operational systems become the responsibility of a systems maintenance group. In the event of significant changes, a particular application can revert from maintenance status to project status. System support activities are functional because they are not oriented toward specific applications.

In addition to conventional systems functions such as forms design, records management, and work measurement, the support activities may include the development of techniques such as critical path methods, simulation, and data base management. Support is provided for communications and the automated office.

As also shown in figure 26.4, the data processing organization usually is oriented toward function rather than projects. An exception would be the dedication of a computer system to a single large project. Other important functional operations are data entry, computer operations, and operations control. Figure 26.5 is a brief summary of typical jobs and duties associated with the organization chart shown in figure 26.4. An important job is that of data base administrator. This job has come into existence because of the increasing number and complexity of the data bases being created to support integrated and distributed data processing systems. The data base administrator is responsible for the integrity of the data base. This person monitors the data base, controls changes to it, maintains its structure, and reorganizes it as required. A data base administrator is essential to the successful implementation and maintenance of the data base management systems (DBMS) discussed in chapter 19.

It is not our purpose in this chapter to examine data processing operations in detail. Rather, it is to develop for the users of these services (including analysts and programmers) an appreciation of the nonroutine nature of so-called "routine" data processing operations. We will, therefore, discuss data processing operations as they affect a user's evaluation of service.

A data processing operation is evaluated by the service it provides to its users. Good service means that the output is correct and that it gets to the users on time. Two factors by which the performance of a data processing center can be measured are performance standards and timeliness.

Data Processing Performance Standards

The Standards Manual   The rules under which analysts, programmers, operators, and other personnel in an information service organization work are its **standards**. They are the reference against which performance can be measured. Standards vary from organization to organization. While it is not essential that standards be the same in all organizations, it is essential that complete and current written standards exist and be understood. Some reasons for the importance of standards are these:

1. The work of analysts and programmers must be understood by other analysts and programmers. This provides back-up.
2. The field of systems analysis, particularly where computers are involved, is changing rapidly. Techniques can be kept current.

Figure 26.5
Typical data processing job descriptions. There are many career opportunities in data processing. Several of those associated with operations are listed and described here.

JOB TITLE	DESCRIPTION
Supervisor of Data Entry	Supervises staff responsible for entry of data for computer processing. Schedules workloads and distributes work assignments.
Supervisor of Operations Control	Supervises library activities, preparation of schedules, production control procedures, and data base maintenance.
Librarian	Stores and issues program documentation and data files kept on disk or tape.
Scheduling Clerk	Prepares and maintains daily, weekly, and monthly schedules for all appointed jobs.
Control Clerk	Checks receipt and acceptability of input data. Establishes controls. Checks output. Maintains error records. Dispatches acceptable output to users.
Data Base Administrator	Maintains the integrity of a data base. Organizes, reorganizes, and controls data base definitions and changes. Controls access to the data base.
Supervisor of Computer Operations	Supervises the operation of all computing equipment. Maintains records of equipment performance. Develops techniques to improve performance.

3. Communications within a department and between departments are improved.
4. New employees can be trained and can become effective sooner if they learn standard procedures.
5. Changes can be implemented more easily when existing standards can be used as references.

Because they affect all members and functions of the information system organization, standards should be established at an appropriate management level. For example, in the information system organization shown in figure 26.4, the establishment and maintenance of standards could be a responsibility of a group that reports to the assistant to the vice-president of information systems. An information systems organization that lacks standards is not "organized," and therefore is poorly equipped to provide lasting user satisfaction.

A *standards manual* is an essential tool. Figure 26.6 lists major topics typically found in standards manuals. The standards policy is a statement to the effect that all work performed in an organization will be in accordance with the content of the manual. Administrative standards relate to organization charts, job descriptions, training of personnel, and administrative information. Systems analysis standards govern the analysts' activities throughout the life cycle of a business system. For example, standards for coding, forms design, charting, and documentation would appear in a manual for information system development.

Programming standards are rules for activities such as computer program flow-charting, language selection, programming techniques, and program documentation. Operating standards relate to computer and peripheral equipment operations. The need to pause to comply with standards sometimes appears to conflict with a need for time-liness. Actually, standards contribute to accuracy, and so lead to faster service because there are fewer errors and complaints.

Timeliness: Response Time, Throughput Time, and Turnaround Time

Response Time   The time that elapses between the release of input data by a user and receipt of computer output is the **response time.** It is the yardstick by which a user measures the timeliness of the output received from a computing center. It is a "door-to-door" measurement, and it is an important measure of usability. Response time includes some elements that are under the control of the data processing manager and some that are not. We can illustrate this by defining two other important "times," throughput time and turnaround time. The relationship among the three components of *timeliness* is illustrated in figure 26.7.

Throughput time is the time required for work to flow through the machine room. *Turnaround time* is the time that elapses between data arrival at the computing center and the availability of output for pickup. The time required to deliver the data to the computer center and that required to deliver the output to a user may be beyond the control of the data processing manager. Yet, these times also contribute to response time.

However, throughput time and turnaround time are very much the responsibility of the data processing manager. An effective manager can do much to keep these times to a minimum. We shall discuss throughput time and turnaround time briefly as they relate to the effectiveness of the data processing manager and his or her supervisor of computer operations.

Throughput Time   Throughput time depends on the efficiency of the computer programs, computing equipment, operating system, and operators. The efficiency of the programs is fixed during the development phase.

The efficiency of the equipment and the operating system is of prime concern to the manager of the data processing department. By measuring productive machine time (for example, production processing and program checkout and test), nonproductive machine time (for example, reruns and set-up time), and maintenance (scheduled and unscheduled), the data processing manager can develop performance standards

Figure 26.6

Contents of a standards manual. Standards manuals contain the rules under which personnel in an information services organization work. Thus, a standards manual for developing information systems must, in addition to a policy statement, contain administrative, systems analysis, programming, and operating standards.

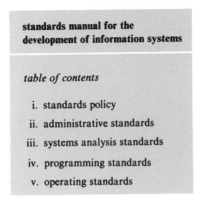

standards manual for the development of information systems

table of contents

i. standards policy

ii. administrative standards

iii. systems analysis standards

iv. programming standards

v. operating standards

Figure 26.7

Data processing performance times. Response time is the overall time between input of data by a user and the user's receipt of computer output. Turnaround time is the total elapsed time within the computer center, including throughput time, which is the actual machine processing time.

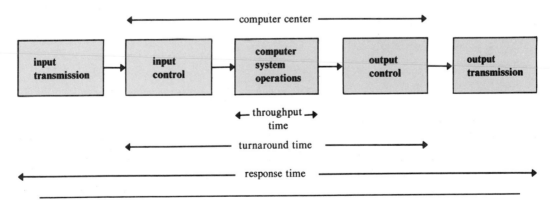

for equipment. In a multi-programming environment (that is, when available machine resources can be used to process more than one job concurrently), the manager should determine the best relationship between job mix and computer configuration. As necessary, changes in equipment can be identified and justified. Examples are changes in main storage capacity; selection of peripheral devices, such as card readers and printers; and requirements for tape or disk storage.

Figure 26.8

Operator's reference manual. The operator's reference manual is an example of the type of installation standard that reduces throughput time and improves overall response time.

---

guide to operator's reference manual

1. *title*   name of computer-based system

2. *purpose*   general description of system

3. *operating procedures*   as appropriate to specific operational duties:

    a. operator inputs: a complete description of all inputs, including:

      1) purpose and use
      2) title of input
      3) input and media
      4) limitations
      5) format and content

    b. operator outputs: a complete description of all outputs, including:

      1) purpose and use
      2) title of output
      3) output media
      4) format and content

    c. file summary: a complete description of all files, including:

      1) file identification
      2) medium
      3) type: master, transaction, and so forth

    d. error and exception handling: procedures for handling hardware and software error conditions

---

Similarly, general installation standards increase the efficiency of operators and thus reduce throughput time. Examples of general installation standards are (1) rules for machine room safety and security; and (2) rules for computer startup and shutdown and for console operation.

Standards for the preparation of operator's reference manuals also are important. We described the general content of the operator's reference manual in figure 22.7 (chapter 22). Figure 26.8 is the same figure. Our purpose in presenting this figure again is to emphasize that specific operating procedures must be developed for each set of related operational duties. Often the entire computer program is not run at once. The sequence and number of computer runs required by an application must be specified in the manual, and the operator must be instructed in a set of duties for each run. (For this reason, the operator's reference manual is sometimes called a "run manual.") The number of runs required depends on both the installation and the application. Typical factors related to the installation include computer main storage capacity, available direct access storage, and number of tape drives. Examples of factors that depend on the application are file maintenance runs and time taken to detect and correct input data errors before proceeding with mainline processing. Figure 26.9 is an

Figure 26.9
Computer run sheet. Computer run sheets, which are contained in the operator's reference manual, provide the detailed input, output, file, and exception handling instructions needed for efficient data processing operations.

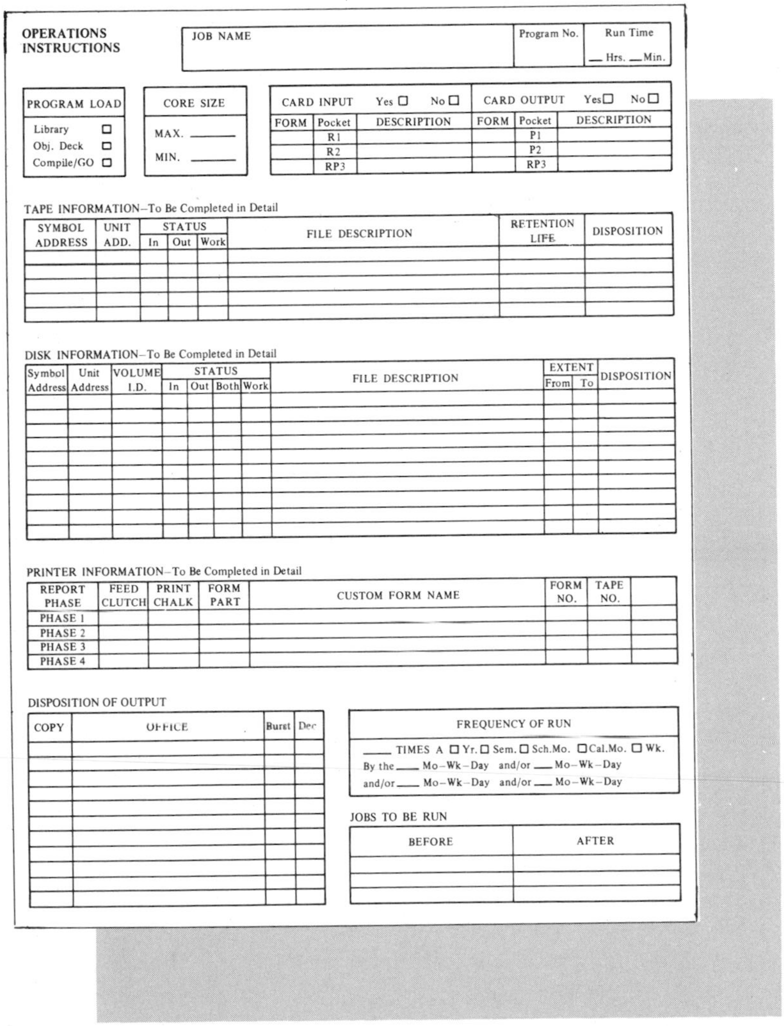

OPERATIONS INSTRUCTIONS

JOB NAME	Program No.	Run Time
		___ Hrs. ___ Min.

PROGRAM LOAD
Library ☐
Obj. Deck ☐
Compile/GO ☐

CORE SIZE
MAX. ______
MIN. ______

CARD INPUT Yes ☐ No ☐			CARD OUTPUT Yes☐ No☐		
FORM	Pocket	DESCRIPTION	FORM	Pocket	DESCRIPTION
	R1			P1	
	R2			P2	
	RP3			RP3	

TAPE INFORMATION–To Be Completed in Detail

SYMBOL ADDRESS	UNIT ADD.	STATUS In	Out	Work	FILE DESCRIPTION	RETENTION LIFE	DISPOSITION

DISK INFORMATION–To Be Completed in Detail

Symbol Address	Unit Address	VOLUME I.D.	STATUS In	Out	Both	Work	FILE DESCRIPTION	EXTENT From	To	DISPOSITION

PRINTER INFORMATION–To Be Completed in Detail

REPORT PHASE	FEED CLUTCH	PRINT CHALK	FORM PART	CUSTOM FORM NAME	FORM NO.	TAPE NO.	
PHASE 1							
PHASE 2							
PHASE 3							
PHASE 4							

DISPOSITION OF OUTPUT

COPY	OFFICE	Burst	Dec

FREQUENCY OF RUN
___ TIMES A ☐ Yr. ☐ Sem. ☐ Sch.Mo. ☐ Cal.Mo. ☐ Wk.
By the ___ Mo–Wk–Day and/or ___ Mo–Wk–Day
and/or ___ Mo–Wk–Day and/or ___ Mo–Wk–Day

JOBS TO BE RUN

BEFORE	AFTER

example of the detailed input, output, file, and exception-handling information that must be provided in the operator's reference manual for every run required to process a complete computer program. Meaningful flowcharts also are an aid to the operator. Figure 26.10 is an example of a form for preparing a "picture" that might be contained in a manual as part of run documentation.

Figure 26.10

File assignment and disposition chart. File assignment and disposition charts are examples of pictorial documentation that assists computer operators in the step-by-step running of computer programs.

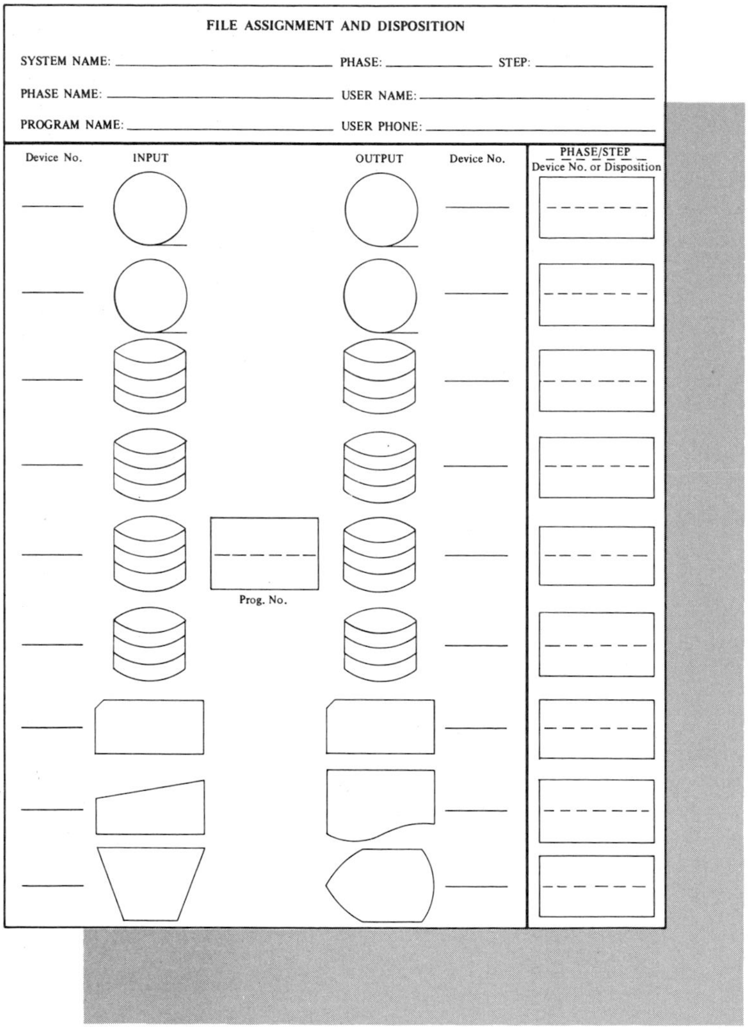

FILE ASSIGNMENT AND DISPOSITION

SYSTEM NAME: _____ PHASE: _____ STEP: _____

PHASE NAME: _____ USER NAME: _____

PROGRAM NAME: _____ USER PHONE: _____

A major purpose of standards is to assure that meaningful and accurate outputs are produced. This means that steps taken to ensure valid output may sometimes increase the throughput time. This is one reason why timeliness and observance of standards must be considered as related factors that, taken together, determine a user's evaluation of computer center service.

Turnaround Time Turnaround time is made up of throughput time plus the time required for jobs to pass through input and output control operations. A typical data processing center may receive work from several plants. The work may be scheduled or unscheduled; the plants may be local or remote. To the extent that remote data entry is via communications networks, turnaround time is reduced. Also, users share some of the responsibility for turnaround time. This is particularly true if remote sites have terminals with a local data-editing or programming capability.

The operations control group is a principal point of contact with the users. It is the point at which inputs are received and outputs are distributed. Often it handles complaints. Therefore, this group must maintain the good user (customer) relations that are essential to the **usability** of the system and the success of the information services organization.

Customer Relations

Good customer relations are important because there will be times when the constructive support and understanding of users are essential to the success of routine operations. Fast response, coupled with standards to ensure a quality product, is not always possible. Most computer installations are not sized to handle peak loads, such as occur at month end and year end. Priority tasks will arise without warning. At these times performance may slip. Also, there always will be unplanned unproductive periods that will cause schedules to slip. Typical causes are machine malfunction and human error. Without customer understanding and support, these periods can be demoralizing and can impair the effectiveness of the entire information services organization.

Security

Management concern about the security of computers often is inflamed by highly publicized incidents of computer-related fraud. The data processing center is expected to provide not only adequate input, processing, and output controls, but also protection against fraud and disaster. This protection should be a shared responsibility. It is the responsibility not only of the data processing manager, but also of top management, insurance companies, security specialists, and users of data processing service. To provide protection against disasters, certain steps should be taken:

1. *Physical location* Select a computer site away from natural hazards. Take steps to reduce known risks. For example, water risk can be reduced by storing data in high locations and by providing drains, pumps, and plastic covers for equipment.
2. *Physical access control* Use badges and controlled entry points. However, the key factor is an alert computer staff who will challenge all strangers.

3. *Fire protection*   Locate the computer center away from fire hazards. Construct the computer area of flame retardant materials. Minimize combustible materials in the computer area. Provide early warning devices, fire detectors, portable extinguishers, and emergency procedures.
4. *Media protection*   Store vital files in a separate room or vault.
5. *Back-up and "fall back" capabilities*   If possible, make arrangements with similarly equipped data processing centers.
6. *Risk insurance*   Evaluate insurance policies to cover data processing losses.

In spite of publicity to the contrary, natural disasters are more frequent causes of catastrophe than is fraud. Nonetheless, steps can, and should, be taken to minimize people-induced security problems. These steps include:

1. Division of duties in data processing.
2. Built-in system controls.
3. File and program change controls.
4. Use of a security specialist.
5. Frequent audits by external auditors.
6. Thorough personnel investigations before hiring.
7. Bonding of staff.
8. Prompt removal of discharged personnel.
9. Protection against voluntary termination by good documentation and cross-training of personnel.

Thus, a well-organized, well-managed, and secure data processing department that uses standards effectively can provide the response time and quality of output that will cause a user to evaluate routine operations as effective. This attitude is conducive to user tolerance and support during exceptional nonroutine situations.

Summary

After the conclusion of a successful acceptance review, the computer-based business information system enters the operation phase. The first stage of the operation phase, called changeover, is a period of transition from the old system to the new system. Changeover is characterized by a series of crises, which usually subside after sixty to ninety days. During the changeover period problems must be handled promptly, and it is useful to maintain a changeover action log.

After changeover is completed the system is turned over to its principal user(s) and enters a stage called routine operation. Routine operation seldom is "routine," since the day-to-day operating environment is subject to unscheduled priorities and emergencies. Routine operation can best be managed if the information services organization reports to an appropriate level. Increasingly, this reporting level is that of a vice-president of information services, due to the complexity of nontraditional information systems that combine data processing, office automation, and communications, and which exhibit a trend toward distributed data processing.

The key to success in managing routine operations is to maintain the usability of the system by providing timely and accurate information to its users. Maintaining good customer relations and data security contribute to the usability of computer-related information systems.

For Review

system changeover	**usability**	**response time**
changeover crisis	computer run sheet	throughput time
changeover action log	Growing Era	turnaround time
user turnover	Refining Era	**standards**
routine operation	Maturing Era	standards manual
timeliness		

For Discussion

1. What is the "changeover crisis"?
2. Describe actions that can be taken to manage changeover.
3. What occurs at user turnover time?
4. Why are so-called routine data processing operations not routine?
5. Discuss the organization of a contemporary information services organization.
6. What standards are contained in a data processing standards manual? Distinguish between them.
7. Discuss the interaction between standards, timeliness, and user satisfaction.
8. Distinguish between throughput time, turnaround time, and response time.
9. What is the purpose of a run sheet? What is its general content?
10. Discuss steps that may be taken to protect the computer center against natural disasters and against fraud.

27
Performance Evaluation and Management of Change

Preview

Business information systems exist in a dynamic and changing environment. After these systems enter the operation phase their performance must be monitored not only to determine whether or not they perform as planned, but also to determine if they should be modified to meet changes in the information needs of the business. System performance evaluation and change management, therefore, are two important stages of the operation phase. They emphasize the relationship between documentation and success in implementing and maintaining effective business information systems.

Objectives

1. You will become aware of the importance of periodic performance evaluation throughout the operation phase.
2. You will learn a technique, based upon the establishment of a performance review board, for managing change.
3. You will be able to relate change control to the life-cycle methodology, and you will reinforce your understanding of the dependence of business information systems upon documentation that is well structured and carefully maintained.

Key Terms

performance review board   A user-oriented board responsible both for the periodic evaluation of the performance of a computer-based business information system and for maintaining its integrity.

change control   The means by which major modifications to computer-based business information systems are managed.

change   A system modification that requires performance review board action.

maintenance   A system modification that does not require performance review board action.

Performance Evaluation

Performance Review Board

A system cannot be forgotten after it has been accepted for routine operation. The dynamic nature of business information systems is an essential aspect of the life-cycle concept. Internal and external factors will cause changes to the operating environment of the system. Typical internal factors that can affect system performance are changes in equipment, work load, programming languages, and personnel. Among the external factors that affect performance are new or revised reporting needs, increases in required output, and changes in schedules. In fact, changes to systems are caused by all of the information generators discussed in chapter 4.

Thus, the life-cycle management process continues throughout the operational life of the system. Computer-based business information systems tend to fall apart rapidly if formal management control is removed, principally because the validity of system documentation is destroyed. One way in which this occurs is the introduction of many small and inadequately documented changes by programmers who deal informally with members of user groups. Although an informal relationship can be healthy, it should not be permitted to destroy the integrity of the system. If laxity in documentation is tolerated, sooner or later a disaster will occur, and the good relationship with the user will disappear. Another type of change, which usually is poorly documented, is the system patch. A *patch* is a "quick fix" programming change that is made under the pressure of an immediate operational need. Most often the patch changes one or more of the computer program modules. At the time the change is made, the intent is to remove the patch and to rewrite the affected routines in the future—but this particular future never arrives, unless the intent is supported by an approved management plan. In an environment in which managed actions continually are secondary to "crisis responses," the original system quickly becomes completely hidden by patches.

One technique for ensuring system integrity is the establishment of a **performance review board (PRB).** Both the user and information systems are represented on the PRB. It is, however, a user-oriented board, which should be headed by the principal user. The PRB is continuously aware of the computer-based system through user organization involvement in the routine operation of the system and through a designated member of the systems maintenance staff. The PRB should not be involved continuously in operational problems, but should respond only to exceptional conditions of a nonroutine nature. In addition, periodic reviews (perhaps quarterly) should be scheduled. The first of these scheduled reviews should take place two to three months after the new system has been installed, to compare actual results with planned results. This review is called the post-installation review. Other PRB actions are triggered by requests for changes to the system. The functions of the PRB related to post-installation and periodic reviews are illustrated in figure 27.1.

Figure 27.1

Performance evaluation. After system changeover is completed and after the computer-related information system enters routine operation, formal performance evaluation occurs. An evaluation report is prepared for review by a performance review board.

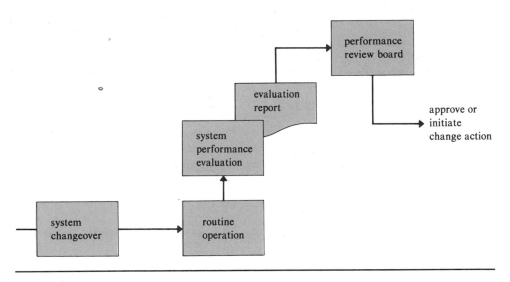

Post-installation Review

System performance evaluation begins with the *post-installation review,* which is intended to determine how well actual performance compares with promised performance. The "promises" were documented as the specific performance objectives, which were in turn derived from the general objectives and anticipated benefits stated in the project directive. This process was described in detail in chapter 12, System Performance Definition.

The post-installation review should not be scheduled until the changeover crisis is over. Usually this is two to three months after the system has been declared operational. The review should be performed by persons who are not directly responsible for system implementation and operation. They can be selected from the user and system staffs or they can be outside auditors or consultants. The first review activity is to gather information related to current system operation. Information-gathering techniques, such as those discussed in chapter 11, Initial Investigation, can be employed. For example, as a member of the review team, you should:

1. Examine the actual system outputs. Compare them with the outputs in the system specification.
2. Use the distribution list to send correspondence inquiring about the operating status of the system and its effectiveness. Ask about any problems that have been encountered or that still exist. Solicit suggestions for improvements.
3. Follow up correspondence by interviews with appropriate users of the system outputs, with operating personnel, and with user management.

After completing its information gathering, the team should compare observed current performance with the specific performance objectives. Some specific objectives with which actual performance can be compared are:

1. elimination of duplicated files
2. reduction or reassignment of personnel
3. cost savings achieved
4. cost avoidance accomplished
5. comparison of past and present error rates

The team then prepares a system performance evaluation report for submission to the PRB. A typical format is shown in figure 27.2. The main elements of the performance evaluation report are these:

1. *Name of system* Identifies the computer-based business system being evaluated.
2. *Specific performance objectives* States the specific performance objectives of the computer-based system.
3. *Method of evaluation* Describes how the system performance was reviewed and evaluated.
4. *Results of evaluation*
   a. Compares measured performance with the specific performance objectives.
   b. Summarizes other factors, such as user and operator satisfaction, intangible benefits, and pertinent observations of the review team.
5. *Recommendations* Recommends actions to the PRB. The range of possible recommendations includes (a) accept the system "as is"; (b) recommend minor modification; (c) recommend major revision; and (d) reject and start over.

For examples of some specific performance objectives that might be evaluated after implementations, let us again consider the ABCO corporation's on-line accounts receivable system (OARS).

Figure 27.3 lists the objectives presented in chapter 12, System Performance Definition. Each of these objectives is "measurable." Each can be evaluated by examining system outputs over several billing cycles and by interviewing users of the system. Thus, actual performance measurement, coupled with an assessment of users' satisfaction and suggestions, results in a meaningful performance evaluation report. This example, by contrast, clearly illustrates the dilemma that management all too often faces in evaluating systems for which specific performance objectives have not been established.

Periodic Review

The post-installation review is followed by *periodic reviews,* which are intended both to ensure that the integrity of the system is maintained and to identify special areas requiring management attention. Two major requirements for system integrity are valid documentation and valid outputs. Management can help to keep documentation valid

Figure 27.2
System performance evaluation report. The system performance evaluation report is prepared as part of the post-installation performance review; it compares actual performance with the specific objectives of the computer-based information system.

```
                    SYSTEM PERFORMANCE EVALUATION REPORT


           1.   Name of system:

           2.   Specific performance objectives:

           3.   Method of evaluation:

           4.   Results of evaluation:

           5.   Recommendations:
```

Figure 27.3
Specific objectives of OARS. This figure lists specific, measurable objectives that can be compared with actual system performance as part of a post-installation review.

```
           specific objectives of on-line
           accounts receivable system (OARS)


       1.  establish billing cycles for each region

       2.  mail customer statements no later than one day after
           the close of the billing cycle

       3.  provide customers with a billing statement no later
           than three days after the close of the billing cycles

       4.  speed up collection, reducing float from an average of four
           days to two days

       5.  establish regional profit centers

       6.  examine customer account balances through on-line
           inquiry during order entry
```

by supporting performance standards that do not accept undocumented changes. Systems and programming staffs often are under pressure to "perform" and not "write," but continual pressure of this kind usually is due to management's lack of understanding of the true cost of poor documentation. The PRB can rectify this situation, in most cases, by allocating the resources necessary to keep the system intact.

Management also can help to maintain the validity of system output. Errors may remain undetected long after changeover has been completed, or they may creep in as a result of changes made to the system. The PRB should request a periodic audit of the accuracy of system outputs.

Audits may be performed "around" or "through" the system. Those performed around the system are external, not an integral part of the system data processing operations. Audits performed through the system are designed into the system; they are computer programs that operate upon system data as it is being processed. This type of audit is becoming more common. Statistical sampling techniques also are coming into use as auditing tools. The results of periodic audits should be reported to the performance review board.

Special areas for management attention may come to light only after the system has operated for an extended period. An example is the performance of the system under unplanned conditions, such as an overload due to unanticipated volumes of data to be processed. Some manifestations of problems are subtle. For example, an increase in circumvention of procedures may be the result of system deficiencies that need to be corrected. As another example, absenteeism can disclose much about the effectiveness of the system. If personnel are misplaced or are subjected to continuing high levels of stress, they will take time off in order to get relief. Reassigning people and rescheduling some of the processing from peak to slack periods may correct this problem.

Throughout all of its reviews, the PRB should be particularly sensitive to users' evaluations of the system. In the final analysis, a system is effective only as long as it continues to be accepted by its users.

Management of Change

Guidelines for System Modification

Computer-based business systems are dynamic. They must be able to accommodate changes in information needs resulting from changes in the business environment. These changes occur not only during the study, design, and development phases of the life cycle of the system, but also throughout its operational life. Provision must be made for *modification* of an operational system, for if it cannot be modified without destroying the integrity of its data base and its outputs, it is a failure. Change can be managed by continuing the life-cycle management process by which the system was created. Inherent in this process are two elements that are essential to the management of change. They are (1) the performance review board, which can make management-level decisions about system modifications; and (2) baseline documentation, which can be referred to, to determine the extent and impact of proposed modifications.

In addition to the conduct of post-installation and periodic reviews, the PRB also must evaluate requests for modification of the operational system. Some requests will be planned; others will be unplanned. Some will be extremely significant, others much less so. Obviously, the PRB should not be involved on a continuing daily basis, but only when requirements for system modification rise above a certain threshold of importance. We can identify this threshold by dividing system modifications into two categories and relating each one to the type of baseline specification it affects, as has been done in figure 27.4. The modification categories shown in this figure are change and maintenance. **Change** is defined as a system modification that requires performance review board action. **Maintenance** is defined as a system modification that does not require performance review board action.

Figure 27.4
Modification categories for computer-based business information systems. System modifications may require ordinary maintenance or action taken by the performance review board. The latter constitutes a system change and generally relates to the type of activity that occurs early in the life cycle.

modification category	baseline specification affected
change (PRB action)	performance specification
	design specification
maintenance (no PRB action)	system specification

The need for PRB action depends upon the impact of a proposed modification, a concept that can be illustrated by again referring to figure 27.4. Let us first consider the two extremes. We recall that a user-oriented performance specification was created at the end of the study phase. In figure 27.4, the first entry in the "baseline specification affected" column refers to the user-oriented performance specification (as this baseline specification was carried forward into the final system specification). This type of modification always involves the users of the system. Therefore, the PRB always should be involved. For example, a change in company credit policy could change the method of billing customers. As another example, a new tax law requiring state withholding tax payroll deductions could affect all of the employees of a company.

At the other extreme, it is not likely that the PRB would be concerned with modifications to the content of the system specification that are not derived from the original performance specification or the original design specification. Examples are the development of a more efficient computer program component and minor changes in hardware and software. We can define this type of modification as technical maintenance rather than as change.

However, we must be careful not to include major programming, hardware, and operating system changes in this definition of technical maintenance. These are changes that should be brought to the attention of the performance review boards for all the ongoing computer-based business systems. The reason is that data processing changes of large magnitude probably cannot be made without errors. Some of these errors will cause strange and sometimes catastrophic things to happen to user outputs. Therefore, it is imperative to have the concurrence and support of all principal users throughout the process of planning and implementing major changes to the computer-based business system environment.

To make sure that the potential impact of technical maintenance modifications is not underestimated because of errors of judgment by programmers or the data processing operations staff, the maintenance analyst should approve *all* modifications made to an operational system. Together with the supervisor of programming, the analyst should make certain that specification and documentation standards are observed regardless of the pressure of work load. Following standards is as necessary for a change in a single instruction as it is for the rewriting of large and complicated parts of the overall computer program component.

In between the extremes shown in figure 27.4 are modifications to the system specification, which are derived from the design specification written at the conclusion of the design phase. The maintenance analyst must judge whether or not the PRB should be involved. Typical situations are those resulting from modifications to the internal system design. Examples are changes in file structure, changes in internal data flow, and minor output revisions. For example, if a report were to be modified slightly by changing the relative position of two output items and by adding a subtotal, the maintenance analyst probably could handle this situation as routine maintenance and deal directly with the users of the report (who probably requested the modification in the first place). However, for any proposed design changes that could have a significant impact on persons supplying input to the system, maintaining the data base, or using the outputs, the maintenance analyst should request the concurrence of the PRB.

We can summarize the foregoing discussion of change and maintenance with the following guidelines for the systems analyst responsible for systems maintenance:

1. Evaluate *all* system modifications.
2. Determine whether the modification is maintenance or change.
3. If it is change, bring the modification to the PRB for a decision.
4. Present a written summary of all maintenance modifications at the next periodic meeting of the PRB.

The importance of establishing guidelines and standards for system modification is underscored when we consider the size of the company investment that the maintenance analyst must safeguard. It is estimated that 25 to 50 percent of all systems and programming effort is spent on maintenance. The continuing annual cost to support a newly operational system is approximately 15 percent of the cost to develop it. Because companies have such a large stake in maintaining computer-based systems, a major value of the life-cycle management process is to provide a framework within which modifications can be accomplished, documented, and approved by management. We call this framework change control.

Change Control
Change control is the means by which major modifications (changes) to a computer-based business information system are managed. By extension, change control also includes the activities and documentation required to preserve the integrity of a system throughout minor modifications (maintenance). Change control is a management process centered on the PRB, which acts as the change control agency. It also relies upon the completeness of the documentation of the computer-based business system. Figure 27.5 is a flowchart that illustrates this process. The process symbols identified by (PRB) refer to action of the performance review board; those identified by (ISO) refer to actions by the information service organization. With reference to figure 27.5:

Figure 27.5
Change management process. Changes can be complex and may involve life-cycle activities similar to those that created the computer-based information system. Therefore, a formal change control process is necessary.

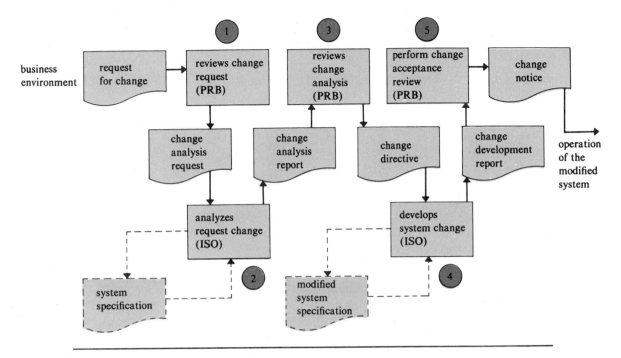

1. The PRB receives all change requests. These may be internal or external in origin. The major internal sources of requests for change are the systems analyst who is responsible for maintaining the system and the operational users of the system. They usually are the persons who can best decide whether a modification is minor and can be handled as maintenance or whether it is major and must be handled as a change. There are many possible external sources of change, for example, a change in the tax laws. Usually external changes must be implemented, but the process is not much different than the process for implementing internally generated changes.

   The PRB reviews all requests for change. Those the board does not reject are transmitted to the information services organization by means of a *change analysis request*. Figure 27.6 outlines its general content. This request initiates a new systems analysis activity. If appropriate, the change analysis request should be accompanied by an information service request (ISR) of the type discussed in chapter 11, Initial Investigation.

2. Upon receipt of the change analysis request, the information services organization assigns a responsible systems analyst, who will determine the effect of the change on the operational system. The effect depends on the nature of the change and on its impact on the system specification. Hence, change analysis is

Figure 27.6
Change analysis request format. A change analysis request is the formal means of
introducing an appeal to change a computer-related information system. The change
request is submitted to the performance review board for consideration.

```
CHANGE ANALYSIS REQUEST


1.  Name of computer-based system:

2.  Description of change:

3.  Reason for change:

4.  Date of need:

5.  Special considerations:
```

an important area in which the necessity for good baseline documentation is
evident. As a result of an investigation, the analyst determines how far back in
its life cycle the system is affected by the change. For example, a need for new
and different user outputs (that is an external performance change) would
relate to the part of the system specification that was derived from the original
performance specification. In this case many performance, design, and
development phase activities might have to be repeated. This process of
analyzing the system specification to determine the effect of the change is
shown by the first dashed-line feedback loop of figure 27.5.

   After completing the analysis of the proposed change, the responsible
analyst prepares a *change analysis report*. Figure 27.7 outlines its general
content. If the change is feasible, the analyst prepares a draft of an information
service request (ISR) to accompany the change analysis report.

3. The PRB receives the change analysis report and reviews it with the systems
   analyst and with other persons who know the system well. If the board decides
   to implement the change, a directive to that effect is prepared and sent to the
   information systems organization. The ISR form is appropriate for this *change
   directive,* as is shown in figure 27.8.

4. Upon receipt of the change directive, the information service organization
   transfers the computer-based system from maintenance to project status. A
   project leader is assigned, and modification of the system is started. The steps
   followed are identical to the life-cycle process for developing a new system. The
   figure showing this process was originally presented in chapter 2; it is repeated
   as figure 27.9. As previously discussed, the point of reentry into the life cycle
   depends on which baseline specification is modified. On some occasions it is
   necessary to start at the beginning and to develop new performance, design, and

Figure 27.7
Change analysis report. The change analysis report, prepared by the information services
organization, is the basis for a decision regarding whether or not to implement a
requested change to a computer-related system.

```
CHANGE ANALYSIS REPORT


1.  Name of computer-based system:

2.  Description of change:

3.  Reason for change:

4.  System modifications:
    a.  external specification:
    b.  internal specification:

5.  Date of availability:

6.  Special considerations:
```

system specifications. The effort to change the ongoing system sometimes is
greater than the effort to develop it in the first place. The requirement for
reentering the life-cycle process to modify the system is shown by the second
dashed-line feedback loop in figure 27.5.

The documented result of these activities is a change development report,
which is analogous to the development phase report prepared when the original
system was developed. The change activities correspond to those previously
derived from the life-cycle process of figure 27.9—including the preparation of
a change study report and a change design report, if required.

5. The change development report is presented to the PRB for a change
   acceptance review. This review is similar to the original acceptance review.
   Again, the PRB requires inputs to prove that personnel are trained, the changed
   system has been tested, and a changeover plan exists.

After a successful acceptance review, the PRB issues a *change notice*.
This notice informs all personnel of the changes to the system and their
effective date. The change notice also acts as a cover sheet to which the
replacement pages for all affected specifications and other system
documentation are attached. Figure 27.10 shows the typical format.

We have described the formal change process for significant system modifications. As we have previously noted, modifications that have taken place as part of system maintenance activities should be presented at the next scheduled periodic PRB
meeting. A formal change notice can then be issued to ensure that these modifications
are included as a part of current system documentation.

Figure 27.8
Change directive. An information service request, called a change directive, is issued by the performance review board to initiate the implementation of a change. The change directive is similar to the project directive, which initiated the original development of the computer-based information system.

INFORMATION SERVICE REQUEST			Page <u>1</u> of <u>1</u>	

JOB TITLE: Change of On-line Accounts Receivable System	NEW ☑ REV. ☐	REQUESTED DATE: 6/15/XX	REQUIRED DATE: 7/15/XX

AUTHORIZATION				
OBJECTIVE: To provide a monthly listing of overcredit accounts	LABOR		OTHER	
	HOURS	AMOUNT	HOURS	AMOUNT
	20	$4,000		

ANTICIPATED BENEFITS:
1. Detection of chronic overdue accounts
2. Revision of credit limits

OUTPUT DESCRIPTION	INPUT DESCRIPTION
TITLE: Overcredit Summary Report	TITLE:
FREQUENCY: monthly QUANTITY: 1	FREQUENCY:
PAGES: 3 max. COPIES: 1/region	QUANTITY:
COMMENTS: Regional Sales Managers; also, CRT display	COMMENTS:
TITLE:	TITLE:
FREQUENCY: QUANTITY:	FREQUENCY:
PAGES: COPIES:	QUANTITY:
COMMENTS:	COMMENTS:

TO BE FILLED OUT BY REQUESTOR

REQUESTED BY: *G. Davis*	DEPARTMENT: 310	TITLE: Head, A/R Dept.	TELEPHONE: X3250
APPROVED BY: *Ben Franklin*	DEPARTMENT: 300	TITLE: Manager, Account. Div.	TELEPHONE: X3208

TO BE FILLED OUT BY INFORMATION SERVICES

FILE NO: ISR-310-1C	ACCEPTED ☑ NOT ACCEPTED ☐		
SIGNATURE: *C. Hampton*	DEPARTMENT: 200	TITLE: Manager, Info. Ser. Div.	TELEPHONE: X2670

REMARKS:

This ISR is accepted as a Change Directive. J. Herring is assigned responsibility for its implementation.

FORM NO: C-6-1	ADDITIONAL INFORMATION: USE REVERSE SIDE OR EXTRA PAGES

Figure 27.9
The life cycle of a computer-based business information system. This is a reminder that the implementation of system change can be a major project, with the tasks to be performed conforming to the life-cycle process for creating business information systems.

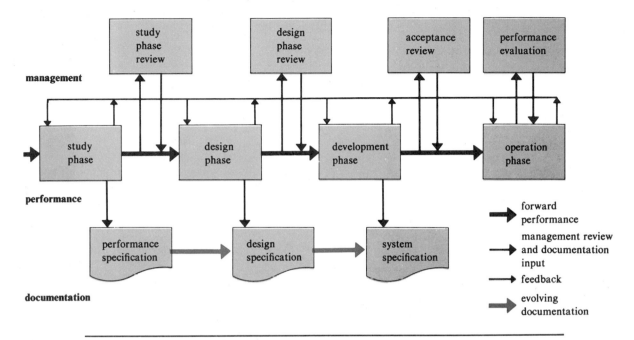

Figure 27.10
System change notice format. After the performance review board has accepted a system change, a system change notice is issued. The system change notice provides the information needed to implement the change.

```
SYSTEM CHANGE NOTICE

1.   Name of computer-based system:

2.   Summary of change:

3.   Effective date of change:

4.   Log of specification/documentation changes:

     specification/documentation       change pages
```

Documentation: The Computer-based System

Expanded Life-cycle Flowchart
In this final chapter, we have returned to the concept of a dynamic life cycle for business information systems. Figure 27.11 summarizes the theme of this text. This figure illustrates the performance, management, and documentation activities that occur

Figure 27.11
Expanded life-cycle flowchart. The expanded life-cycle flowchart is made up of flowcharts for each of the four life-cycle phases: study, design, development, and operation. The expanded life-cycle flowchart is a reference flowchart that summarizes the theme of this text.

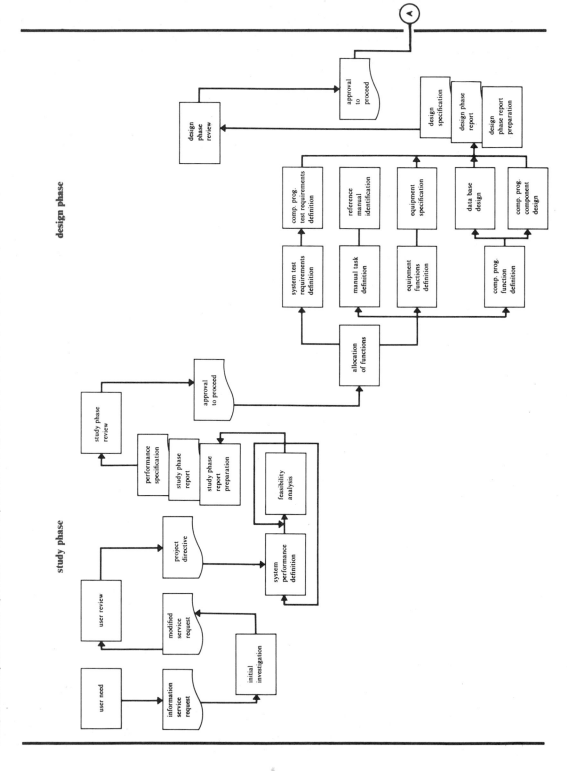

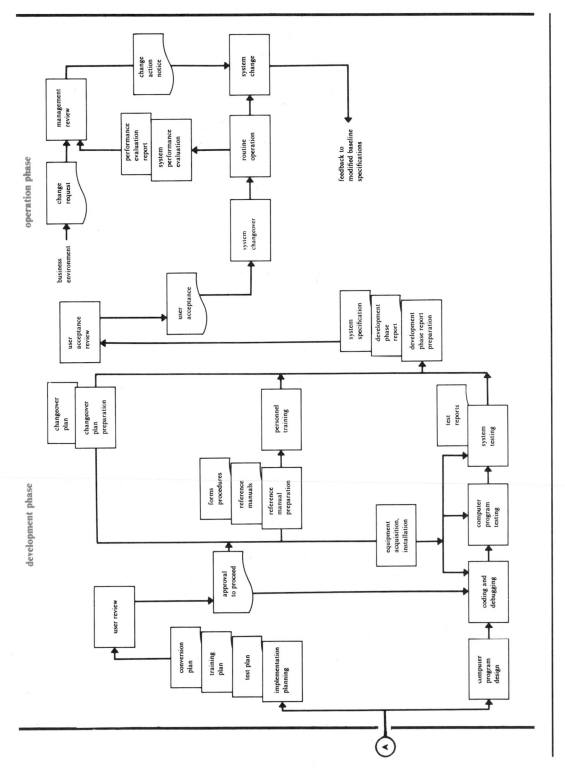

throughout all four phases of the life cycle. Further, it emphasizes feedback, which can occur at any time in the life cycle in order to accommodate change. This is particularly evident throughout the operation phase of a business information system, where the likelihood of change is great.

Structured Documentation

This chapter also emphasizes the importance of documentation structured to support the life-cycle methodology. Figure 27.12 shows the key documents that make possible the management and control of the life-cycle process.

The importance of the three *baseline documents,* the performance specification, design specification, and system specification, is emphasized by the lines and arrows extending forward in time. Other essential documents also are shown. The hierarchical structure depicted in this figure should remind us that, in a very real sense, documentation *is* the computer-based business information system.

Summary

Computer-related business information systems must be reviewed after they are installed and changeover is completed to be certain that they perform as planned. They also must be reviewed periodically thereafter both to insure that the integrity of the system and its documentation are maintained and to respond to changes due to the dynamics of the business environment. One method for performing these reviews and for managing change is to establish a user-oriented performance review board (PRB). All system modifications are not significant enough to require prior approval by the PRB. We distinguish between maintenance tasks, which do not require PRB approval, and changes, which do require approval. All modifications do affect baseline specifications; however, changes usually relate more to fundamental decisions made in preparing the performance specification and design specification than do maintenance tasks.

Change control is the means by which major modifications (changes) to a computer-based business information system are managed. The change control process involves the PRB, which reviews all change requests. The PRB may request further analysis from the information service organization prior to rejecting, modifying, or implementing the change request. When a decision is made to change the system, the PRB issues a change directive to the information service organization, and the life-cycle is reentered at a point that depends upon the extent of the modification. Sometimes the effort required to change a system is greater than that required to develop it initially.

Change management emphasizes the importance of the life-cycle methodology, which is based upon a documentation structure that is continuously and carefully maintained. In a very real sense, documentation *is* the computer-based business information system.

Figure 27.12
Documentation structure for computer-based business information systems. Structured relationships among documents essential to management of the life-cycle process for creating computer-based business information systems are summarized here.

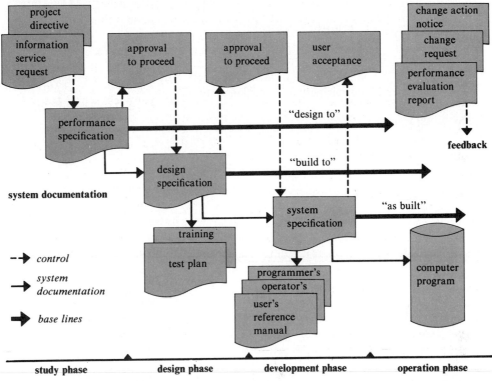

For Review

performance review board (PRB)
post-installation review
periodic review
modification
change

maintenance
change control
patch
change analysis request

change analysis report
change directive
change notice
baseline document

For Discussion

1. What is the purpose of the performance review board? Who should be on this board?
2. Under what circumstances are undocumented changes acceptable?
3. What is the purpose of the post-installation review? When should it be scheduled?
4. Why are periodic reviews necessary?
5. What are the differences between modification, change, and maintenance?
6. Describe the change management process.
7. What is the meaning of the statement "In a very real sense, documentation *is* the computer-based business information system"?

Glossary

Acceptance review   The user review of a computer-based information system project for the purpose of accepting the system and initiating the operation phase. The development phase review.

Achievement index   One of three performance indices. It is the ratio of actual achievement to planned achievement. *See also* **Cost index, Status index.**

Activity   The application of time and resources to achieve an objective. A basic element of critical path networks.

Algorithm   A set of rules or instructions used to accomplish a task.

Allocation of functions   The design phase process of identifying the functions that the system must perform, such as manual tasks, equipment functions, or computer program functions.

Alphabetic code   A code that describes items by the use of letter and number combinations. The two categories of alphabetic code are mnemonic and alphabetic derivation.

Alphabetic derivation code   A code consisting of characters taken or derived from the name or description of the coded item according to a set of rules.

Applications Package   A complete business system, including computer programming and all documentation. Usually applies to purchased systems designed and developed out-of-house.

Automated office   An office that takes advantage of extensive automation, including data processing, word processing, and communications technologies.

Bar chart   A chart constructed using vertical or horizontal bars to represent quantities and to show relationships between the plotted data.

Baseline document   A major reference document containing specifications to which changes can be referred. *See also* **Baseline specification.**

Baseline specification   A reference for system maintenance and change. The principal baseline specifications are the performance specification, the design specification, and the system specification.

Basic symbols   Flowcharting symbols that are common to both program and system flowcharts.

Best system   The system that meets the performance requirements at the least cost.

Block   *See* **Physical record.**

Blocking factor   The number of logical records in each physical record or block.

Block sequence code   A sequence code in which a series of consecutive numbers is divided into blocks, with each block reserved for identifying a group of items with a common characteristic.

Bottom-up computer program development   A development technique where the most detailed subroutines are developed first and then combined with higher-level routines until the program is completed.

Boxed style   A form style in which data items are allocated a specific amount of space. This space is identified by a name or a brief description.

Business   A combination of personnel, material, facilities, and equipment to accomplish specific objectives and to achieve defined goals. A system of systems.

Business information system   A system that uses resources to convert data into the information needed to accomplish the purposes of the business.

Candidate evaluation matrix   A table that lists evaluation criteria and rates alternative systems in terms of these criteria.

Candidate system matrix   A table that lists functions to be performed and alternative systems for performing them.

Card design aid   A worksheet useful in the trial-and-error design process of determining the layout of punched cards.

Carriage control   The technique of controlling the vertical positioning of the paper in a computer printer.

Cathode Ray Tube   *See* **CRT.**

Change   A system modification that requires performance review board action.

Change analysis report   A report prepared by the analyst after completing the analysis of a proposed system change.

Change analysis request   A request to initiate the analysis of an existing system for possible changes. The changes must have been approved by the performance review board (PRB).

Change control   The means by which major modifications to computer-based business information systems are managed.

Change directive   The authorization for the modification of an existing system that is sent to the information systems organization from the performance review board (PRB).

Change notice   A notice sent by the performance review board to all personnel that identifies changes to a system and the effective date of those changes.

Changeover   The process of changing over from the old system to the new system; the transistion from the development phase to the operation phase.

Changeover action log   A form used to keep track of actions, responsibilities, assignments, and completion dates during a system changeover.

Changeover crisis   The "crisis-like" period during which an organization changes from an "old" system to a new one.

Changeover plan   The plan of how the change from the existing system to a new system will be made.

Channel 1   The carriage control reference to the first printed line on a page.

Channel 12   The carriage control reference to the last printed line on a page.

Chart   A graphical or pictorial expression of relationships or movement.

Code   A group of characters used to identify an item of data.

Code dictionary   A listing of codes and their corresponding data items.

Code plan The identification of the particular characteristics of an item that need to be contained in the code.

Coding The process of writing instructions in a programming language such as COBOL, PL/I, or RPG.

Communication The process of transferring information from one point to another.

Computer program The instructions followed by a computer to accomplish a particular task.

Computer program function A system task to be performed by a computer, such as master file load, master file update, input data editing, and so forth.

Computer run sheet A form containing detailed input, output, file, and exception-handling information that must be provided in the operator's reference manual for every run required to process a complete computer program.

Constraint A condition, such as time or money, that limits the solutions a systems analyst may consider.

Control The actions taken to bring the difference between an actual output and a desired output within an acceptable range.

Conversion The process of performing all of the operations that result directly in the turnover of the new system to the user.

Conversion plan The plan that provides for the conversion of procedures, programs, and files preparatory to actual changeover from the old system to the new one.

Correspondence and questionnaires A method of determining the currency of procedures; a fact-finding technique used during an initial investigation.

Cost index One of three performance indices. It is the ratio of actual costs to planned costs. *See also* **Achievement index, Status index.**

Critical path The path in a critical path network, such as PERT, along which the sum of all activity times, from the first to the last event, is the greatest.

Critical path method Use of a network—often computer generated—to display the relationship between tasks and schedules. An example is the program evaluation and review technique (PERT).

Critical path network *See* **Critical path method.**

CRT Cathode ray tube; a television-like input/output station that can display information.

Cumulative documentation The concept that describes the increases in the level of documentation detail and volume of documentation as a project moves through the life cycle.

Cut forms Forms that do not require special equipment to manufacture or to use the form. Typically, they are single sheet forms.

Cylinder The surface area covered by all read-write heads in one position of the access mechanism on a magnetic disk or magnetic drum.

Data Base Administrator (DBA) The authority that controls the data base management system (DBMS) by controlling the data base schema and subschema.

Data Base Management System (DBMS) Software that allows data descriptions to be independent from computer programs. It provides the capability of describing logical relationships between files to facilitate efficient maintenance and access to the data base.

Data carrier A term that refers to media used to communicate or store data. Examples are records, reports, magnetic files, and CRT screen displays.

Data collection   Gathering and organizing all of the documentation related to data carriers; a fact-finding technique used during an initial investigation.

Data element analysis   A process for understanding the meanings of data names and codes; a fact-analysis technique used during an initial investigation.

Data element list   A form used to list information describing each data element; accompanies an output specification.

Data flow diagram   A network that uses special symbols to describe the flows of data and the processes that change, or transform, data throughout a system.

DBMS   *See* **Data Base Management System.**

Debugging   The process of testing a computer program for errors and correcting any errors found.

Decision Support System (DSS)   An information system designed to provide managers with information to support their decision-making processes; an advanced management information system.

Decision table   A table used to describe logical rules.

Decision tree   A network-type chart that is the logical equivalent of a decision table.

Design phase   The life-cycle phase in which the detailed design of the recommended system is accomplished.

Design phase report   A report prepared at the end of the design phase; it is an extension of the study phase report and it summarizes the results of the design phase activities.

Design phase review   A review held with the user organization at the conclusion of the design phase to present the results of the design phase activities and to determine whether or not to proceed with the development phase.

Design specification   A baseline specification that serves as a "blueprint" for the construction of a computer-based business information system.

Development phase   The life cycle in which the system is constructed according to the design phase specification.

Development phase report   A report prepared at the end of the development phase; it is an extension of the design phase report and it summarizes the results of the development phase activities.

Development phase review   A review held with the user organization at the conclusion of the development phase to determine whether or not to enter the operation phase.

Direct files   Files in which records may be accessed randomly by calculating the record address.

Display layout sheet   A form used to design CRT screen layouts. The form is divided into 24 lines of 80 characters each to simulate the possible display positions on a screen.

Display station   *See* **CRT.**

Distributed Data Processing (DDP)   The ability to locate processing power wherever it is needed.

DSS   *See* **Decision support system.**

Early Era   1940–1955. A period of significant changes in computers and their applications.

EBCDIC   Extended Binary Coded Decimal Interchange Code. A computer coding system for recording data in memory and on magnetic media.

Edit characters   Characters such as dollar signs, commas, and decimal points that are to be inserted into numeric values.

Event   A point in time at which an activity begins or ends. A basic element of critical path networks.

Expanded process-oriented system flowchart
A process-oriented flowchart that has been expanded in detail until each of the processing functions can be identified.

External performance   A term that refers to the performance of a computer-based information system external to the computer program component.

Fact analysis   Activities related to the analysis of data collected during the fact-finding stage of an initial investigation.

Fact finding   Activities related to the collection of facts during the initial investigation of the business information system.

Feasibility analysis   A procedure for identifying candidate systems and selecting the most feasible system.

Feedback   The process of comparing an actual output with a desired output for the purpose of improving the performance of a system.

File   The collection of logically related records, that is, records of a single type such as employee records, payroll records, and so forth.

Flowchart   A pictorial representation that uses predefined symbols to either describe data flow and processing in a business system or the logic of a computer program.

Forms control   The coordination of the forms needs of multiple departments or users to control the costs of forms.

Functional file   A file of forms by subject, operation, or function.

Gantt-type chart   A horizontal bar chart, typically used to show a project schedule and to report progress on that schedule.

Goal   A broadly stated purpose of a business.

"Gozinto"   A term that refers to the expansion of a specification into successive levels of detail; derived from "goes into".

Group classification code   A code that designates major, intermediate, and minor data classifications by successively lower orders of digits.

Growing Era 1955–1965.   A period of significant changes in computers and their applications.

Hardcopy   A relatively permanent output; for example, a printed report.

Hardware   The physical components of a computer system.

Hardware end-product   An end-product that is described primarily by its physical attributes, which can be observed and measured as the product moves through the several stages of its development.

Hierarchy chart   A chart that shows the hierarchy, from top to bottom, of functions to be performed. It is part of the HIPO planning and design technique.

HIPO charts   A set of charts made up of a hierarchy chart (H) plus input-processing-output (IPO) charts, which emphasizes the functions of a system or computer program.

Immediate replacement   A changeover method requiring immediate use of the new system; at the same time, the use of the existing system is discontinued.

Implementation   The process of bringing a developed system into operational use and turning it over to the user.

Implementation plan   A plan for implementing a system; it includes test plans, training plans, an equipment acquisition plan, and a conversion plan.

Incremental commitment   A term that refers to management committment to project expenditures. Committment is rekindled at scheduled study, design, and development phase reviews.

Indexed-sequential file   A file created in sequential order, as are sequential files; additionally, a set of address indexes is created in order to look up and randomly access records.

Information   Refined data; a useful system output.

Information flow   The network of administrative and operational documentation.

Information generator   A business information need, either external or internal in origin.

Information-oriented system flowchart   A flowchart that uses a grid structure to trace the flow of data through a system.

Information resource management
Management of the integrated information resources of a corporation, including data processing, communications, and office automation.

Information Service Request (ISR)   A written request for information services support.

Initial Investigation   An investigation conducted by the systems analyst to clarify the problem and define it in detail that is mutually understood by the user and the analyst; results in a project directive.

Input-output analysis   A general term for analysis techniques that is based upon the perception of a system as a process that converts inputs into outputs. Examples are: data flow diagrams, information-oriented system flowcharts, process-oriented system flowcharts, and IPO charts; also, a fact-analysis technique used during an initial investigation.

Integrated Circuit (IC)   A miniaturized electronic circuit, produced by application of microelectronics technology.

Interface   A term that refers to data elements that are shared by more than one system.

Internal Performance   A term that refers to the performance of a computer-based information system internal to the computer program component.

IPO (Input, Process, Output) chart   A detail-level chart that lists the inputs, processing steps, and outputs of each functional module of a hierarchy chart. It is part of the HIPO planning and design technique.

Key-to-disk   A device that records data on magnetic disk from input keystrokes; provides relatively high data rate input to a computer.

Key-to-tape   A device that records data on magnetic tape from input keystrokes; provides relatively high data rate input to a computer.

Life cycle   A concept that describes the relationships between the four phases of a business information system.

Life-cycle management   The management of any, or all, of the phases of the life cycle of a computer-based business information system. Often, a systems analyst acts as a life-cycle manager.

Life-cycle methodology   A four-phase, systems-analysis process for solving business information system problems.

Life-cycle phase   One of the four life-cycle phases: study, design, development, and operation.

Limited Information Service Request   An initial Information Service Request (ISR), often incomplete, which usually initiates an initial investigation.

Linear programming   A management science technique, which often makes use of a computer; it involves use of a mathematical model to find the optimum combination of available resources for achieving a given result.

Line chart   A chart constructed by connecting a set of plotted points; also called a graph.

Link pin   An organizational position characteristic of a superior-subordinate relationship; it is a position between an upper and a lower element of the organization.

Logical Record   The unit of data to be processed at one time, that is, the payroll data for one employee.

Maintenance   A system modification that does not require performance review board action.

Management Information System (MIS) An information system designed to provide managers with the information needed for planning, control, and decision making.

Manual   Printed and assembled pages of instructional material.

Manual system   A system designed around a pattern of manual operations.

Matrix   *See* **table.**

Maturing Era   1980–   . A period of significant changes in computers and their applications; the Information Era.

Microcomputer   A small, low cost computer produced by application of microelectronics technology.

Milestone   A key activity to be completed in accordance with a project plan.

MIS   *See* **Management information system.**

Mnemonic code   Letter and number combinations obtained from descriptions of the coded item that serve as a memory aid.

Modification   A general term that includes both maintenance and change to an operational system. *See* **Maintenance, change.**

Modified Information Service Request   An Information Service Request (ISR) prepared by the systems analyst after completion of the initial investigation.

Narrative   A form of writing that tells a story.

Network   Refers to critical path techniques, such as PERT, for project management; also, the term may be applied to graphs with many branches, such as data flow diagrams and decision trees.

Nonprocedural language   A "user-friendly" language that does not require knowledge of computer logic or procedures.

Nontraditional computer-oriented system A computer-oriented system broader in scope than data processing.

Numerical file   A file of forms in form number sequence.

Objective   A concrete, specific accomplishment necessary to the achievement of a goal.

Observation   Observation of the operation of an on-going system; a fact-finding technique used during an initial investigation.

Open style   A style of form consisting largely of headings and open areas in which data can be entered.

Operation phase   The life-cycle phase in which changeover from the old system to the new system occurs and in which the system is installed, operated, and maintained.

Optical bar-code reader   An optical reader that detects combinations of marks by which data is coded; for example, the Universal Product Code (UPC).

Optical Character Reader (OCR)   An optical reader that detects special symbols; for example, embossed credit cards.

Optical mark reader   An optical reader that accepts data in the form of pencil marks on paper.

Optical reader   A class of input device that captures data directly, using optical techniques. Examples are: optical bar-code readers, Optical Character Readers (OCR), and optical mark readers.

Organization chart   A flowchart that identifies organizational elements of a business and displays areas of responsibility and lines of authority.

Organization function list   A document that describes the major activities performed by each organization shown on an organization chart.

Output data source analysis   A technique for estimating the storage requirements of an information system by listing all of the output data elements and determining their source.

Output specification   A form used to describe computer-generated output; accompanied by a data element list.

Parallel operation   A changeover method that is based upon a period of operation of both the old and the new system.

Patch   A "quick fix" programming change; a major cause of poor documentation and high maintenance costs.

Payback analysis   An analysis performed to determine the length of time for the investment in a new system to be recovered as a result of cost saving or cost avoidance.

Performance Review Board (PRB)   A user-oriented board responsible for both the periodic evaluation of the performance of a computer-based business information system and for maintaining its integrity.

Performance specification   A baseline specification that describes what the computer-based system is to do.

Periodic review   *See* **System performance evaluation.**

Personal interview   A fact-finding technique used during an initial investigation.

PERT   Program Evaluation and Review Technique. A critical path network technique.

Phased changeover   A changeover method based upon the incremental replacement of the old system by the new system; it is a compromise between parallel operation and immediate replacement.

Physical record   The unit of data to be inputted into or outputted from the computer's memory at one time. This may be one or more logical records. The physical record is also referred to as a block.

Pie chart   A type of chart that uses "slices" of a pie to present relationships as percentages of a circular area.

Policy   A broad written guideline for conduct or action.

Post-installation review   The first system performance review; it occurs in the operation phase after the changeover crisis has passed.

Presentation   The oral communication of plans or results made in order to influence people and to obtain decisions.

Principal user   The person who, in practice, will accept or reject the computer-based business information system.

Print chart   A form used to design computer printer outputs. Each line on the form is divided into print positions to allow for the detailed design of titles, column headings, detail lines, and so forth.

Procedure   Specific statements that tell how policies are to be carried out.

Procedure analysis flowchart   A chart that uses special symbols to record the details of manual procedures in pictorial form for analysis and improvement of those procedures.

Process-oriented system flowchart   A flowchart that shows the data processing operations that convert inputs into outputs.

Product flow The flow of raw materials into finished goods.

Production enterprise A business enterprise with a product as its goal; for example, manufacturing.

Program *See* **Computer program.** Also, sometimes used as a synonym for a project; *see* **PERT.**

Programming symbols Flowcharting symbols used to represent computer program logic and related operations.

Project cost report A line chart that is used to report actual project cost versus planned cost.

Project directive The final version of an Information Service Request; the written contract between the user and the information services organization.

Project plan and status report A horizontal bar chart that is used to plan milestone events for a project and to report progress in achieving those milestones. A Gantt-type chart.

Pseudocode The use of a pseudo programming language to describe the logic of a computer program. It is a form of structured English.

Punched card A card that records data coded in the form of combinations of holes (punches). The data is entered into the card by an operator of a key punch or data recorder machine. Subsequently, it is entered into a computer by a card reader.

Recurring data analysis Analysis of documents in order to identify unnecessary data duplication and redundant files; a fact-analysis technique used during an initial investigation.

Reference manual An important part of the documentation of a computer-related business information system. The three principal reference manuals are: programmers reference manual, users reference manual, and operators reference manual.

Refining Era 1965–1980. A period of significant changes in computers and their applications.

Report A formal communication of results and conclusions due to a particular set of actions; it summarizes work that has been performed.

Report use analysis An analysis of report usage in order to identify unnecessary reports and to improve the usefulness of reports; a fact-analysis technique used in an initial investigation.

Research Investigation of sources of background information related to a business information system; a fact-finding technique used during an initial investigation.

Resource Facilities, personnel, material, or equipment used by a system.

Response time The elapsed time between the release of input data by a user and the receipt of computer output.

Routine operation The stage in the operation phase that follows system changeover.

Self-checking code A code that uses check digits to test the validity of the data.

Sequence code A coding system in which codes are assigned in sequence. *See also* **Simple sequence code** and **Block sequence code.**

Sequential file A file created by writing records on the storage medium in sequence, according to a control field within each logical record.

Service enterprise A business enterprise with a service as its goal; for example, education.

Significant digit code A numeric code in which the numbers describe a measurable physical quantity.

Simple sequence code The assignment of consecutive numbers to a group of items as they occur.

Simulation A technique for the design of systems using computer models for simulating performance and answering "what if" questions.

Skipping The process of moving a form rapidly through the printer carriage to a designated position.

Softcopy A relatively transient output; for example, a CRT screen display.

Software The collection of programs that facilitates the use of a computer.

Software end-product An end-product associated with computer based systems. As contrasted with a hardware end-product, it is information, and it does not possess attributes that can readily be observed and measured from concept to end-product.

Spacing The process of moving a computer-printed form one, two, or three lines at a time.

Specialty form A form that requires special equipment for its construction, usually obtained from a company that specializes in forms manufacture.

Specification A reference document that contains basic detailed data.

Specific objective A specific, measurable performance outcome for a system.

Staff position An organizational position that is service in nature; as contrasted with a line position, it does not represent direct authority.

Standards The rules under which personnel in an information services organization work.

Status index One of three performance indexes. It is the ratio between the achievement index and the cost index and is the most significant single measure of the status of a project.

Step chart A type of chart that conveys a pattern of motion by connecting points to form a "staircase".

Structured design and development A top-down process for the design and development of computer-based information systems. It proceeds from high-level system modules to successively lower-level modules.

Structured english A method for displaying a logical process in an outline format.

Structured walk-through A technical review to assist people working on a project; used to discover errors in the logic of a computer program or other system components.

Stub A term that describes a lower-level computer program module that has been identified, but that has not yet been written; the use of stubs is part of a top-down computer program development process.

Study phase The life-cycle phase in which a problem is defined and a system is recommended as a solution.

Study phase report A comprehensive report prepared for the user-sponsor of the system and presented at the conclusion of the study phase activities.

Study phase review A review for the purpose of presenting the results of the study phase activities and for determining future action.

Style The design of the body of a form, that is, boxed or open.

Subsystem A component of a system that is, itself, a system.

Superior-subordinate relationship An organizational relationship in which authority flows downward and responsibility flow upward.

System A combination of resources working together to convert inputs into outputs.

System candidates Alternative systems from which the best system is selected as the result of a feasibility analysis.

System change *See* **Change.**

System changeover The period of transition from the old system to the new system. *See also* **Changeover.**

System flowchart A flowchart used to analyze existing systems, synthesize new systems, and communicate with others.

System of systems A reference to a complex system, such as a computer-related business information system, that is made up of smaller elements, called subsystems, which also are systems.

System performance definition A process involving the statement of general constraints, identification of specific objectives, and description of outputs.

System performance evaluation An evaluation of the performance of a computer-based information system that occurs during the operation phase. The evaluation is repeated periodically.

Systems analysis A structured process for identifying and solving problems; the performance, management, and documentation of the activities related to the four phases of the life cycle of a business information system.

System analyst An individual who performs systems analysis during any, or all, of the life-cycle phases of a business information system. A life-cycle manager.

System specification A baseline specification that contains all of the essential system documentation; it is a complete technical specification.

System symbols Flowcharting symbols that represent equipment or specialized input, output, or processing operations.

System team Representatives of areas affected by a computer-based business information system; they participate in the design and development of the system, usually under the leadership of a systems analyst.

Table A chart made up of intersecting horizontal and vertical lines to form rows and columns; also called a grid chart or matrix.

Technical writing A document written with the purpose of communicating facts.

Throughput time The time required for work to flow through the machine room of a computing center; a component of response time.

Top down A structured method for the design and development of computer-based systems; in successive phases, major modules are expanded into additional, increasingly detailed modules.

Top-down computer program development A structured technique that starts with a general description of the system and expands it into successively greater levels of detail.

Track The magnetic surface (on magnetic disk storage devices) covered by one read-write head with the access arms in one position.

Traditional computer-oriented system A computer-oriented system with data processing as the focal point.

Turnaround time The time that elapses between the arrival of data at a computing center and the availability of the output.

Type   The classification of a form by the complexity of its manufacturing process, that is, cut or specialty.

Usability   The value of an information system as perceived by its users.

User turnover   A term that refers to the turnover of a computer-related information system to its principal user by the systems team; it occurs when the changeover stage of the operation phase is completed.

Weighted candidate evaluation matrix   A table that weights the candidate evaluation matrix entries by their importance and applies a rating number; it is a means of calculating comparative total scores for each candidate.

Photo Credits

Index

study phase   design phase

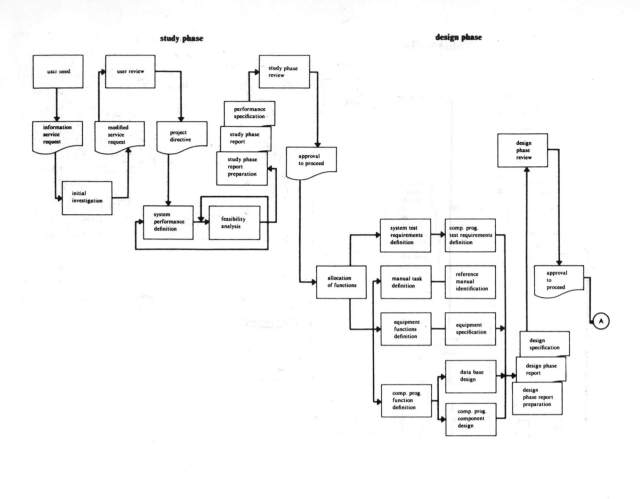